Q 660

49⁰⁰/5⁹⁸

D1786577

Powell's Ratite Management, Medicine, &

$49.00 / 5.98   PC

Science & Natural History   117537

RATITE
Management, Medicine, and Surgery

RATITE
Management, Medicine, and Surgery

EDITED BY
Thomas N. Tully, Jr., DVM MS
&
Simon M. Shane, BVSc. Ph.D. MBL. dip.ACPV

KRIEGER PUBLISHING COMPANY
MALABAR, FLORIDA
1996

Original Edition 1996

Printed and Published by
**KRIEGER PUBLISHING COMPANY
KRIEGER DRIVE
MALABAR, FLORIDA 32950**

Copyright © 1996 by Krieger Publishing Company

All rights reserved. No part of this book may be reproduced in any form or by any means, electronic or mechanical, including information storage and retrieval sustems without permission in writing from the publisher.
No liability is assumed with respect to the use of the information contained herein.
Printed in the United States of America.

> **FROM A DECLARATION OF PRINCIPLES JOINTLY ADOPTED BY A COMMITTEE OF THE AMERICAN BAR ASSOCIATION AND A COMMITTEE OF PUBLISHERS:**
> This publication is designed to provide accurate and authoritative information in regard to the subject matter covered. It is sold with the understanding that the publisher is not engaged in rendering legal, accounting, or other professional service. If legal advice or other expert assistance is required, the services of a competent professional person should be sought.

Library of Congress Cataloging-In-Publication Data

Ratite management, medicine, and surgery / edited by Thomas N. Tully, Jr. and Simon M. Shane.
    p.   cm.
    Includes index.
    ISBN 0-89464-874-8 (hardcover : alk. paper)
    1. Ratites—Diseases. 2. Ratites—Surgery. 3. Ratite farming.
  I. Tully, Thomas N. II. Shane, Simon M.
SF995.5.R37 1996
636.6—dc20   95-48173
                                                                 CIP

10 9 8 7 6 5 4 3 2

Contents

List of Tables		vii
List of Figures		ix
List of Appendixes		xiii
Contributors		xiv
Foreword—*John A. Crawford*		xv
Preface		xvii

CHAPTER

1	Clinical Anatomy of Ratites *Murray E. Fowler*	1
2	Ratite Nutrition *C. Roselina Angel, Sheila E. Scheideler, and Jerry L. Sell*	11
3	Biosecurity and Control of Disease *Simon M. Shane and Lori Minteer*	31
4	Restraint and Handling of the Ostrich *John R. Wade*	37
5	Restraint and Handling of the Emu *David Mouser*	41
6	Reproduction *Karen D. Hicks-Alldredge*	47
7	Hatchery Management in Ostrich Production *James S. Stewart*	59
8	Hatchery Management in Emu Production *Simon M. Shane*	69
9	Anesthesiology of Ratites *Janyce L. Cornick-Seahorn*	79
	Color Section appears between pages 94 and 95	
10	Surgical Conditions of Ratites *Mark R. Crabill and Clifford M. Honnas*	95
11	Clinical Hematology and Chemistry of Ratites *Alan M. Fudge*	105
12	Parasites of Ratites *Thomas M. Craig and P. Lea Diamond*	115

13	Infectious Diseases *Simon M. Shane and Thomas N. Tully, Jr.*	127
14	Developmental Problems in Young Ratites *Brian L. Speer*	147
15	Therapeutics *Thomas N. Tully, Jr.*	155
16	Guide to Examination and Health Certification of Ratites *Amy M. Raines and Simon M. Shane*	165
17	Jurisprudence *Simon M. Shane*	173
Index		179

List of Tables

TABLE		
1.1.	Characteristics, Speed, and Distribution of Ratites	1
1.2.	Comparative Length of the Intestines of Ratites	4
1.3.	Sex Differentiation in Ratites	8
2.1.	Comparative Length of Ratite Intestines	12
2.2.	Protein and Amino Acid Requirements for Maintenance and Growth of Ostriches	14
2.3.	Definition of Energy Terms	15
2.4.	TME_n Values of Various Feedstuffs for Adult Ostriches	16
2.5.	Mineral Requirements of Ratites	20
2.6.	Classification of Ingredients by Contribution to Diets for Ratites	21
2.7.	Characteristics of Ingredients	22
2.8.	Nutrient Composition (Excluding Amino Acids) of Feedstuffs Commonly Used for Ratites	24
2.9.	Amino Acid Composition of Feedstuffs Commonly Used for Ratites	26
2.10.	Nutrient Composition of Forage Sources Commonly Used for Ratites	26
2.11.	Element Composition of Mineral Sources Commonly Used for Ratites	28
2.12.	Predicted Energy and Amino Acid Requirement of Ostriches	28
2.13.	Nutrient Recommendations for South African Ostriches	28
2.14.	Estimated Recommendations for Energy Intake in Laying Ostriches	28
2.15.	Estimated Amino Acid Requirements for Mature Ostrich Hens	29
2.16.	Macromineral Recommendations for South African Ostriches	29
2.17.	Trace Element and Vitamin Supplementation used in South African Ostrich Diets	29
2.18.	Growth Parameters of South African Ostrich	29
2.19.	Nutrient Ranges in Practical Diets for Ratites	30

7.1.	Egg Production and Hatchability of South African Ostriches, 1994	60
7.2.	Repeatability of Reproductive Parameters in South African Ostriches	60
9.1.	Normal Physiologic Values for Ratites	80
9.2.	Dosages of Anesthetic Agents Used in Ratites	82
9.3.	Suggested Concentrations for Inhalation Agents Used in Ratites	85
9.4.	Endotracheal Tube Selection for Ratites	85
11.1.	Literature Values for Ostriches	106
11.2.	Literature Values for Emus	107
12.1.	Ectoparasites of Ratites	115
12.2.	Helminths of Ratites	117
12.3.	Protozoans of Ratites	122

List of Figures

NOTE: "cl" (instead of page number) means refer to color section.

FIGURE		
1.1	Diagram of the structures of the foot of an ostrich	2
1.2	Sternum of ratites	3
1.3	Thoracic girdle and sternum of the ostrich	3
1.4	Pelvic girdle of an adult ostrich	4
1.5	Diagrams of the stomach of ratites	4
1.6	Diagrammatic illustration of a left lateral view of the thoraco-abdominal viscera of an ostrich	5
1.7	Diagrammatic illustration of the thoraco-abdominal viscera of an ostrich in a dorsoventral view	5
1.8	Diagram comparing ostrich and emu digestive tracts	6
1.9	Diagram of right lateral view of the cloaca of an ostrich	6
1.10	The tracheal cleft and diverticulum of an adult emu	7
1.11	Diagram of left lateral view of the cloaca and phallus of an adult male ostrich	8
1.12	Diagram of left lateral view of a retracted and erect phallus of a male emu or rhea	9
1.13	Erect phallus of an adult male ostrich	9
3.1	Effective biosecurity requires fences, secure gates and constant supervision	33
3.2	Emu breeder in well-drained pen	36
3.3	Ostrich egg collection from a clean nest	36
3.4	Impervious surfaces in stainless steel work areas	36
5.1	Six-foot chain link breeder pens on large farm	42
5.2	The circular "carousel" breeder pen arrangement	42
5.3	Restraint technique which allows acceptable control of bird	43
5.4	"Flipping" an emu to allow vent sexing of young adult birds and examination of the genitalia and cloaca in adult birds	44
5.5	The bird is immediately picked up and rolled up on its back.	44
5.6	The subject is gently lowered to the ground.	45
6.1	The single (left) ovary of a 2-year-old ostrich hen prior to onset of production	47

6.2	Paired testes of mature male ostrich located equidistantly from the midline in the body cavity	48
6.3	Phallus of a mature ostrich, protruding from cloaca	48
6.4	Phallus of a mature emu, everted from cloaca	49
6.5	Male ostrich in typical courtship position (kanteling)	50
6.6	Young ostriches held in pen with mature male "caretaker"	51
6.7	Stimulating the neck of a male emu to initiate release of semen	52
6.8	Cone used to collect semen	52
6.9	Pressure on phallus with cone to release semen	53
6.10	Micrograph of emu semen showing spermatozoa	53
6.11a	Diagram of female reproductive tract	54
6.11b	Diagram of avian egg showing principal structures	54
6.12	Presence of egg yolk material in the body cavity of an ostrich due to internal laying	54
6.13	Prolapse of the phallus in an ostrich	56
6.14	Partial prolapse of the phallus of an ostrich	56
6.15	Radiograph of a kiwi showing egg retention	57
7.1	Bacterial infection of an ostrich egg	cl
7.2	Modern incubator with adequate air exchange capacity and installations to maintain temperature and humidity	63
7.3	Normal position of the ratite embryo at the time of pipping and malposition Type II	64
7.4	Ostrich egg at the time of external pipping	cl
8.1	Typical emu nest in pen	69
8.2	Chart to document moisture loss during incubation	71
8.3	Oxygen requirements of the developing embryo	72
8.4	Infrared imaging system to determine size of air cell and viability of embryo	72
8.5	Simple psychometric chart showing relationship of temperature and humidity	75
8.6	Sophisticated psychometric chart, including parameters such as enthalpy	77
9.1	Hood used on juvenile and adult ostriches for restraint	80
9.2	Juvenile ostrich restrained on a nonslip surface for induction of anesthesia via mask delivery	83
9.3	Adult ostrich positioned for intubation	86
9.4	Adult ostrich placed in lateral recumbency	87
9.5	Catheter positioned in the jugular vein of an ostrich	88
9.6	Fourteen-gauge, 13-cm catheter positioned in the right jugular vein of an emu	88

List of Figures

9.7	Normal electrocardiographic and arterial blood pressure on an anesthetized ostrich	89
9.8	Twenty-two-gauge catheter positioned in an artery for direct arterial blood pressure measurement and collection of arterial blood	89
9.9	Blood pressure cuff connected to an oscillometric blood pressure monitoring device for indirect measurement of blood pressure	91
9.10	The Ratite Rap	91
9.11	The Ratite Rap positioned on an ostrich to facilitate a safe anesthetic recovery	92
9.12	Ratites placed in sternal recumbency during recovery from anesthesia	92
10.1	Foreign body removed from the proventriculus of an ostrich	95
10.2	Lateral radiograph of the abdomen of an ostrich showing the presence of nails and rock within the proventriculus	96
10.3	Proper positioning of the ratite for a left paramedian approach to the abdomen	96
10.4	The cranial landmark for a left paramedian approach to access the proventriculus	96
10.5	Blunt separation of the air sac with the fingers to access the abaxial surface of the proventriculus	97
10.6	Fluid is aspirated from the proventriculus through a stab incision	97
10.7	Lengthening of the proventriculotomy incision	97
10.8	Placement of a tube into the proventriculus through an esophagotomy for postoperative feeding of debilitated patients	98
10.9	Esophagotomy tube is secured to adjacent skin	98
10.10	Lateral abdominal radiograph of an emu with egg retention	99
10.11	Exteriorization of the uterine segment containing the egg in preparation for hysterotomy	99
10.12	An incision is made directly over the retained egg	99
10.13	Closure of the hysterotomy with a simple continuous pattern oversewn with a continuous Cushing pattern	100
10.14	Laceration into the oral cavity and pharnyx of an ostrich	100
10.15	Longitudinal neck laceration in an ostrich	101
12.1	*Argus persicus*, soft-shelled tick	cl
12.2	*Struthiolipeurus struthionis*, louse	cl
12.3	*Libyostrongylus douglassii*	cl
12.4	*Dicheilonema* sp.	cl
12.5	*Baylissascaris* sp. in cerebral tissue of emu	cl

12.6	*Chandlerella quiscali*	cl
12.7	*Syngamus trachea*, gapeworm	cl
12.8	Ovum of *Syngamus trachea* on fecal flotation	cl
12.9	*Balantidium* sp. in fecal suspension	cl
13.1	Dead embryo in shell showing omphalitis due to congenital infection	cl
13.2	Lesions of tuberculosis in a juvenile emu	cl
13.3	Site for intradermal mycobacterium sensitivity test in ostriches	cl
13.4	Severe aspergillosis in emu chick showing numerous pulmonary granulomas	cl
13.5	Characteristic hemorrhages associated with eastern equine encephalitis in emus	cl
13.6	Superficial pox lesion	cl
13.7	Severe chronic pox lesions in an ostrich chick	cl
14.1	Enlarged yolk sac and unhealed navel in an ostrich chick	cl
14.2	Surgical removal of the retained yolk sac in an ostrich chick	cl
14.3	"Tumbler" emu chick showing retraction of the head and incoordination	cl
14.4	Angular leg deformity in a young emu chick	cl
14.5	Angular limb deformity in a juvenile ostrich	cl
14.6	Radiograph showing osteochondritis dissicans	151
14.7	Untreated rolled toe in a juvenile ostrich	cl
14.8	Splinting of rolled toe in young ostrich chick	cl
14.9	Impaction of the proventriculus and ventriculus with fibrous material	cl
14.10	Prolapse of the cloaca in an emu chick	cl
16.1	Efficient emu breeding unit	165
16.2	Large ostrich pens, to provide exercise	166
16.3	Automatic water receptacles	166
16.4	Working pen adjacent to barn and mobile chute for handling, examining, and treating emus	166
16.5	Microchip reader is essential to positively identify individual ratites for certification	167
16.6	Leg bands used for farm identification of individual birds	167
16.7	Examination of emus requires close observation within the group	167
16.8	Collection of a blood sample from the jugular vein	168
16.9	Radiograph showing foreign body in ventriculus	169
16.10	Obtaining a cloacal culture for microbiological examination	169
16.11	Intradermal *Mycobacterium* sensitivity test	169
16.12	Handling biological material with zoonotic potential requires appropriate precautions	170

List of Appendixes

APPENDIX
7.1	Identifying Incubation Problems in Ostrich	67
8.1	The Application of the Psychrometric Chart in Incubation	74
15.1	Ratite Formulary	158
16.1	Veterinary Certificate for Insurance of Ratites	172

Contributors

C. Roselina Angel, Ph.D.
Nutritional Consultant, Research and Development
Purina Mills Inc.
St. Louis, MO 63166

Janyce L. Cornick-Seahorn, D.V.M., M.S., Dip. ACVIM, ACVR
Associate Professor
Louisiana State University, School of Veterinary Medicine
South Stadium Road
Baton Rouge, LA 70803

Mark R. Crabill, D.V.M.
Large Animal Resident
Texas A&M University, College of Veterinary Medicine
College Station, TX 77843-4475

Thomas M. Craig, D.V.M., Ph.D.
Professor
Texas A&M University, School of Veterinary Medicine
College Station, TX 77843

John A. Crawford
Managing Editor, *The Ostrich News*
P.O. Box 860
Caché, OK 73527

P. Lea Diamond, M.S., D.V.M
Professor
Texas A&M University, School of Veterinary Medicine
College Station, TX 77843

Murray E. Fowler, D.V.M., Dip. ACZM
Professor of Veterinary Medicine
School of Veterinary Medicine
University of California at Davis
Davis, CA 95616

Alan M. Fudge, D.V.M., Dip. AVBP (Avian practice)
Director, Avian Medical Center of Sacramento
California Avian Laboratory
6114 Greenback Lane
Citrus Heights, CA 95621

Karen D. Hicks-Alldredge, D.V.M.
Director, Mesquite Veterinary Clinic
Midland, TX 79707

Clifford M. Honnas, D.V.M., Dip. ACVS
Assistant Professor
Texas A&M University, College of Veterinary Medicine
College Station, TX 77843-4475

Lori Minteer, D.V.M.
Ratite Veterinarian
Jupiter, FL 33478

David Mouser, D.V.M.
Veterinary Practitioner
Madisonville, TX 77867

Amy M. Raines, D.V.M.
Director, Boondocks Ratite Hospital
Oklahoma City, OK 73165

Sheila E. Scheideler, Ph.D.
Associate Professor
University of Nebraska, IANR
Lincoln, NE 68583-0908

Jerry L. Sell, Ph.D.
Professor
Iowa State University
Ames, IA 50011

Simon M. Shane, B.V.Sc., Ph.D., M.B.L., Dip. ACPV
Professor
Louisiana State University, School of Veterinary Medicine
Baton Rouge, LA 70803

Brian L. Speer, D.V.M., DipAVBP (Avian practica)
Director, Oakley Veterinary & Bird Hospital
Oakley, CA 94561

James S. Stewart, M.S., D.V.M.
Ostrich Consultation Services
San Ramon, CA 94583

Thomas N. Tully, Jr., D.V.M., M.S.
Assistant Professor
Louisiana State University, School of Veterinary Medicine
South Stadium Road
Baton Rouge, LA 70803

John R. Wade, D.V.M.
Director, Pacesetter Ostrich Farm
Folsom, LA 70437

Foreword

John A. Crawford

During the late 1980s, a small number of U.S. entrepreneurs recognized the potential of the ostrich, emu, and rhea as producers of low fat, red meat. Production from domesticated ostrich at that time was limited to the Klein Karoo, a high density production area in South Africa. The emu was available only to aboriginal tribes in Australia; and production from the rhea, once a source of meat in South America, had vanished due to widespread wildlife protection programs. No serious effort at domestication of either emus or rheas had been initiated prior to 1988.

As the managing editor of *The Ostrich News*, I was able to observe and chronicle the initiation and expansion of the industry during the late 1980s. During this decade, production operations spread throughout the United States, Canada, Mexico, and other countries around the world. Producers are now found in every state, and each of the three species of ratites has adapted to environments as diverse as Florida and Montana.

The early entrepreneurs have been joined by farmers, ranchers, and small homesteaders who have rapidly increased the population of ratites in the United States. Through the efforts of diligent, dedicated, and hard working producers, there is currently a stable production base of ostrich, emus, and rheas to support an emerging commercial market for low fat, red meat products.

During this initial phase of multiplication of stock, producers identified other valuable products in addition to meat. Each of the birds produces a unique hide which can provide valuable and durable leather products. Emus produce oil which is incorporated in cosmetics.

The success in building a production base has drawn the attention of manufacturing and marketing enterprises. New consumer products have emerged that apply the inherent qualities of the meat, leather, and oils harvested from these birds.

Ostrich, emus, and rheas have proven to be profitable additions to American agriculture. Each has demonstrated a capacity to grow and thrive on marginal lands in diverse climates with the potential to generate profit for the producer at commercial harvest prices. Producers know that domestication, current management practices, health care, nutrition, and breeding must improve to enhance performance.

Today, producers, nutritionists, geneticists, veterinarians, and scientists are cooperating to improve production parameters and contain costs consistent with an evolving produts-oriented commercial market. These efforts will provide answers to existing and future problems.

This text is a positive step in spreading reliable and verifiable information on ratites, raising the standard of management, and improving health care. The contributions of the 17 authors will increase the opportunity for commercial producers to be more successful and profitable.

Preface

This text represents a compilation of the knowledge and experience of veterinary educators and practitioners involved in ratite health and production. Despite a century of commercial production of ostriches in southern Africa, it is only during the past decade that structured scientific research and accepted clinical practices have been applied to ratites. The industry is on the threshold of expansion, having progressed from the stage of multiplication and speculative investment to emergence of an industry committed to the sale of valuable end-products.

Information in this text will apprise practitioners of the latest developments in the field of ratite medicine and will provide owners and farm managers with the background to enhance management of flocks of ostriches, emus, and rheas. The collation of knowledge and appraisal of published reports in the text will improve communication among the participants in the industry. The authors hope the material provided will dispel misinformation and bias which characterize any new and potentially profitable enterprise. Despite the unfounded promises at promotional meetings and the nonreviewed articles which appear in the less responsible ratite periodicals, there are no simple solutions to many of the production problems which exist. Some diseases and metabolic-environmental interactions will become more severe as the industry progresses to large production units.

Although the text is divided into topic-related chapters, it is emphasized that many of the problems facing producers involve overlays of nutrition, management, disease, environment, and genetics. Evaluating the relative contribution of these factors to a specific problem and developing appropriate corrective strategy requires sound knowledge, the ability to derive conclusions from observations, and to develop cost-effective solutions. Many of the principles contained in this text have been adapted from the poultry industry, which has achieved success through horizontal and vertical integration, and application of technology to supply the world's population with meat and eggs. It will be necessary for successful ratite producers and their advisors and health professionals to emulate the structure, organization, and high level of technical development of U.S. poultry producers who generate an ex-farm income approaching $20 billion annually.

The contributors to this text trust that the information provided will contribute to enhanced productivity and profitability. It is the editors' wish that advances in basic and clinical research in ratite medicine and management will complement the information which has been documented, adding to our collective knowledge. The editors would like to thank Gail Lungaro, Harry M. Cowgill, Michael L. Broussard, Kristine J. Raab, Jenise A. Coyne, and Joshua R. Pinkston for their help in preparing this manuscript for publication. The editors welcome suggestions and comments for inclusion in subsequent editions of this text.

Chapter 1

Clinical Anatomy of Ratites

Murray E. Fowler

INTRODUCTION

Ratites (ostrich, emu, cassowary, rhea, and kiwi) are flightless birds descended from flighted ancestors. They are classified in four different orders and five families. They share many of the evolutionary adaptations of other birds, but some anatomical characteristics are unique and will be noted in this chapter. Veterinarians dealing with management and propagation of ratites should have a basic understanding of ratite anatomy in order to safely handle these birds and to understand how to collect laboratory samples, administer medication, evaluate radiographs, perform surgery, and identify organs at necropsy.

GENERAL CHARACTERISTICS

Size and other general characteristics are listed in Table 1.1. The sutures of the ratite skull remain open throughout life. Male ratites have a protrusible phallus.

All the ratites have elongated necks and relatively long, heavily muscled legs adapted to running. All use their legs and feet in both defense and offense by thrusting forward with disabling or lethal effect from the larger species. An ostrich or cassowary is capable of causing severe injury, even disemboweling with one slashing blow. It is unwise to approach or work in front of these birds.

ORGAN SYSTEM ANATOMY

Integumentary System

Leather has been made from the skin of ostriches for centuries, and emu pelts are processed into garments by reattaching the feathers. Ratite feathers have no barbules to stabilize the filaments attached to the central shaft; they therefore resemble hair. The feathers of the emu and cassowary have a double shaft.

TABLE 1.1. Characteristics, Speed, and Distribution of Ratites

	Size				Body Color		No. of toes	Speed	Native Distribution
	Height Meters		Weight kg						
	♂	♀	♂	♀	♂	♀			
Ostrich	3.0	2.5	160	120	Black & White	Gray	2	64 kph/40 mph	Africa, So. of Sahara
Rhea, greater	1.7	1.5	25		Gray		Monomorphic 3		S.E. Brazil, Bolivia, Paraguay, Uruguay, N.E. Argentina
Rhea, Darwin's	1.1	1.1			Gray		Monomorphic 3		High plateau of Peru, Bolivia, Chile and grass land Patagonia
Emu	1.5	1.8	36–38	55	Grayish brown to		Monomorphic 3	48 kph/30 mph	Throughout Australia has a booming except east of dividing range
Cassowary	1.8	2.0	85		Glossy black		Monomorphic 3	48 kph/30 mph	Extreme N.E. Australia, and New Guinea tropical rain
Kiwi	.35	.55	0.67–4.0		Tan-Brown, may be mottled		Monomorphic 3	Can outrun a human or dog, 40–48 kph/25–30 mph	New Zealand

Feathers are not distributed uniformly over the surface of the skin, but are restricted to feather tracts (pterylae). The massive thighs of ostriches are devoid of feathers, but the legs of other ratites are feathered down to the tarsometatarsus. Callosities (dermal thickenings) are found at strategic locations of wear or pressure. Ratites may sit fully sternally recumbent or poised on the tarsometatarsus. When resting on the tarsometatarsus, the proximal end of the metatarsal bone is protected by a plantar metatarsal pad (1 cm thick × 5 × 12 in ostrich). Prominent callosities in the ostrich are located over the most ventral aspect of the sternum (1 × 8 × 11 cm) and over the bony prominence produced by the ventrad and craniad projection of the pubic bones (1 × 4 × 9 cm). These callosities bear the bird's weight when in sternal recumbency. Rheas and emus have only the sternal callosity (approximately 4 × 9 cm).

Skin on the plantar surface of the digits is modified for arid environments. In the ostrich, the dermal pad is thick and the epidermal surface is covered with tightly packed vertical rods (0.8 cm long) of cornified tissue (Figure 1.1). In other ratites, the plantar pads are simple callosities. There is a second pad over the tarsometatarsal-phalangeal articulation. Additional cushioning is provided by paired, tubular, deep plantar fat bodies enclosed by a fibrous capsule. This tissue is similar to the digital cushion of horses or the bulb of the heel of goats and sheep.

The large digit of the ostrich has a prominent blunted toenail. The toenail on the small digit is less well formed. Toenails of other ratites are sharper. The medial toenail of the cassowary is a spike 12 cm long, especially adapted for fighting.

Unique to the cassowary is a bony, horny casque on the forehead, perhaps an adaptation to movement through foliage in its dense rain-forest habitat.

Muscular System

Except for the substantial thigh muscles, there is a dearth of tissue for intramuscular injections. The major pectoral muscles are either absent or vestigial. Ostriches in particular use the wings in behavioral displays but the muscles involved are diminutive. The ventral midline area of the abdominal wall consists of the aponeuroses of the abdominal muscles. There is no muscle tissue for 19 cm on either side of the linea alba in the ostrich. In both the rhea and the emu, the rectus abdominus muscle forms a thin sheet on either side of the linea alba, extending for about 13 cm caudally from the sternum. Otherwise, muscle fibers are found only dorsally, and the ventral abdomen is supported by a tunic.

An incision made along the linea alba penetrates the skin, subcutaneous fat (minimal) and a dense fibrous

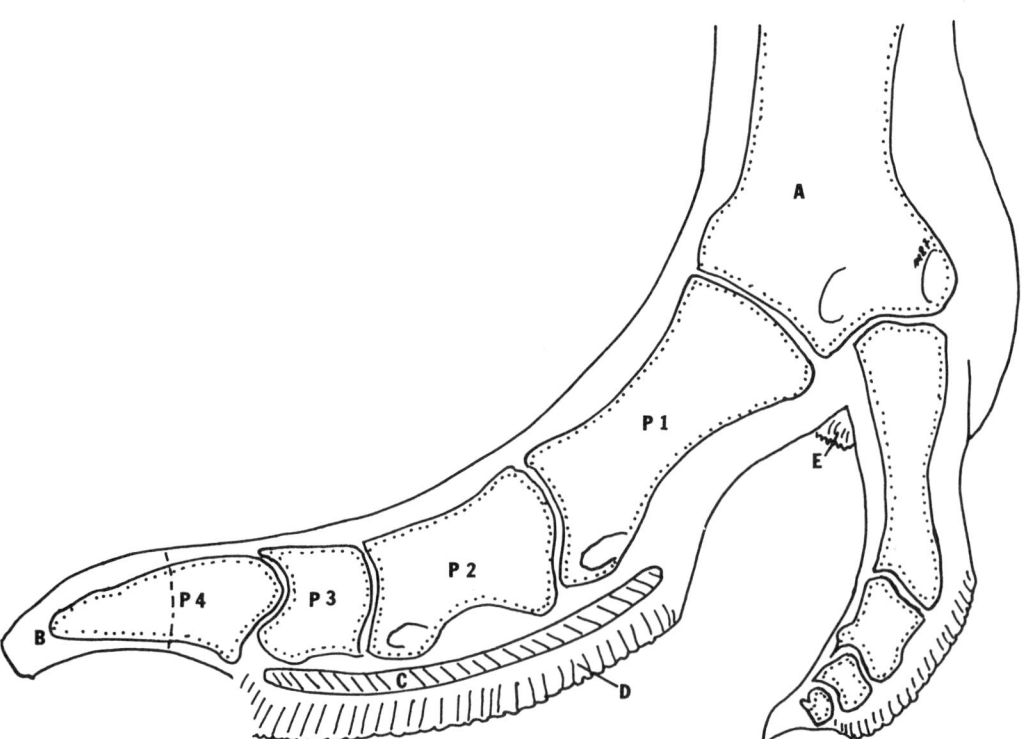

Figure 1.1 Diagram of the structures of the foot of an ostrich. A. tarsometatarsal bone, B. toenail, C. digital cushion, D. phalangeal pad, E. metatarsal phalangeal pad and P-1–4 phalangeal bones.

Clinical Anatomy of Ratites

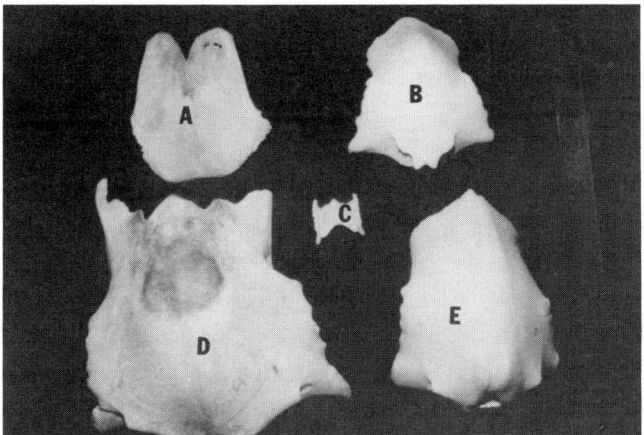

Figure 1.2 Sternum of ratites. A. rhea, B. emu, C. kiwi, D. ostrich and E. cassowary.

abdominal tunic. Next is a retroperitoneal fat layer that may be 2–8 cm thick. This layer is most pronounced in emus and is a source of oil. When a laparotomy is performed, the bulk of this adipose tissue should be peeled away prior to closing the body wall. Finally, the parietal peritoneum is penetrated.

Skeletal System [3,4]

The group name ratite (Latin ratis = raft) is derived from the unique sternum that is concave dorsally and convex ventrally, somewhat like a soup plate (Figure 1.2). There is no keel and the ventral surface is devoid of muscles. The thoracic girdle is modified compared to other avian orders since there is no need for flight. In the ostrich, the scapula, coracoid, and clavicle are fused in the adult bird and attached to the cranial sternum (Figure 1.3). The ostrich and rhea have relatively large wings, while those of the emu and cassowary are rudimentary.

The pelvic girdle is characterized by ilia that form an inverted osseous shield over the top of the fused vertebrae (synsacrum). In the ostrich, the ischial and pubic bones project caudad to fuse and then turn ventrad and craniad to form a pubic symphysis (Figure 1.4). In other ratites, the ischial and pubic bones are separate and there is no pubic symphysis.

The patella is absent in ratites. In its place, in the ostrich, there may be a small bone in the tendon of insertion of the muscle on the cnemial crest of the tibiotarsus. This crest provides extra leverage for quick, forceful forward movement of the leg [3,4] in running or swimming.

The tibiotarsal-tarsometatarsal articulation may be confused with the stifle. In the ostrich and emu, one of the tarsal bones remains unfused to the contiguous bones and its location gives the appearance of a patella. The free tarsal bone is absent in other ratites. The

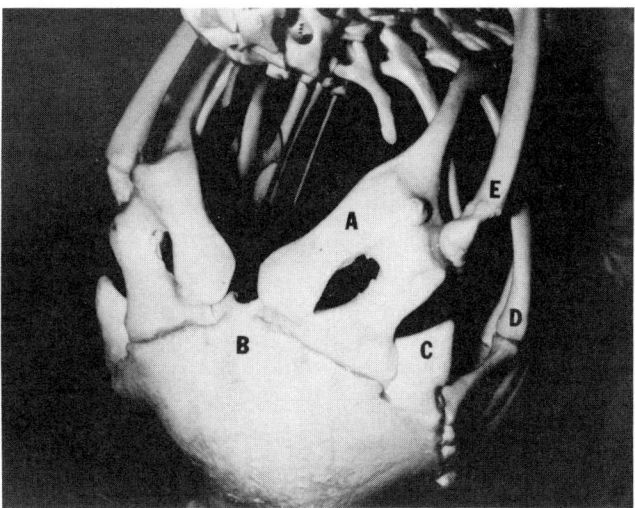

Figure 1.3 Thoracic girdle and sternum of the ostrich. A. fused coracoid and scapula, B. sternum, C. cartilaginous extension of the sternum, D. rib and E. humerus.

ostrich has two digits (3 and 4), and other ratites have three (2, 3, and 4). There are four phalanges on each digit, differentiating ratites from other birds.

Radiographic Characteristics

The thoracic girdle, sternum and pelvic areas are complex. The synsacrum shields all the dorsal structures. In the ostrich, the pubic bones are fused to form a pubic symphysis, but in other ratites the pubis is open as in other birds. Because ratites are flightless, there is little need for reducing the mass of bones by pneumatization. An exception is the femur of the ostrich. Radiographs of the femorotibial articulation (stifle) are difficult to evaluate. A proper exposure for the femur will be inadequate for the tibia and vice versa, thus at least two different exposures are necessary.

Digestive System

The digestive system is basically similar to that of other plant-eating birds [2–4]. Within the ratites there are differences based primarily on the forages available in their native habitat which influenced the evolution of digestive strategy to process feed. Only anatomical characteristics important for surgery or understanding feeding and nutrition are discussed in this review. Source references may be consulted for additional details.

The esophagus generally traverses the right side of the neck, but is moveable. The esophagus has a markedly expansible diameter and when contracted, contains numerous longitudinal rugae. The crop is absent in all ratites. The esophagus enters the proventriculus within the thoracic cavity.

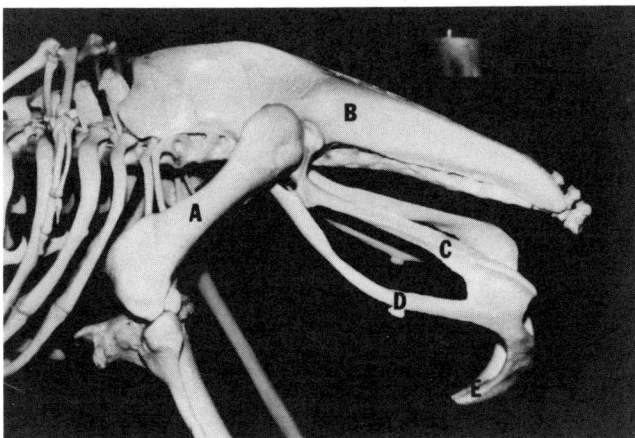

Figure 1.4 Pelvic girdle of an adult ostrich. A. femur, B. synsacrum, C. ischium, D. pubic bone and E. pubic symphysis.

TABLE 1.2. Comparative Length of the Intestines of Ratities

	Ostrich	Emu	Rhea
Small intestine	36	94	62
Caeca	7	2	21
Rectum (colon)	57	4	17

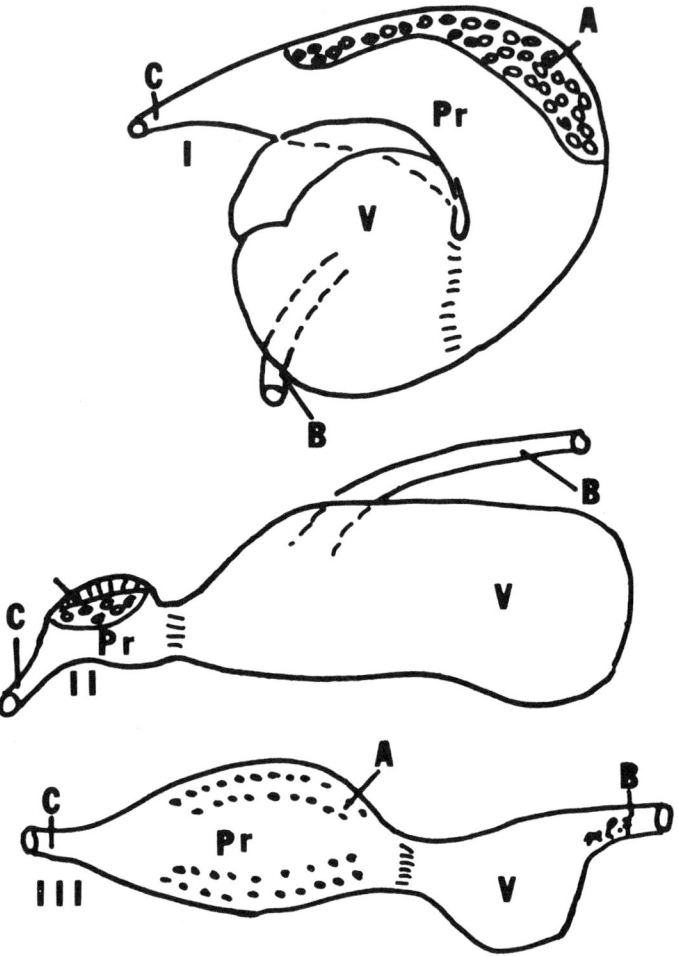

Figure 1.5 Diagrams of the stomach of ratites. I. Ostrich, II. rhea and III. emu and cassowary. A. glandular area of the proventriculus, B. duodenum, C. esophagus, Pr. proventriculus and V. ventriculus.

Comparative dimensions of the gastrointestinal tract are documented in Table 1.2. The proventriculus of the ostrich is a large, dilated, thin-walled structure (Figure 1.5-I). In contrast with most other birds in which the entire inner surface of the proventriculus produces digestive enzymes, the secretory function is restricted to a patch on the greater curvature. The distal extremity of the proventriculus of the ostrich passes dorsal to the ventriculus (gizzard) and empties into this organ on the caudal aspect (Figure 1.5). This arrangement does not occur in any other species of birds. The opening between the proventriculus and ventriculus is large, making it possible to extract ventricular foreign bodies via an incision in the proventriculus.

The ventriculus of the ostrich is a thick-walled structure similar to the ventriculus of seed-eating birds. The ventriculus is situated slightly to the left of the midline at the caudal border of the sternum (Figures 1.6 & 1.7). Though both the proventriculus and ventriculus may normally contain small stones, impaction is common, particularly in juveniles.

The dark, tough lining of the ventriculus (koilin) and proventriculus of birds is formed by protein secreted from the glands combined with entrapped desquamated cells and cellular debris. The green or brown color arises from refluxed bile pigments from the duodenum. This normal membrane should not be identified as an abnormal structure at necropsy.

In the rhea, the proventriculus is a small dilated structure cranial to the ventriculus. The secretory area is limited to a thickened, dorsally located patch (Figure 1.5-II). The ventriculus is elongated. The stomach of the rhea is situated slightly to the left of the midline, dorsal to the sternal notch, but more caudal to the sternum than in the ostrich. In the emu and cassowary, the proventriculus is large and spindle-shaped (Figure 1.5-III), and the ventriculus is slightly larger and less heavily muscled. The stomach of the cassowary lacks a koilin membrane. The pyloric orifice to the small intestine is on the right side of the ventriculus in all species.

In the ostrich, the folds of the small intestine occupy the left mid to caudal abdomen (Figure 1.7). In the rhea, the small intestine is relatively short and straight. The ileum is easily located within the mesenteric attachment between the paired caeca in the ostrich and rhea. The ileum enters the large intestine at an

Clinical Anatomy of Ratites

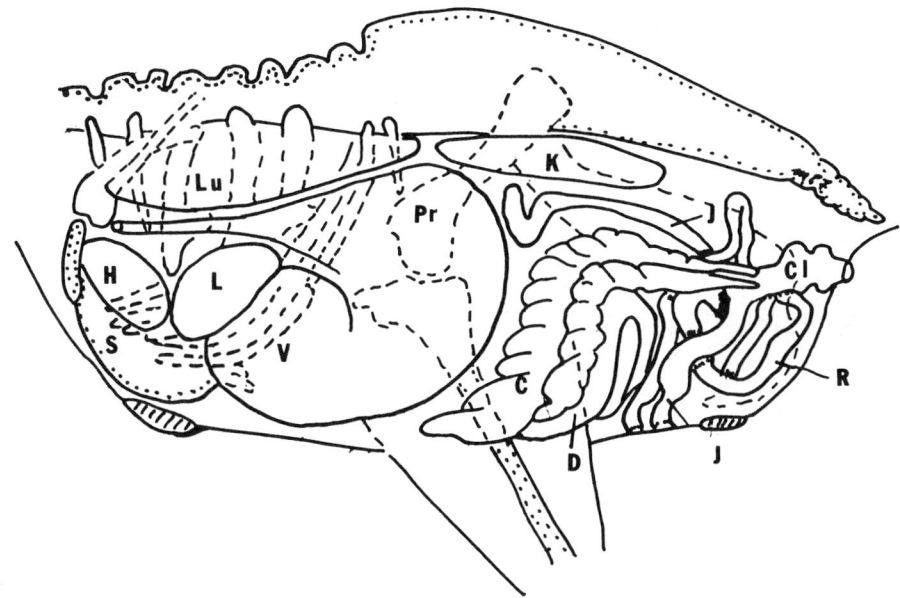

Figure 1.6 Diagrammatic illustration of a left lateral view of the thoraco-abdominal viscera of an ostrich. Lu. lung, H. heart, S. sternum, L. liver, V. ventriculus, Pr. proventriculus, K. kidney, J. jejunum, C. cecum, D. duodenum, R. rectum and Cl. cloaca.

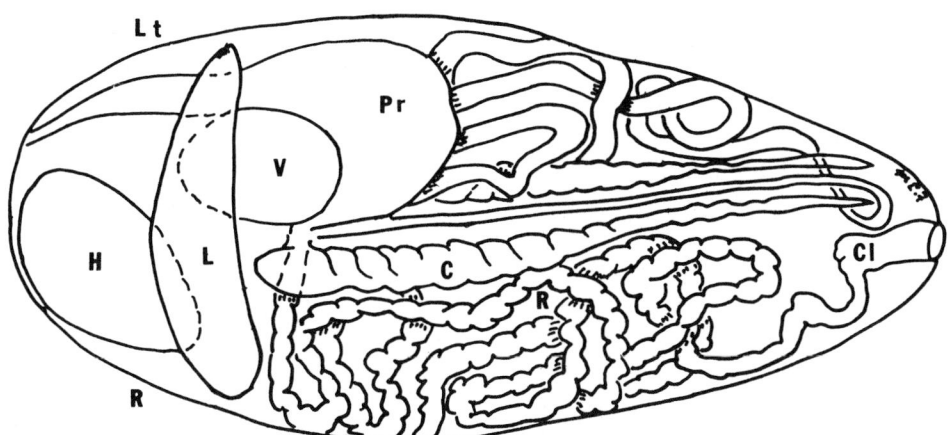

Figure 1.7 Diagrammatic illustration of the thoraco-abdominal viscera of an ostrich in a dorsoventral view. H. heart, L. liver, V. ventriculus, Pr. proventriculus, C. cecae, I. ileum, D. duodenum, J. jejunum, R. rectum and Cl. cloaca.

ileocecorectal junction. The small intestine of the emu occupies most of the abdomen caudal to the ventriculus. The comparison of ostrich and emu digestive systems is illustrated in Figure 1.8. The comparative length of the small intestine in emus (Table 1.2) may indicate an evolutionary adaptation to more nutritive forage.

The caeca are paired in ratites. In the ostrich, the elongated caeca are visible upon initial incision of the ventral midline abdominal wall and extend diagonally from right to left in a caudal direction. Spiral folds within the lumen of the caeca of the ostrich and rhea produce a sacculated appearance. The caeca of the emu and cassowary are vestigial and nonfunctional (Figure 1.8).

The rectum (large intestine) of the ostrich is voluminous and occupies the caudal right abdomen (Figure 1.7). A long, voluminous rectum may be necessary to digest fibrous feed and absorb fluid. The rectum enters the cloaca via a rectal pouch into the coprodeum (Figure 1.9). The rectal pouch is separated from the coprodeum by a rectocoprodeal fold. The coprodeum is a large dilated sac, which may be covered by a dark tough membrane, similar to koilin. The coprodeum and urodeum are partially separated by a coprourodeal fold. The urodeum is short but

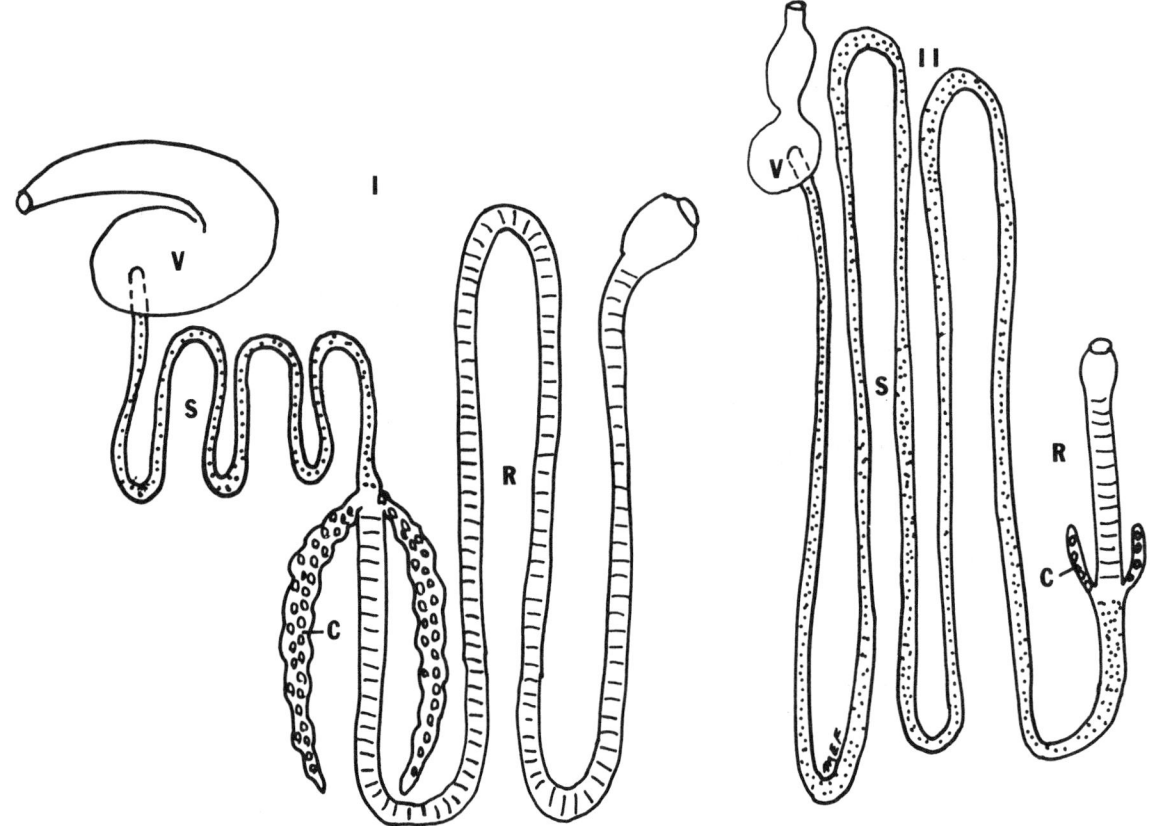

Figure 1.8 Diagram comparing ostrich and emu digestive tracts. I. Ostrich, II. emu, V. ventriculus, S. small intestines, C. caeca and R. rectum.

within this compartment are orifices for the ureters, oviduct of the female and vas deferens of the male. A uroproctodeal fold partially separates the urodeum from the proctodeum through which all excretions pass via the vent to the exterior. The bursa of Fabricius is located on the dorsum of the proctodeum, but much of the structure is distributed as gut-associated lymphoid tissue. The general structure of the ostrich cloaca is illustrated in Figure 1.9 [4].

The liver is cranial to the ventriculus and caudal to the transverse membrane (primitive diaphragm). There is no gallbladder in the ostrich, but one is present in the rhea and the emu [10].

Respiratory System [3,4]

The larynx of ratites is well developed. These birds are easily intubated because the glottis is large and readily accessible when the mouth is held open and the tongue pulled forward. The trachea has complete but flexible cartilaginous tracheal rings. The size of an endotracheal tube necessary for anesthesia is easily determined by palpation of the trachea. An 18 mm (ID) endotracheal tube is suitable for an adult ostrich.

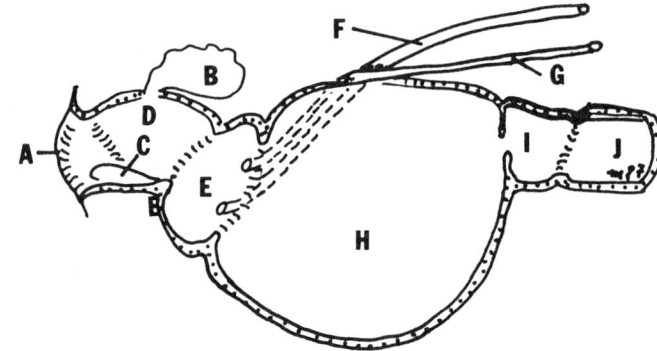

Figure 1.9 Diagram of right lateral view of the cloaca of an ostrich. A. vent, B. bursa of Fabricius, C. genital eminence, D. proctodeum, E. urodeum, F. ureter, G. genital duct, H. coprodeum, I. rectal pouch and J. rectum.

In the emu, the complete cartilaginous rings are interrupted by a 6–8 cm long cleft on the ventral surface of the trachea 10–15 cm cranial to the thoracic inlet (Figure 1.10) [4]. In the chick, a thin membrane covers the cleft, but as the bird matures, an expandable pouch (30 cm long) forms cranial to the cleft. When air is directed into the pouch, the skin of the neck expands

Clinical Anatomy of Ratites

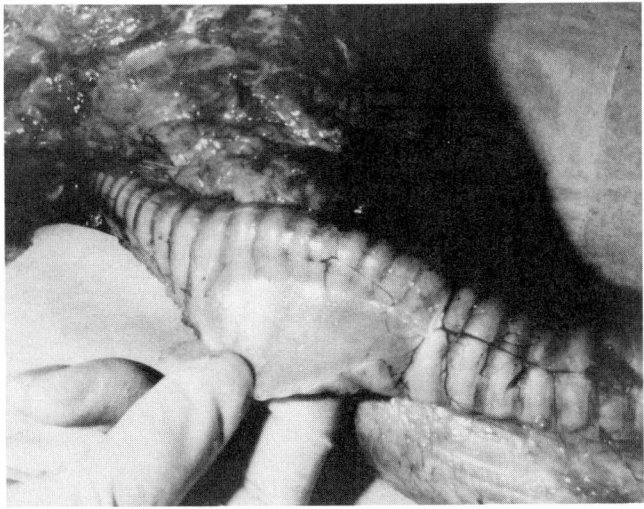

Figure 1.10 The tracheal cleft and diverticulum of an adult emu. Head to the right.

and a booming sound is produced by the female and a growling sound by the male. In mature emus, the presence of the expandable pouch may complicate inhalation anesthesia. If positive pressure is used to inflate the airsacs and lungs, air will inflate the pouch. This can be prevented by wrapping the lower neck with an elastic bandage (Vetrap, Animal Care Products—3M, St. Paul, MN 55144), taking care to avoid compression of vessels. Lungs and air sac anatomy are similar to flighted birds [2]. The femur is the only pneumatized bone in the ostrich and emu [4].

The respiratory rate of an adult ostrich in a mild ambient temperature varies from 6–12/minute. If the bird is subjected to thermal stress, panting may increase the rate to 40–60/minute. Studies have shown that an ostrich can maintain a body temperature of 38–40°C, even when ambient temperature reaches 50°C [35]. Body temperature is maintained by evaporative (adiabatic) cooling from the trachea, air sacs and the gular area of the pharynx. Ostriches do not become alkalotic when panting because air is shunted through the primary bronchi and air sacs, bypassing the lungs.

Circulatory System [1,5]

Locating superficial veins is important when obtaining blood samples and for intravenous medication. The jugular vein is a primary site, and similar to other avian species, the right jugular vein is larger than the left. Other sites for blood collection include the brachial and medial metatarsal veins. The brachial vein is easily identified on the featherless craniomedial aspect of the humerus [1]. This vein is suitable for insertion of an intravenous catheter. The medial metatarsal vein is readily accessible only in the sedated or immobilized bird.

Ratites have a renal portal system similar to other birds and reptiles [4,5]. Though the precise physiologic function of the system is unknown, it is established that the flow of venous blood through the kidney is under the control of the autonomic nervous system. There is considerable controversy concerning the clinical significance of the renal portal system. Caution should be exercised when administering nephrotoxic drugs such as aminoglycoside antibiotics which may reach the kidney in high concentration if administered into the muscles of the leg. Drugs excreted through the kidney tubules such as Ketamine (Vetalar, Parke Davis) may be excreted before reaching the general circulation, precluding the desired effect.

The shape and location of the spleen is important for evaluating radiographs, at surgery and at necropsy. In the ostrich the spleen is oval and elongated in cross section (1.5 × 3 × 9 cm). The spleen is situated on the right side dorsal and lateral to the proventriculus. The location is similar for the rhea, but the shape resembles a bent cylinder, 1 × 5 cm. The spleen of the emu is cylindrical. The spleen of the cassowary is flattened and shaped like an irregular polygon [3,4].

Reproductive System [3,4]

As in other avian species, only the left ovary develops. In the mature ostrich a flattened cluster of ova (varying in diameter from 1 to 8 cm) is present in the sexually active hen. The fan-shaped infundibulum continues as a thin-walled oviduct, approximately 118 cm long and 3 cm in diameter, in the ostrich. The uterus is a thick-walled expanded segment of the genital tract. The genital duct continues on to the urodeum of the cloaca. The paired testes are tan in the ostrich, rhea (1 × 1 × 4 cm) and cassowary, but black in the emu. The testes enlarge during the breeding season.

The comparative anatomy of the reproductive systems of the four ratite species is summarized in Table 1.3.

Ratites have an intromittent organ commonly called a phallus [2-4]. Although this structure is analogous to the mammalian penis, the organs are not homologous. There is no urethra in the avian phallus, and it does not have the urinary function of the mammalian penis.

Ostriches and kiwis have an intromittent phallus with no internal cavity (Figure 1.11). When relaxed, it is folded on the ventrum of the proctodeum of the cloaca. A phallic sulcus is located on the dorsum of the phallus. The rhea, emu, and cassowary have an

TABLE 1.3. Sex Differentiation in Ratites

SPECIES	CHICK		ADULT	
	Male	Female	Male	Female
Ostrich	Phallus round in cross section. Phallic sulcus dorsal, 1–4 cm long.	Phallus flat in cross section, 0.5–1 cm long.	Black plumage, Larger than ♀. Phallus, relaxed 20 cm, erect 40 cm, curved.	Gray to brown plumage. Phallus on genital mound, flattened.
Emu	Phallus. Hollow tube 5–10 mm long. Spirals as bird ages.	Genital mound, slight phallus.	Shorter than ♀. Lighter plumage. Phallus spiralled, hollow tube similar to rhea.	Darker plumage than ♂. Usually the "boomer."
Rhea	Phallus similar to emu, but more elongated.	Genital mound, no phallus.	Phallus spiralled, 3–7 cm relaxed, 8–12 cm erect. Round cross section. Hollow tube and phallic sulcus.	Genital mound.
Cassowary	Phallus 5–10 mm triangular.	Small genital mound.	Triangular shaped phallus pointed caudally.	Usually much larger than ♂.

intromittent phallus with a cavity and a partially inverted sleeve that everts during erection (Figure 1.12). The tubular structure gives the appearance of a urethra, but it does not communicate with the urinary tract. A phallic sulcus directs the semen into the cloaca of the female.

Reproductive Organs of Adult Ratites

Ostrich [3,4].

The phallus of the adult male ostrich is irregularly round in cross section, and lies folded in a wide pocket on the floor of the proctodeum when flaccid (20 cm long).

The phallus is so bulky that it may occupy most of the proctodeum and protrude from the vent in order to allow defecation and urination. When erect, the bright red phallus (40 cm long) projects from the cloaca in a ventrocranial curve, with the phallic sulcus on the dorsum at the base (Figure 1.13). Because of the asymmetry of the fibrolymphatic bodies, the erect phallus tends to deviate to the left. The exact mechanism for stimulation of erection is unknown.

The female ostrich has a diminutive phallus (3 cm long) which projects from a genital mound or eminence on the floor of the proctodeum. There may be a minimal groove on the dorsal surface. The female phallus is differentiated from the male phallus by being flattened in cross section and is straight.

Kiwi.

The retracted adult male phallus lies on the floor of the proctodeum and is triangular in shape with the apex pointed caudally.

Rhea, Emu, and Cassowary.

The flaccid male phallus projects slightly from the ventral wall of the proctodeum. The remainder of the

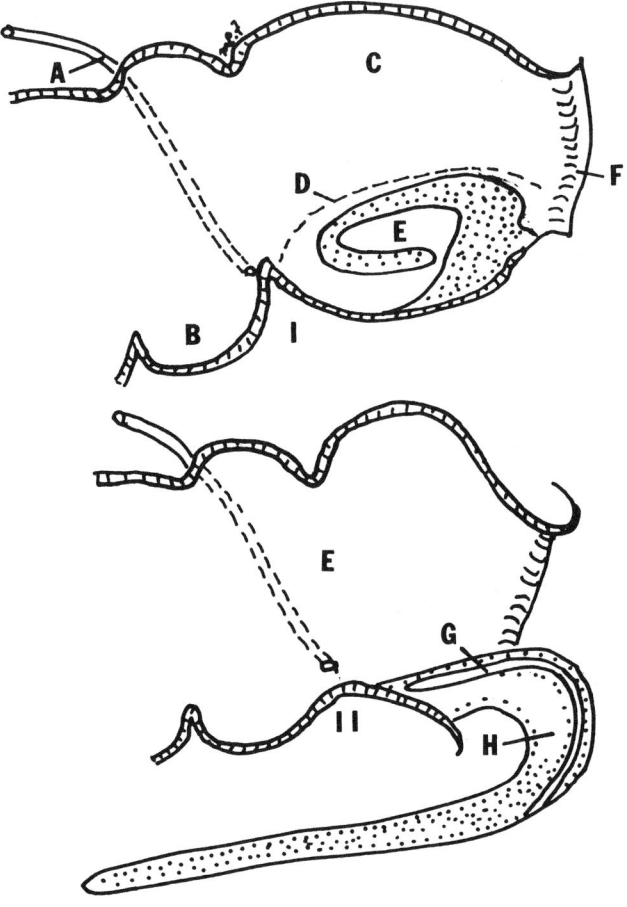

Figure 1.11 Diagram of left lateral view of the cloaca and phallus of an adult male ostrich. A. vas deferens, B. urodeum, C. proctodeum, D. uncovered crypt on the floor of the proctodeum, E. retracted phallus, F. vent, G. dorsal sulcus, and H. erect phallus.

phallus (a blind hollow tube, like the finger of a glove) lies within a pocket beneath the mucosa of the proctodeum (Figure 1.12). Approximately half of this blind cavity everts during erection to extend the phallus; however, an attachment at the base of the blind pouch prevents full eversion. The phallic sulcus lies on the dorsum of the organ. The everted, but relaxed phallus of a rhea is 1.4 × 7 cm long. This phallic anatomy makes vent sexing more difficult in the emu because it is not as easy to insert a finger to extrude the phallus as in the ostrich. Instead, the finger must be inserted deeply into the proctodeum and pressure exerted on the floor to force extrusion of the phallus as the finger is withdrawn. The adult female emu has a slight prominence on the genital mound.

Phallus of Chicks and Juvenile Ratites

Experienced clinicians will become skilled at identifying the sex of immature ratites by observation or palpation of the cloaca. However, even the experienced clinician may occasionally misidentify the sex.

Ostrich.

In the female juvenile, the phallus is flattened, with only a trace of a groove on the dorsal surface. The male phallus is irregularly rounded in cross section with a prominent groove on the dorsal surface. The dimensions of the phallus of a 7-month-old 53-kg juvenile was 1.0 × 1.5 × 3.0 cm. The crura of the phallic bodies were prominent on the floor of the proctodeum.

Emu.

The female has a slight protuberance on the ventral proctodeum. In two 5-week-old 2.95-kg males, the phallus was 2 mm in diameter, 5 mm long, slightly spiraled, and round in cross section. The phallic sulcus is barely visible.

Rhea.

The rhea has a phallus similar in size and shape to the emu.

Cassowary.

The female has no phallus. The juvenile male phallus is similar in shape to the adult but diminutive in size.

Additional reproductive data for ratites are listed in Table 1.3.

Organs of Special Sense

Ratites have excellent vision. Ostriches have the largest vertebrate eye relative to body size. The sense of hearing is acute, and the external ear orifice is located caudal to the eye.

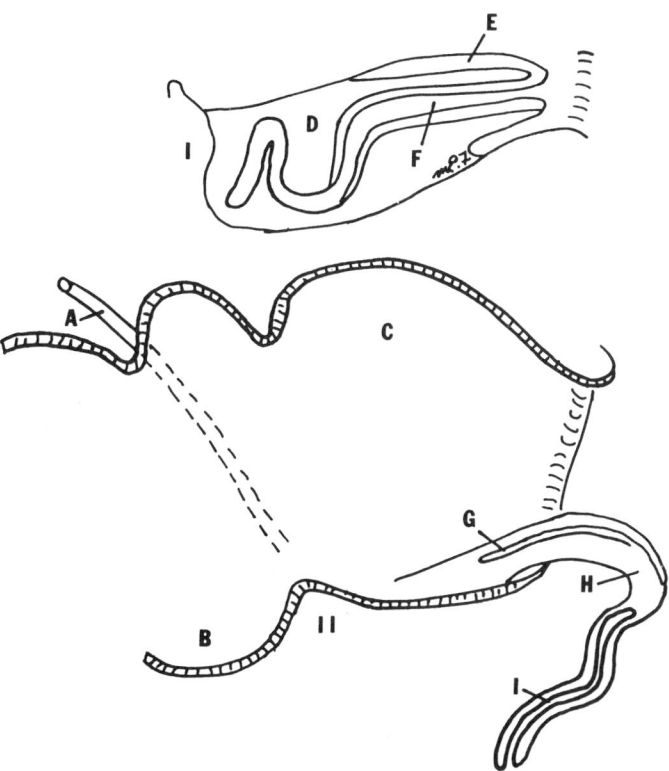

Figure 1.12 Diagram of left lateral view of a retracted and erect phallus of a male emu or rhea. Note: The top drawing represents the phallus within the pouch. A. vas deferens, B. urodeum, C. proctodeum, D. pocket to contain phallus, E. erectile wall of phallus, F. inverted hollow tube of phallus, G. phallic sulcus, H. erectile tissue, and I. erect phallus with blind hollow tube.

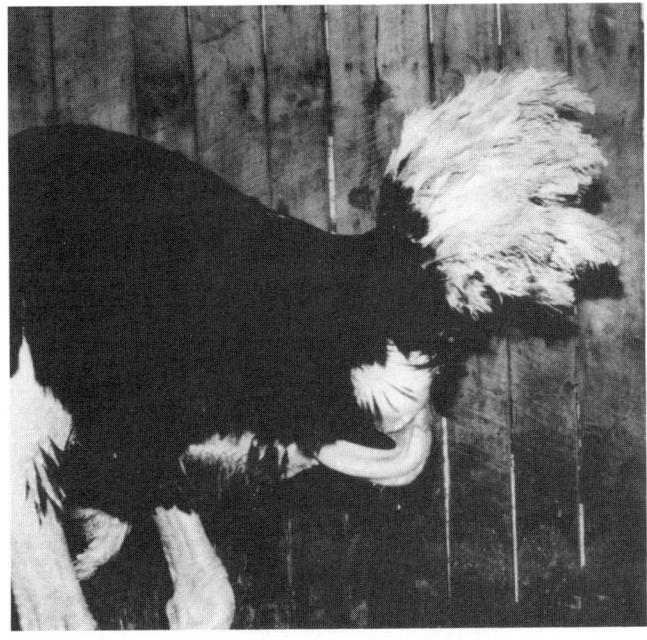

Figure 1.13 Erect phallus of an adult male ostrich.

Nervous System

The nervous system of the ratite has no unique characteristics. While ostriches can lie in sternal recumbency for days, lateral recumbency for an hour may result in peroneal nerve paralysis. Heavy muscles over the proximal tibiotarsal bone produce a significant bulge, and the weight of the bird pressing on this bulge may cause direct pressure on the peroneal nerve or produce ischemic edema. Adequate padding in the form of an air mattress or foam pad should be used when adult or large juvenile birds are subjected to general anesthesia.

ACKNOWLEDGMENT

Permission has been granted by the American Association of Zoo Veterinarians for liberal use of illustrations and textual material from a paper by the author on the Comparative Clinical Anatomy of Ratites which appeared in the *Journal of Zoo and Wildlife Medicine* 1991;22(2):204–227.

REFERENCES

1. Bezuidenhout AJ and Coetzer DJ. The major blood vessels of the wing of the ostrich. *Onderstepoort Jour Vet Res* 1986;53:201–203.
2. Bezuidenhout AJ. The topography of the thoraco-abdominal viscera in the ostrich (*Struthio camelus*). *Onderstepoort Jour Vet Res* 1986;53:111–117.
3. Cho P, Brown R, Anderson M. Comparative gross anatomy of ratites. *Zoo Biology* 1984;3:133–144.
4. Fowler ME. Comparative Clinical Anatomy of Ratites, *J Zoo Wildl Med* 1991;22:204–227.
5. Oelafsen BW. The renal portal valves of the ostrich, *Struthio camelus*. *So African J Sci* 1977;73:57–58.

Chapter 2

Ratite Nutrition

C. Roselina Angel, Sheila E. Scheideler, and Jerry L. Sell

INTRODUCTION

This chapter summarizes key areas related to nutrition of ratites. Information on ingredients and nutrients, the gastrointestinal tract, and neonatal development are provided as a basis for understanding the practical aspects of feeding of ostriches, emus, and rheas. Given the limited information on ratite nutrition available in the scientific literature, data for other relevant species are adapted and used where applicable. Present knowledge of nutrient requirements, based on published information, is included for the ostrich, emu, and rhea, the ratites of commercial importance. The information in this chapter should be considered a summary of basic nutrition and its application to ratites.

POSTHATCH DEVELOPMENT OF THE GASTROINTESTINAL TRACT AND EARLY NUTRITION

Little information is available concerning the development of the gastrointestinal tract of ratites at the time of hatching. It seems reasonable that their status would generally correspond to that of other precocial avians such as chickens and turkeys, although the chronology of post-hatch development may differ among these species. An overview of information on chickens and turkeys is therefore provided, with the assumption that it is applicable to ratites.

The physical and functional development of the gastrointestinal tract of chickens and turkeys is advanced at the time of hatching. The glandular tissue of the true stomach (proventriculus) and gizzard is fully functional in terms of secretory activity. The intestinal tract is developed at hatch with the epithelial tissue lining the lumen arranged in many folds with relatively long villi. The surface area available for intestinal digestion and absorption is extensive, and development of smooth muscle to support motility and movement of ingesta is evident. Several functional components of the gastrointestinal tract and the pancreas are not fully developed in chicks and poults at hatch. The pancreatic enzymes amylase, trypsin, and lipase, responsible for efficient digestion of dietary starches, proteins, and fats, respectively, show low activity at hatch. In the instance of amylase and trypsin, activity increases substantially during the first 10 days after hatching. Lipase activity increases more slowly, requiring 21 or more days to reach optimum. Concurrent with relatively low pancreatic lipase activity, the secretion of bile into the intestinal tract of chicks and poults seems to be inadequate to support efficient digestion of fat.

Activities of digestive enzymes associated with the intestinal wall are relatively high at hatch but decrease markedly in chicks and poults placed in a conventional environment. The primary reason for decrease activities of intestinal digestive enzymes, such as maltase, sucrase and the aminopeptidases, is related to erosion of the intestinal epithelium following exposure to microorganisms in the environment. In chicks and poults, reduced activities of intestinal enzymes normally lasts through 5 to 9 days posthatch with a rapid increase thereafter concurrent with intestinal repair.

As a consequence of the relatively low activities or decreases in activities of digestive enzymes during the first 7 to 21 days after hatching, utilization of dietary nutrients by chicks and poults is impaired. The digestibility coefficient of dietary protein for young turkeys has been shown to increase from approximately 28% at 4 days to 53% at 7 days posthatch. Fat digestion in chicks was reported to be only 68% at 6 days but was 83% at 14 days posthatch. Data obtained from one experiment with ostriches showed that nutrient utilization by ratites is also inefficient during early development. Digestibilities of neutral detergent fiber (NDF) and dietary fat by 3-week-old ostriches were 6 and 44%, respectively. By 6 weeks posthatch, the digestibility of these dietary components increased to 27 and 74%, respectively. It is important to note that the digestibility of NDF by ostriches continued to increase through 17 wks of age, reaching 58%. Overall utilization of dietary energy increased proportionately. Utilization of fibrous components of ostrich diets, represented by NDF, depends on microbial fermentation in the large intestine and ceca. The establishment of an appropriate flora in the intestinal tract is an important part of "maturation" in ostriches and other ratites that depends on products of microbial fermentation of complex carbohydrates for energy.

In general, physical and functional immaturity of the gastrointestinal tract during the early posthatch period makes young ratites particularly vulnerable to numerous maladies, including nutrient deficiencies and diseases of microbial origin. Adequate nutrition and management are critical during this time.

THE GASTROINTESTINAL TRACT OF RATITES

Certain similarities exist between the gastrointestinal tract of chickens and the three commercially significant ratite species. Ratites do not have a crop which is a food-storage organ in poultry. The large proventriculus and gizzard in ostriches, and to a lesser extent in the emu, and the very large gizzard in the rhea, may store food.

The proventriculus or true stomach is the first digestive organ in ratites. Acid, as well as a pepsin precursor, are secreted into the lumen. Acidic conditions activate pepsin and the initial stages of protein digestion begin. The pH of the proventriculus and gizzard in the emu were measured at 2.8 and 2.5, respectively [1]. The ostrich has a larger proventriculus in proportion to body size than the chicken, emu, and rhea. The rhea has the smallest proventriculus in relation to body weight. High levels (158.8 mM) of volatile fatty acids in the proventriculus and gizzard (139.3 mM) of ostriches indicate that foregut fermentative digestion occurs in ostriches [2]. The significance of this fermentation and the contribution to energy requirements has not been determined.

The avian ventriculus (gizzard) prepares feed for digestion by grinding ingested particles to a size small enough to pass through to the duodenum (proximal small intestine). The gizzard in the ostrich is thick walled and has well-developed muscles. The extent of muscle development is partly related to the presence of hard particulate material in the lumen that promotes muscle contractions and results in greater muscle development. The gizzard in emus is similar in size to the proventriculus. The gizzard wall is thinner and less muscular in emus than in ostriches. The gizzard in the rhea is very large in comparison to the proventriculus and the musculature is not as well developed as in the ostrich.

TABLE 2.1. Comparative Length of Ratite Intestines

Length	Ostrich[1] cm	%[1]	Emu[2] cm	%	Rhea[3] cm	%
Small Intestine	512	36	315	88	132	63
Cecum	94	7	12	3	46	22
Colon	800	57	29	8	31	14

Sources:
[1] Fowler [3]
[2] Herd and Dawson [1]
[3] Cho et al. [4]

As in the domestic fowl, the duodenum in ratites forms a loop with the pancreas lying between the descending and ascending segments. In the ostrich, a section comprising 20% of the ascending portion of the loop deviates from the pancreas to form a secondary loop of unknown significance. Digestive enzymes (trypsin, lipase, amylase) have been identified and characterized in the ostrich, but not quantified. These enzymes enter the intestinal lumen at the distal portion of the duodenum. Most of the digestion and absorption occurs in the small intestine. In ostriches, the small intestine constitutes 36% of the total length of the intestinal portion of the gastrointestinal tract (Table 2.1). In contrast, the small intestine of emus, rheas and chickens comprises 88, 63 and 90% of the intestinal tract, respectively.

The length of the intestine differs among ratites (Table 2.1). The emu has the shortest intestine in proportion to body size. The total length of the emu intestinal tract in proportion to body weight is approximately 30% of the length in the chicken. The ceca in ratites are paired. In proportion to body size, the rhea has the longest ceca, the ostrich has the longest colon, and the emu has the shortest relative length of both colon and ceca. The proximal portion of the colon in ostriches is sacculated and has a larger internal diameter than the distal colon. Fermentation of fiber occurs in the proximal colon with the distal colon being active in water resorption.

Passage rate or retention time varies with age and species. In ostriches, the retention time in young birds (6.8 kg) is 38 hours and in older birds (48 kg), 48 hours [5]. In contrast, the passage rate in emus varies between 3 hours and 48 hours, depending on the composition of the ingesta. In the emu, a selective separation of large particles, occurs first in the proventriculus and gizzard, and subsequently, the larger fibrous particles are selectively retained in the distal ileum [6].

WATER

An adequate supply of water of acceptable quality is essential for the health and well-being of ratites. Water is the nutrient required in greated amounts and a restriction of water intake results in a marked reduction in growth, poor feed efficiency, and imparied reproduction [7]. Poultry are adversely affected by as little as 12 hours of water deprivation. Ostriches, 4 to 6 months old, deprived of water for 24 hours reduced feed intake by 45%. A further decline of 67% was observed during 24 to 48 hours of water deprivation, and body weight decreased 31% by the end of 48 hours [8].

The importance of water is indicated by its prevalence in the body. Water constitutes 50 to 85% of body weight, depending on age and body condition of animals, and is a major component of every cell of the body.

Water possesses many chemical and physical properties that make it a very effective participant in and facilitator of virtually all body processes.

1. Water is an excellent solvent, has a high dielectric constant, and low viscosity. Consequently, water solubilizes many substances and transports them within the body. Processes dependent on the properties of water include nutrient absorption from the gastrointestinal tract, transport of nutrients within the body, and elimination of waste products of metabolism in the urine and feces. Water also provides the medium and chemical entities required for numerous chemical reactions such as digestion of carbohydrates, fats, and protein. For example the digestive breakdown of proteins into their constituent amino acids requires one molecule of water for each amino acid released.

2. Water has a high specific heat and a change in state from liquid to vapor requires input of energy in the form of heat. The combination of these two properties facilitates cooling of the body. This is especially important for ratites, which depend on increased respiration rate to dissipate excess body heat by evaporation of water from the lungs and upper respiratory tract.

3. Water provides a medium in which ionic homeostasis essential for life, can be maintained.

4. Water contributes to cell turgidity and body structure, and function.

Ratites should receive adequate quantities of water. Water is obtained from metabolism of nutrients (metabolic water), from the diet and from drinking water, which is quantitatively the most important source. Drinking water must be readily available at all times to ensure adequate consumption. The amounts of water needed will depend on a number of factors, including body size and activity of the birds, ambient temperature and humidity, composition and texture of the feed, and health status. Although research data on water intake of ratites are limited, available information shows that ostriches consume about 2 to 3 times as much water as feed, on a weight basis. A study on water and feed intake in ostriches from hatching to 350 days of age confirmed that the ratio of water-to-dry matter intake ranged from 1.8 to 2.6, with an average of 2.3 [9]. This range of water-to-feed intake is similar to commercial poultry (2 to 3.5).

Ambient temperature and composition of the diet are major factors contributing to a wide range in water intake. Water consumption increases markedly with elevation in ambient temperature. In chickens, water intake may increase twofold as ambient temperature increases from 25°C to 35°C. A similar effect of high temperatures on ratites would be expected, and an adequate supply of water under high temperature conditions is critical to performance or survival.

Water consumption usually increases as salt (sodium chloride) or fiber content of diets increases. Increased water intake facilitates the excretion of sodium and chloride when salt intake is excessive, whereas extra water intake is needed to mositen ingesta and enhance passage of digesta when greater amounts of fiber are consumed.

Disease can affect consumption of water. Intake usually decreases with onset of a disease and often precedes clinical signs. A sudden reduction in water intake by ratites may indicate the early onset of a disease.

It is noteworthy that water intoxication occurs in young poultry following overconsumption of water after a 36-to 48-hour period of deprivation. The potential exists for water intoxication to occur in ratites, emphasizing the importance of a consistent and adequate water supply, especially for birds under 21 days of age.

Water quality is an important aspect of ratite nutrition. Specific guidelines concerning water quality for poultry, such as those listed in the National Research Council publication, *Nutrient Requirements of Poultry* [10], should be regarded as a guide for ratites. Water with fewer than 3,000 mg total dissolved solids (TDS)/liter is acceptable for all avian species. Water containing more than 3,000 mg TDS/liter may prove detrimental to health and productivity, especially if the concentration exceeds 5,000 mg/liter. Composition of the TDS also influences the acceptability of water. Sulfate concentrations above 500 mg/liter should be considered undesirable for young poultry. Nitrites and

TABLE 2.2. Estimated Protein and Amino Acid Requirements (% of Diet) for Maintenance and Growth of Ostriches [11]

Age	LW	PROT	LYS	MET	CYS	MET+CYS	ARG	THR	VAL	ISL	LEU	HIS	PHE	TYR
	kg													
30	4	23.9	1.06	.31	.28	.59	.98	.65	.79	.87	1.45	.36	.85	.44
60	11	27.2	1.25	.36	.33	.69	1.15	.76	.93	1.03	1.70	.43	1.0	.51
90	19.5	22.4	1.08	.32	.28	.60	1.01	.66	.82	.90	1.47	.38	.87	.45
120	28.5	20.7	1.06	.32	.27	.59	.99	.64	.81	.88	1.43	.38	.85	.43
150	39.5	17.4	.91	.27	.23	.50	.85	.55	.70	.76	1.23	.33	.73	.39
180	52.1	16.8	.90	.27	.23	.50	.85	.55	.69	.76	1.22	.33	.72	.39
210	63.4	14.8	.85	.26	.21	.47	.80	.51	.66	.72	1.14	.31	.68	.37
240	73.3	13.5	.82	.25	.20	.45	.78	.50	.64	.70	1.10	.31	.66	.36
270	82.4	13.0	.83	.26	.20	.46	.79	.50	.65	.71	1.11	.31	.66	.37
300	91.0	12.8	.84	.26	.20	.46	.81	.51	.67	.72	1.12	.32	.67	.38
330	96.3	8.5	.63	.20	.15	.35	.61	.38	.51	.54	.84	.24	.50	.29
360	99.6	7.4	.59	.19	.13	.32	.57	.35	.48	.51	.78	.23	.47	.27
390	103.5	7.4	.59	.19	.14	.33	.58	.36	.48	.52	.79	.33	.47	.27
420	107.0	7.5	.61	.20	.14	.34	.59	.37	.49	.53	.81	.24	.48	.28
450	110.0	7.3	.61	.20	.14	.34	.59	.37	.50	.53	.80	.24	.48	.28
480	112.3	6.9	.60	.19	.13	.32	.59	.36	.49	.52	.79	.24	.48	.28
510	114.2	6.7	.60	.19	.13	.32	.59	.36	.49	.52	.79	.24	.47	.28
540	116.0	6.7	.60	.20	.13	.33	.59	.36	.50	.53	.80	.24	.48	.28
570	118.6	7.4	.64	.21	.14	.35	.62	.38	.52	.56	.84	.25	.51	.30
600	120.3	6.8	.62	.20	.14	.34	.61	.37	.51	.54	.82	.25	.49	.29

nitrates should not exceed upper limits of 10 and 100 mg/liter, respectively.

PROTEINS AND AMINO ACIDS

The term "protein" designates the crude protein content of feedstuffs and diets. Crude protein is defined as the nitrogen content multiplied by the factor 6.25, assuming that the nitrogen content of proteins averages 16%, by weight. Proteins are composed of long chains of amino acids linked together. The physical and chemical properties of proteins vary greatly, depending on the total number and linkage sequences of amino acids as well as the proportion of the different amino acids in the protein.

The dietary crude protein specification reflects the requirements for amino acids. Amino acids are components of all tissues, including muscle, bone, skin, feathers, and eggs. Amino acids also are required for synthesis of biologically active substances including hormones and enzymes, for transport of nutrients in the body, and to maintain ionic balance of body fluids. A deficiency in dietary protein or amino acids results in reduced growth in young animals and decreased reproductive performance of mature animals, impaired feed efficiency, poor feather development, lethargy and reduced resistance to disease.

Protein of body tissue contain as many as 22 amino acids. Eleven of these amino acids must be supplied in the dietary protein or as amino acid supplements for chickens and turkeys because they cannot synthesize these amino acid at rates sufficient for maintenance and growth. These 11 amino acids, termed essential amino acids, are [10]

Arginine Leucine Threonine
Glycine Lysine Tryptophan
Histidine Methionine Valine
Isoleucine Phenylalanine

A part of the dietary requirements for glycine, methionine, and phenylalanine can be satisfied by serine, cystine, and tyrosine, respectively. Dietary glycine is not required by chickens and turkeys after 3 to 4 weeks of age.

Little information is available on the essential amino acid requirements of ratites. A commonality exists among a large number of animals with respect to essential amino acids, although absolute requirements for each amino acid differ among species. Data on amino composition of 7-month-old ostriches and feed consumption during different age intervals from hatch to maturity were used by South African researchers to estimate essential amino acid requirements of ostriches (Table 2.2) [11]. These requirements should be viewed as tentative because of assumptions made in deriving them. In the absence of information on essential amino acid requirements of emu and rheas, values used for broiler chickens are assumed to reflect the needs of these ratites.

Amino acid requirements of young poultry are relatively high during early growth. As birds grow toward physical maturity, dietary amino acid requirements decrease because of changes in growth rate and metabolic needs in relation to feed intake.

A blend of dietary protein sources and supplemental methionine and lysine can be used to formulate diets to meet essential amino acid requirements. Sufficient protein should be included in the diet to satisfy most of the metabolic needs for nonessential amino acids, which otherwise would have to be synthesized from essential amino acids. In practical terms, diets are formulated to meet essential amino acid and protein requirements in the most economical way.

The principal sources of dietary protein are cereal grains (corn, barley, sorghum, wheat, and oats), alfalfa meal, and ingredients with relatively high protein content (soybean meal, animal by-product meals, and fermentation by-products). Appropriate blending of these ingredients, within the constraints of economic and dietary considerations, results in formulas that fulfill essential amino acid requirements. Frequently, mixtures of protein sources will not supply sufficient methionine or lysine, and supplemental amino acid sources are needed to balance the diet. Supplemental sources of these amino acids are available commercially and are well utilized by avian species.

The feeding quality of a protein-contributing ingredient depends on amino acid composition and digestibility. Protein sources vary in digestibility, which influences amino acid availability. Proteins sources derived, in part, from animal tissues, such as feathers or hooves, are generally less digestible than proteins derived from most plant sources. Improper processing involving either under- or overheating of any major protein source, including soybean meal, corn gluten meal, meat and bone meal, or blood meal, will reduce protein digestibility and the availability of amino acids. Data are presented in *Nutrient Requirements of Poultry* [10], listing the amino acid composition of feed ingredients and amino acid digestibilities for poultry.

Feeds are usually formulated to contain predetermined protein and amino acid levels. Any factor that affects feed intake will influence the quantity of protein and the amino acids consumed. Major factors influencing feed intake include ambient temperature, dietary energy content, physical texture of the diet, accessibility to feed, and health status. To maintain adequate, but not excessive, protein and amino acid intake, adjustments may be required in protein and amino acid percentages of the diet to compensate for changes in feed intake.

ENERGY

Energy represents the capacity of dietary carbohydrates, fats, and amino acids to support all life processes through oxidation. Numerous terms are used to describe dietary energy, as depicted in Table 2.3.

Terms used to describe energy in animal nutrition are based on an initial measurable quantity, gross energy

TABLE 2.3. Definition of Energy Terms

Gross Energy (GE)	Energy of combustion
Digestible Energy (DE)	GE feed + GE feces
Metabolizable Energy (ME)	GE feed + (GE feces + GE Urine + GE gaseous products of digestion)
Net Energy (NE)	ME—Heat Increment
Calorie (cal)	Heat required to raise the temperature of 1 g of water from 16.5 to 17.5 C, equivalent to 4.184 joules
Kilocalorie (kcal)	1,000 cal
Joule (J)	One joule is 10^7 ergs. One erg is the amount of energy needed to accelerate a 1 g mass by 1 cm/s. Used worldwide except in the Americas.
To convert 1 J to 1 cal	Multiply joules by 0.239

(GE). This is the heat of combustion or the heat released when a substance is completely oxidized (burned) to carbon dioxide and water. The gross energy of an ingredient is the maximum energy in a substance and is measured using an adiabatic bomb calorimeter. The gross energy of an ingredient depends on the proportion of the component nutrients, fat, carbohydrate, and protein. The gross energy content of the major nutrient categories are carbohydrates, 3.7 (glucose) 4.2 (starch); protein, 5.6; and fat, 9.4 kcal/g. Minerals, vitamins, and water do not contribute to dietary energy.

Although gross energy is the maximum energy in a substance, not all of this energy is available. Apparent digestible energy (DE) is the gross energy of a feed minus the gross energy in feces. Most birds excrete feces and urine together through the cloaca, making it difficult to measure digestible energy without surgical alteration of experimental subjects. Some separation of urine and feces occurs in ostriches due to voiding at different times. It is difficult to completely separate feces from urine in ratites, and thus digestible energy is not used as a basis for feed formulation for avians.

Apparent metabolizable energy (ME) is defined as the gross energy of the feed minus the gross energy of the feces, urine and gaseous products of digestion. In poultry, the gasses produced during digestion are considered to be negligible. In ratites, gaseous substances, such as volatile fatty acids (VFA) and methane are produced by microbial fermentation of fibrous compounds in the lower digestive tract [2]. In emus, this takes place in the distal ileum. The proximal large intestine and ceca are the site of VFA production in ostriches. The large ceca and proximal large intestine are the sites of cellulolysis in rheas. The energy from gaseous products of digestion in ratites may be of importance although this has not yet been determined. Current methodology (total collection or a marker system) used to determine energy values of ingredients or diets fed to ratites has assumed that the gaseous components are negligible.

The metabolizable energy of a feed can be calculated, on the basis of certain assumptions. Assuming 100% digestibility of fat, protein, and nitrogen-free extract, the concentration of these can be multiplied by 8.7, 4.4, and 4.0 kcal/g, respectively, resulting in an approximate metabolizable energy value for an ingredient. The nitrogen-free extract of an ingredient can be calculated by subtracting the proportion of water (moisture), ash, crude fiber, protein, and fat contained in an ingredient from unity. It is important to note that digestibility varies considerably among ingredients, and thus known digestibility values should be used to avoid errors. This system to predict metabolizable energy has limited use in poultry. The predictive value for dietary energy in ratites is even more questionable because of microbial digestion of dietary fiber. Other indirect methods for estimating metabolizable energy are available, but have limitations [12].

Net energy (NE) is the metabolizable energy less the heat increment. Heat increments represents the heat released during digestion and fermentation of feeds, absorption of nutrients, product formatioon, and formation and excretion of wastes. Net energy is used for body maintenance, physical activity and production of body tissues or eggs. The efficiency with which metaboligable energy is used for maintenance, activity, growth, or egg production differs; thus, there is no specific net energy value for an ingredient or feed.

The ability of commercial poultry to ferment fiber is very limited and fiber does not contribute to the metabolizable energy value of an ingredient for chickens and turkeys. The apparent digestion of fiber (neutral detergent fiber) in chickens varies between 6 and 9%. In adult emus, the digestibility of neutral detergent fiber is as high as 45% [1] and in ostriches may attain 63% [2]. The greater capacity of the emu and ostrich to ferment fiber will have an effect on the amount of metabolizable energy these species can derive from ingredients containing varying levels of fiber. For example, the metabolizable energy of alfalfa for poultry is 1200 kcal/kg [10]. The determined metabolizable energy of alfalfa for ostriches is 2127 kcal/kg [13]. Poultry metabolizable energy values have been used for formulation of ratite feeds due to a lack of specific metabolizable energy values for ingredients incorporated in ratite diets. When using poultry metabolizable energy values for juvenile and mature ratites, the actual energy of the diet will be underestimated, especially when the diet contains high levels of fiber. It is also important to note that the actual energy derived from an ingredient will probably differ between ostriches and emus, due to the higher capacity for fiber digestion in

TABLE 2.4. Nitrogen corrected metabolizable energy (TME_n) Values of Various Feedstuffs for Adult Ostriches[a]

Ingredients*	TME_n values for ostriches	
	MJ	Kcal/lb
Yellow corn	15.06	1,636
Alfalfa	8.91	967
Malting barley	13.93	1,513
Oats	12.27	1,333
Wheat bran	11.91	1,300
Sunflower oilcake meal	10.79	1,172
Soybean oilcake meal	13.44	1,460
Saltbush hay (*Atriplex nummularia*)	7.09	770
Meat & bone meal (ostrich origin)	12.81	1,392
Fish meal	15.13	1,644

*90% dry matter basis.

[a]Cilliers [11]

ostriches. It is estimated that microbial degradation of dietary fiber (including cellulose and hemicellulose) may contribute up to 75% of the metabolizable energy requirement for maintenance in the growing ostrich [14].

Studies in South Africa have determined the true metabolizable energy (TME_n) for local feedstuffs. The TME_n values which incorporate a correction factor for nitrogen retention are depicted in Table 2.4. Values for TME_n were validated by continuous 5-day feeding trials using 7-month-old ostriches [11].

Assuming that a diet is adequate in essential nutrients, commercial poultry consume feed to meet energy requirements. The ratio of energy to other essential nutrients is the basis for formulation of diets for chickens and turkeys. This approach should be used with caution in ratites, and the obvious limitations regarding cellulose digestion should be considered. Deficiencies or excesses of essential nutrients can influence feed consumption, affecting the relationship of energy to essential nutrients. If a diet has a gross excess of a specific nutrient, feed consumption may decrease in proportion to the potential deleterious or toxic effect of the nutrient. Similarly, a nutrient deficiency may lead to an increase or a decrease in feed consumption, depending on the nutrient and the severity of the deficiency.

The relationship between energy and feed consumption in immature ratites is assumed to be similar to that of poultry, but this has not been proven. Observations based on free choice diets in ostriches, emus, and rheas have shown a preference based on energy intake in the absence of obvious deficiencies or excesses of nutrients. This implies that high energy diets will be selected in preference to formulations with lower metabolizable energy values, but data is not available on potential intake changes.

FATS

Fats are classified chemically as simple lipids and are fatty acids bound to a glycerol molecule by ester linkages. Fats in animal feeds are derived from vegetable and animal sources. In general, fats differ in their relative proportion of fatty acids. Vegetable fats contain high concentrations of unsaturated fatty acids, compared to the more saturated fatty acids in most animal fats. Fats incorporated into feed usually contain long-chain (12–22 carbon) fatty acids.

Fats added to animal diets in supplemental form include yellow grease (a by-product of the cooking industry), animal-vegetable fat blends, vegetable oils, or animal fats. Animal-vegetable fat blends contain by-products of the rendering and vegetable oil refining industry. The Association of American Feed Control Officials has defined several categories of feed-grade fats. The definition stipulate the fat components and limits of nonfat materials, such as moisture, insoluble, and unsaponifiables, permitted in these fats. Fats included in ratite feed should always be stabilized with an effective antioxidant at a suitable level to guard against peroxidation.

Feed-grade fats increase energy concentration in diets because they provide about 2.25 times as much energy for metabolism per unit weight compared with carbohydrates and proteins. In addition to supplying energy, supplemental fat improves absorption of lipid-soluble nutrients, reduces dustiness of feed, and prolongs the operating life of feed mixing and handling equipment.

Several factors affect the contribution of fats to dietary energy available to poultry. These include the degree of unsaturation of the fatty acids and age of the birds. Poultry utilize unsaturated fats from vegetable oils more efficiently than saturated fats such as beef tallow. This difference in utilization is especially evident with poultry less than 3 weeks of age because of a limited supply of the bile and pancreatic enzymes needed for fat digestion. The results of many studies demonstrate that the ability of young poultry to use fat, particularly saturated fats, improves from 1 to 21 days of age. Research has shown that young ostriches share this characteristic with poultry [16].

Fatty acids derived from dietary fat can be used by poultry in several ways other than as energy sources. Some of the dietary fatty acids are incorporated as constituents of cell membranes, or they may be stored as adipose tissue, which constitutes an energy reserve for the body. In egg-producing hens, dietary fatty acids are deposited in the egg yolk lipids. The fatty acid composition of dietary fats can influence the fatty acid composition of body tissues and egg yolk.

Fats are important dietary constituents because they contribute to meeting essential fatty acid requirements. Linoleic acid is the only essential fatty acid for which a specific dietary requirement has been demonstrated for chickens. A dietary concentration of 1% linoleic acid is considered adequate in commercial poultry. In contrast, dehydrated alfalfa meal and meat by-product meals contribute less linoleic acid than corn. In many diets, supplemental fats, including animal-vegetable blends, are important sources of dietary linoleic acid.

CARBOHYDRATES AND FIBER

Carbohydrates contain carbon, hydrogen, and oxygen and are classified as monosaccharides, oligosaccharides, and polysaccharides. Monosaccharides are single-sugar molecules containing five (pentose) or six (hexose) carbon atoms. Ribose and xylose are pentoses. Glucose, fructose, and galactose are hexoses. Disaccharides contain two monosaccharides bonded together. Examples include sucrose, lactose, and maltose. Other oligosaccharides are composed of three sugar molecules, such as raffinose, and four sugar molecules, such as stachyose.

Polysaccharides comprise chains of sugar molecules and can be divided into pentosans (xylan), hexosans (dextrin, cellulose, starch, glycogen), and mixed polysaccharides (hemicellulose, pectins, and gums). Monosaccharides and disaccharides, and some polysaccharides (starch and dextrin) are easily digested by all avian species. Cellulose is the main constituent of plant cell walls. Generally, it is not digested by commercial poultry species and may reduce the digestibility of other nutrients contained in forages. Hemicellulose contains many different sugars and is more easily digested than cellulose but less easily digested than starch.

Fiber comprises the nonstarch polysaccharides plus lignin and is resistant to the digestive enzymes of any higher vertebrate animal, including ratites. There is controversy concerning the specific definition of "fiber" and methods of analysis and measurement. "Crude fiber" refers to the organic residue remaining after sequential digestion with dilute acid and alkali following a standard procedure. The broad term "crude fiber" does not identify any specific components and does not relate to the relative digestibility of components. At best, crude fiber gives a rough estimate of the indigestible material in feedstuffs.

A widely used system to evaluate fiber separates plant components based on their solubility in detergents [15]. The cell contents are solubilized in neutral detergent and are known as "neutral detergent solubles" (NDS). The residue after treatment is termed the "neutral detergent fiber" (NDF) and comprises the cell wall constituents.

Neutral detergent removes the cell contents (proteins, lipids, starches, sugars, free amino acids, or NDS) and leaves the cell wall constituents (cellulose, hemicellulose, and lignin). The residue remaining after treatment with acid detergent solution is called "acid detergent fiber" (ADF) and comprises cellulose and lignin. The neutral detergent fiber component (cellulose, hemicellulose, and lignin) is not digested by commercial poultry. In contrast, ostriches are capable of digesting hemicellulose (66%) and cellulose (31%) [17], and emus can digest hemicellulose (57%) and cellulose (19%) in high (36%) NDF-containing diets [1].

Components of fiber cannot be digested by the intestinal enzymes secreted by ratites. Cellulolysis is dependent on a viable population of microorganisms (including bacteria, protozoa and yeasts) within the lumen of the gastrointestinal tract. A site of microbial fermentation and passage rate important is in digestion of fiber. Any factor that alters the composition of microbial flora, such as abrupt changes in diet, stress and antimicrobial medication, will affect the capacity of the bird to digest fiber. Microflora present in the digestive tract are influenced by diet. A high-fiber diet stimulates cellulolytic microorganisms. If a sudden change is made from a high-fiber to a low-fiber diet, the cellulolytic bacteria and protozoa population will fall with a concomitant increase in the populations of starch digestors. Abrupt dietary changes may result in proliferation of potentially pathogenic intestinal bacteria such as *Clostridium* spp., precipitating necrotic enteritis. The establishment of a specific microflora in the intestinal tract is a relatively slow process. Shortly after hatch, microbial populations are slowly established as the chick is exposed to microorganisms in the environment. The capacity to digest fiber increases with age. The apparent NDF digestibility in ostriches was 6.5, 27.9 and 58.0 at 3, 6 and 17 weeks of age, respectively [16].

Fermentation of fiber in the ostrich commences in the proventriculus and gizzard but most occur in the ceca and colon. Volatile fatty acid concentrations were found to be high in the gizzard and proventriculus decreased in the small intestine and high in the hindgut. The bacterial counts in samples obtained from the ceca and colon were similar to values obtained from typical rumen cultures. Counts in the colon were 15% to 40% higher than in the ceca [5].

Despite apparent limitations to digestion of fiber in the emu due to vestigial ceca and a relatively short colon, the emu selectively retains fibrous dietary components in the distal ileum. This allows some fermentation of fiber to occur [6]. Based on anatomical considerations, the rhea is primarily a cecal fermenter with some fermentation occurring in the colon.

VITAMINS

Vitamins can be categorized as fat-soluble (Vitamins A, D, E, and K) and water-soluble (B-complex and C). Fat soluble vitamins are absorbed in the intestine in lipid micelle complexes. Efficiency of absorption can be affected by type and amount of dietary fat and by the ability of the animal to digest and absorb fat. Fat-soluble vitamins can be stored in body tissues to a certain extent compared to the minimal storage of water-soluble vitamins. Most vitamins required by ratites should be added as supplements, as cereals and forages are inadequate sources of critical nutrients. There is no evidence at this time that vitamins administered in water are more efficiently absorbed compared to feed. Oxidation is a major disadvantage of water supplementation and the potential exists for over- or under-dosing creating deficiencies.

Vitamin A is required by all birds for integrity of mucosal and epithelial membranes and immunological functions. Sources of vitamin A vary in biological activity, with retinol being the most active form, followed by B-carotene and retinoic acid. Because vitamin A in the retinol form is unstable in feeds, vitamin A in commercial supplements is usually added in the more stable forms, retinyl acetate or retinyl palmitate. Oversupplementation with vitamin A will depress absorption of vitamins D and E. A deficiency of vitamin A will result in increased susceptibility to infection, conjunctivitis, and blindness in immature birds and reproductive failure and nephritis in adults.

Vitamin D_3 is an essential nutrient that functions as a hormone in its active form (should be 1,25-dihydroxycholecalciferol) and must be supplemented in avian diets as vitamin D_3. After absorption, it is hydroxylated in the liver and kidney into the active form. The kidney and liver of the recipient must be functional to support efficient hydroxylation. Vitamin D_3 stimulates the synthesis of a calcium binding protein in the intestinal tract and in the shell gland of the oviduct. Calcium binding protein transports calcium and phosphorous across the gastrointestinal tract and calcium across cellular membranes of the shell gland. Vitamin D_3 is also active in regulating osteoblast activity which is responsible for growth and integrity of bone and deposition of intramedullary bone in the laying hen.

Vitamin E is required for a range of functions in avian tissues, including antioxidant action at the subcellular level. The efficiency of Vitamin E absorption in the gastrointestinal tract in newly hatched birds is limited, and embryonic stores may be depleted before the bird attains efficient dietary vitamin E absorption. It is important to adequately supplement breeder rations with vitamin E to sustain fertility and promote survival of chicks. Natural vitamin E (tocopherol) is unstable when exposed to the environment, and destruction is accelerated by pelleting, storage at high temperature, or rancidity. Commercial sources of vitamin E are usually supplied as esters such as tocopheryl acetate encapsulated in a gelatin-coated particle to maintain potency. Vitamin E deficiency depresses reproduction in adults and may result in muscular dystrophy or encephalomalacia in progeny. Ingestion of free radicals present in diets as a result of peroxidation of nonstabilized fats will destroy vitamin E and can lead to encephalomalacia and immunosuppression in commercial avian species including ratites.

Vitamin K is an essential component of blood clotting proteins. It is necessary to supplement natural Vitamin K in ratite rations to prevent bleeding and bruising. Natural Vitamin K is not readily available from ingredients and relatively inexpensive synthetic additives, such as menadione sodium bisulfite and menadione dimethylpyrimidol, are incorporated in premixes. Ratites and other avian species can absorb and convert these water-soluble analogues into metabolically active Vitamin K.

The B vitamins supplemented in most ratite rations should include thiamin (B_1), riboflavin (B_2), pantothenic acid, niacin, choline, B_{12}, pyridoxine (B_6), biotin, and folacin. A regular daily intake of these nutrients is necessary to avoid deficiencies. The B-complex vitamins are essential in many metabolic processes and are not stored in the body. Dietary supplementation is usually adequate without additional supplementation in water. Deficiencies can result in lowered egg production, embryonic mortality and malformations, and lowered chick viability. Choline, as with other B vitamins, is involved in many metabolic processes and interacts with other essential nutrients such as manganese, zinc, niacin, folic acid, biotin, and methionine. The B vitamins associated with hatchability and embryonic mortalities include pantothenic acid, riboflavin, and folic acid. It is important that an appropriate quantity of each B-vitamin be supplied in all breeder, starter, and grower rations for ratites.

Vitamin C is not specifically required in ratite rations, as most avian species can synthesize this nutrient. Supplemental Vitamin C must conform to high standards of biological activity and must be handled, stored, and added to feeds to maintain biopotency.

It is important to note that reproductive failure, including low egg production, decreased fertility, embryonic death, and malformations, may be attributed to a large number of causes. Genetics, disease, mismanagement, toxicity, and environmental factors function alone and interact with nutrition to depress efficiency of production. Generally, adequately formulated, fat-stabilized diets from a reputable supplier, purchased within the manufacturer's stated shelf life under appropriate storage conditions (cool and dry) should contain sufficient levels vitamins. Vitamin supplementation in drinking water or super-addition to commercial diets will not improve performance problems due to non-nutritional causes but may in fact precipitate imbalances. Thorough and detailed investigation of problems relating to fertility, and viability of chicks by a competent veterinarian and nutritionist is suggested before resorting to supplementation with specific nutrients in water or by injection.

MINERALS

All species of animals require a range of macrominerals (calcium, phosphorus, potassium, sodium, and chloride) and microminerals (magnesium, manganese, zinc, iron, copper, molybdenum, selenium, iodine, cobalt, and chromium). These elements serve important structural and metabolic functions and are crucial to the maintenance, growth, and reproduction of domestic livestock, including ratites. Although little research has been conducted on the actual requirement of ratites for specific minerals, generalizations from other avian species can be made in relation to the need and function of macro- and microminerals. Table 2.5 provides dietary guidelines and toxic levels of minerals in ratite diets. This table should be regarded as a basis for further studies.

Sodium (Na), potassium (K), and chloride (Cl) are essential in all animals to maintain osmotic balance in extracellular (sodium and potassium) and intracellular tissues (potassium). All animals have inherent homeostatic mechanisms to regulate cations and anions. Most avian species are fairly tolerant of excessive dietary salt. Salt added to water is extremely toxic to all avian species due to the relative inefficiency of the mesonephric kidney. Fecal moisture content will increase when the osmotic balance is disturbed by excessive dietary sodium or potassium. Sodium and chloride are supplemented in nearly all rations as salt, and natural potassium present in cereals and forages satisfies nutrient requirements. Deficiencies in the major cations will result in reduced growth, reduced skeletal density, and abnormal appetite manifested by excessive pica and geophagia. Dietary excess of salt will result in a high intake of water and wet droppings from polyurea. Excess salt may be erroneously mixed into rations in small nonautomated feed mills. This problem may be detected by tasting feed and confirmed by sodium or chloride analysis.

Calcium (Ca) and phosphorus (P) are the two major macrominerals required for skeletal integrity. Calcium is essential for reproduction in birds, being specifically required for endocrine control of ovulation and forma-

TABLE 2.5. Mineral Requirements of Ratites

Nutrient[a]	Ostriches	Emus	Tolerance Levels [18]
Calcium, %			
Starter/Grower	>1.0	>1.0	>2.5:1 ratio of Ca:P
Breeding	>2.0	>2.0	5–6% in reproduction
Phosphorus, %			
Total/Nonphytate	>.6(.5)	>.6(.5)	
Sodium, %	>.15	>.15	1.0% in water
Chlorine, %	>.12	>.12	4.0% dietary NaCl
Potassium, %	>.3	>.3	
Magnesium, %	>.07	>.07	0.60%
Manganese, mg/kg	>100	>60	2,000 mg/kg
Zinc, mg/kg	>80	>70	2,000 mg/kg
Iron, mg/kg	>80	>80	1,000 mg/kg
Copper, mg/kg	>20	>20	500 mg/kg
Selenium, mg/kg	>.20	>.20	5 mg/kg (may be lower for ratites)
Iodine, mg/kg	>0.4	>0.4	50 mg/kg

[1]The data contained in this table is based on nutrient information available from scientific publications, and from nutrient levels commonly found in diets worldwide. Performance of birds on these diets is not documented. These nutrient levels should be used as a guide and must not be the only source of information in formulating diets. Care must be taken to avoid nutrient imbalances or adverse nutrient interactions.

tion of eggs. In addition, acid-base balance, blood clotting, neuromuscular function, and other processes at the cellular and tissue level require adequate calcium intake. Some ingredients such as poultry meal, meat and bone meal, or alfalfa have relatively high levels of calcium. Most ratite diets require calcium supplementation in an inorganic form, such as limestone or oyster shell. Hens in production have an increased demand for calcium and should receive diets with calcium levels consistent with maintenance and production requirements. There is evidence that the calcium levels in diets for laying ostriches (2-3%) are deleterious to spermiogenesis in males. Competitive inhibition of zinc absorption may depress male fertility, according to recent studies conducted in South Africa.

Phosphorus is essential for skeletal integrity, muscle activity, nutrient metabolism, and acid-base balance. Phosphorus is abundant in cereals and forages but is often chelated to phytic acid and is not available to monogastric animals. Ratite diets are supplemented with a minimum level of nonphytate phosphorus from inorganic sources such as dicalcium phosphate. The calcium:nonphytate ratio in starter, grower, and maintenance ratite rations should approximate 2:1 and should increase to 3-4:1 ratio in layer rations. An imbalance in calcium or phosphorus will lead to a dietary deficiency manifested as abnormal bone development, poor shell calcification, and, in immature birds, depressed growth. Oversupplementation of calcium or phosphorous will depress the absorption and utilization of microminerals such as manganese, zinc, or copper.

Magnesium is an essential micromineral required for enzyme function, egg production and other metabolic activities. Magnesium is usually adequate in alfalfa meal and other forages and supplementation is not needed in ratite rations formulated from natural ingredients. Oversupplementation of magnesium in ratite diets will cause diarrhea, feed refusal, and an inhibitory effect on utilization of dietary calcium. Magnesium is stored in the bone of most species.

Manganese is an essential mineral for cartilage formation in ratites. It is required as a cofactor in several enzymes involved in amino acid and other metabolic functions. Manganese is not adequately supplied by cereals and forages included in ratite rations and must be supplemented in either organic or inorganic forms. South African diets include manganese at a level of at least 200 ppm in ostrich chick grower rations [19]. Excessive levels of calcium and phosphorous should be avoided because of the inhibitory effect on absorption and utilization of manganese, copper, zinc, and other microminerals. Analysis of tissues and serum shows that ostriches have lower manganese values compared to other avian species [20]. Dietary manganese deficiency results in chondrodystrophy in immature poultry and embryonic malformations in the progeny of deficient hens. Manganese deficiency has not been documented in ratites.

Congenital and environmentally induced leg abnormalities (osteodysplasia) are common in all young ratites due to rapid growth of the tibiotarsal and tarsometatarsal bones. Rotation of either of these bones will produce a valgus or varus deformation of the hock joint. This may progress to unilateral slipped tendon (perosis). This condition, usually of genetic or environmental origin, should be distinguished from a deficiency of manganese or other micronutrients which affect formation of cartilage in the epiphyseal end plates of long bones.

Zinc is an essential micromineral usually requiring supplementation in the inorganic form in ratite rations. Zinc is essential for formation of bone, skin, and feathers. Zinc deficiency results in shortening of leg bones, enlargement of the hock joints, dermatitis, and poor feathering in poultry. Excess calcium in ostrich breeder rations will reduce zinc absorption by males, which may result in decreased fertility. Zinc toxicity can occur when galvanized waterers are used over long periods of time. Excess zinc intake or supplementation will interfere with the absorption of other essential microminerals such as copper and molybdenum.

Iron and copper are essential trace elements which are components of hemoglobin and are also required for normal feather development. Iron and copper are usually supplemented in ratite rations in the inorganic form. Copper also plays an important role in formation of elastin for cartilage, bone, and blood vessels. Copper deficiency has been implicated as a cause of aortic rupture in ratites, although this condition may have a genetic basis or be associated with a genetic-nutritional interaction. Several minerals can interfere with absorption and transport of copper in ratites. These include calcium, phosphorus, zinc, iron, and molybdenum.

Molybdenum is a required trace element which functions as a cofactor of the enzyme xanthine dehydrogenase. In the Rocky Mountain and Sierra Nevada watersheds, high levels of soil and water molybdenum may interfere with availability of copper.

Selenium is a critical micromineral required for the enzyme glutathione peroxidase which maintains the integrity of cell membranes. Selenium functions with vitamin E as a biological antioxidant but cannot completely replace vitamin E in other functions such as maintaining the integrity of muscles and blood vessel. Selenium is usually supplemented in most ratite rations, but levels are regulated by the FDA for all animal feeds. It is emphasized that the Se requirement is in the region of 0.2 ppm. Excessive selenium supplementation will lead to toxicity, manifested as embryonic mortality and malformations. Cases of degenerative myopathy have been reported in ratites [21]. Supplementation with vitamin E rather than selenium is recommended due to possible toxicity. Selenium should never be added to a balanced commercial ration or injected into chicks or breeders without establishing that a deficiency exists following feed and tissue studies. Parenteral administration of Se to attempt to reverse congenital leg abnormalities in ratite chicks or to improve fertility or egg production usually results in toxicity.

Iodine is an essential component of thyroxine and triiodothyronine, required for normal thyroid gland function in all animals. Iodine is usually incorporated in the trace mineral premix in commercial ratite rations. An iodine deficiency will cause enlargement of the thyroid glands, alter metabolic rate, and decrease hatchability.

Other trace elements are required by most species of animals but are usually present in adequate amounts in the feedstuffs incorporated in conventional diets. It is unnecessary to supplement diets with elements such as chromium. Additional information on trace elements roles and metabolic functions is available in standard texts on nutrition [10,18,22,23]. Mineral nutrition is influenced by interactions which occur in absorption and metabolism of micronutrients.

TABLE 2.6. Classification of Ingredients by Contribution to Diets for Ratites

Sources of protein	Sources of energy	Sources of fiber
Soybean meal	Corn	Alfalfa hay
Alfalfa	Corn byproducts	Wheat middlings
Meat meal	Milo	Grass pastures
Fish meal	Wheat*	Legume pastures
Poultry meat meal	Barley*	Sunflower meal
Cottonseed meal	Oats*	Safflower meal
Legume forages	Rye	Corn cobs
Sunflower meal		Wheat bran
Safflower meal		Rice hulls
Soybean meal		Soybean hulls
Blood meal		Oat hulls

*Enzyme supplementation enhances nutritional value.

Suspected deficiencies should be investigated, applying ingredient, feed, serum, and tissue assays before resorting to uncontrolled supplementation with a wide range of minerals.

INGREDIENTS

Incorporation of ingredients into ratite diets in the United States is based on availability, cost, and documentation on ingredient utilization. Information on ingredient utilization by ratites is limited. Currently, practice is based on principles of poultry nutrition, modified by a knowledge of the anatomy, natural diet, and performance characteristics of ratites. The use of specific ingredients in practical diets without any apparent deleterious effects in South Africa (ostriches) and Australia (emus) has guided nutritionists in the United States. The most commonly used ingredients (Table 2.6) in ratite diets and some general information on these ingredients is presented in Table 2.7. Tables 2.8 through 2.11 depict nutrient information for the principal ingredients used in ratite diets.

Ingredient selection for ratite diets is done on the basis of cost, availability, suitability, and protein, energy, and fiber content. Vitamin and mineral composition of the ingredients are of secondary consideration since they will be supplied primarily by supplemental sources in the diet. Quality, cost, palatability, availability, digestibility, and the presence of antimetabolites or toxic factors in ingredients influence inclusion rate in diets. Knowledge of the harvest year and origin of the ingredient is important in the process of selection. In seasons with excessive rainfall, especially at the time of harvest, mold proliferation may lead to mycotoxicosis. Cereals should be dried to 14% maximum moisture.

The region where the ingredient was harvested may indicate possible low or high levels of specific minerals. Regions in the northwest United States have a

TABLE 2.7. Characteristics of Ingredients

Ingredient	Considerations/Characteristics
Soybean meal	Limited in sulfur amino acids but high in lysine; good quality and consistency; widely available; may contain inhibitors unless properly processed.
Alfalfa	Variable in quality, availability of protein; supplies calcium but low in available phosphorus.
Meat meal	Limited in methionine and tryptophan; variable quality; good availability; high in calcium and phosphorus; quality control of sources is important to prevent *Salmonella* microbial contamination and low availability of amino acids; must be antioxidant stabilized to prevent peroxidation.
Fish meal	Variable quality and high cost. If poorly processed may have suboptimal amino acid availability, peroxidation of fats, and possible production of gizzerosine, a toxic by-product from overheating, contains unidentified growth factors and is a good source of essential nutrients.
Peanut meal	High probability of toxins derived from *Aspergillus* and should therefore be used with caution; limited in methionine, tryptophan, and very low in lysine; variable quality; meal regionally available but may contain inhibitors if improperly processed; high-fat meals should be stabilized with an antioxidant.
Cottonseed meal	Cottonseed meal should not be used in starter and breeder diets; methionine and lysine content may be low; regional in availability, only low gossypol meals should be selected. Gossypol is toxic in avian species, reducing growth and affecting egg quality. Only low-fat cottonseed meal should be used due to the presence in the fat of sterculic acid which reduces growth, affects egg quality and body composition.
Sunflower meal	Limiting in lysine; inverse relationship between protein and fiber levels; regional and seasonal availability. Good source of fiber.
Blood meal	High protein source limited in isoleucine; very variable protein quality (protein digestibility and amino acid availability), spray-dried blood meal has more consistent protein quality than vat-dried blood meal. May be contaminated with *Salmonella* spp.
Corn	Limiting in lysine and tryptophan; variable protein and fat content; good source of energy; only low moisture corn (under 14%) should be used to prevent mycotoxicosis.
Milo/Sorghum varieties	Limited in lysine and sulfur amino acids; variable quality, regional in availability; low tannin cultivars should be used.
Wheat	Limited in lysine and sulfur amino acids; very variable protein content; should be used in pelleted rations or cracked or rolled before use due to stickiness/pasting (due to high gluten content) that may influence palatability; low moisture wheats should be used. Addition of enzymes to hydrolyse arabinoxylans is recommended.
Oats	Limited in total sulfur amino acids and lysine; whole oats may contain from 25 to 50 percent hulls. Oats from which the hulls have been removed are called oat groats and are an excellent energy source.
Barley	Limited in lysine and total sulfur amino acids; as protein level increases lysine decreases, thus in high protein barleys, lysine becomes more limiting. Supplementary enzymes addition is recommended.
Rye	Limited in lysine and total sulfur amino acids. Because of the risk of ergot toxicity and antimetabolics, rye should be used with care in ratite rations.

high soil selenium content that results in high concentration of this element in grains, pastures and oil seeds. It is not uncommon to find copper deficiency in animals reared in areas with high molybdenum in the soil due to an antagonism between molybdenum and copper.

The protein quality of cereals tends to be low primarily due to a relative low in lysine level. Combining cereal grains with protein supplements that supply essential amino acids contributes to a balanced intake. This may be important in ratites because the lysine requirements in young birds may be higher than in poultry [11]. Corn and milo (sorghum) tend to have the lowest level of lysine, compared to other cereals. Cracking, rolling, crimping, or flaking of grains may improve digestibility especially in young birds. In general, whole grains should not be used for emus since kernels are not digested but passed in feces. Cereals are low in calcium and moderate in total phosphorus. The availability of phosphorus from plant sources is low (20 to 40%) due to phytate binding.

The hulls of grains and some oil seeds are used primarily as bulk ingredients. Oat and rice hulls are poorly digested and are mainly used as bulk. Care must be taken when using peanut meal and hulls ratite rations because of the danger of aspergillosis from *Aspergillus flavus*. Although soybean hulls are a good source of digestible fiber, care must be taken to determine that the product has been heat processed to ensure destruction of trypsin inhibitor and of the antinutritional factors present in pieces of raw soybeans present in the hulls.

Pastures can contribute to the dietary requirements of ostriches and rheas and, to a lesser extent, emus. Appropriate management of pastures is necessary to ensure a mixture of grasses and legumes which will result in the best combination of nutrients, resistance to trampling, and desirable yields. Grasses which are usually low in protein have a high fiber content in contrast to legumes that have a high protein content. A satisfactory mixture of grasses and legumes will result in a high quality pasture. Species and cultivars should be selected in relation to climate, temperature, humidity, precipitation, and soil type. The county extension service or a forage specialist should be consulted to determine the best cultivars and fertilization and maintenance requirements.

Alfalfa is the most commonly used legume for ratites. This crop or ingredient is a good source of protein, fiber, minerals, and vitamins. Care must be taken to balance calcium and phosphorus intake when using alfalfa pasture or free choice alfalfa hay in a feeding program. The calcium level in alfalfa is high and available phosphorus is low. Some of the calcium in alfalfa is in an oxalate form, which may be unavailable to ratites. The calcium to phosphorus ratio in alfalfa is not conducive to adequate bone growth. Most legumes contain a high level of calcium and a low level of phosphorus in contrast to grasses.

The following quality considerations are relevant to the purchase and storage of feeds and ingredients:

- purchase ingredients or feeds from a reputable supplier against known specifications
- avoid moldy ingredients
- do not purchase grains with a moisture content above 14%
- store feeds and ingredients at low humidity and moderate temperature in insulated areas
- prevent direct contact of bags and ingredients with floors by using pallets
- provide adequate ventilation in storage areas
- control insects, rodents and wild birds that serve as reservoirs and vectors of disease

Producers should consult with feed suppliers to ensure that acceptable quality control and good manufacturing practices are followed. Feed should be delivered in new paper containers clearly designated as to type of diet and date of manufacture. Bags should be correctly labeled, specifying proximate analysis and additives.

NUTRIENT REQUIREMENTS

Scientific knowledge of the nutrient requirements of ratites is developing. Existing information is based on the nutritional requirements of domestic animals supplemented by reports from South Africa [11, 14, 27]. Limited information on nutritional requirements of emus is available from Australia [24]. South African scientists have reported on extensive research on cape ostriches.

Nutrient Requirements of Emus

Scientific information available on the energy and protein requirements of emus has been published following studies on a limited number of adult birds in Australia [28]. Four adult emus ranging in weight between 28 kg and 48 kg had a maintenance energy requirement of 67.9 cal ME/kg$^{.75}$ (body weight)/day. The maintenance nitrogen requirement for these birds was found to be 0.09 g/kg$^{.75}$ (body weight)/day or 0.56 g protein/kg$^{.75}$ (body weight)/day. This means that a 45.5 kg (100 lb) emu would require 1,303 kcal of metabolizable energy/day and 10.9 g of protein/day for maintenance. The apparent nitrogen requirement is low compared with other avian species. This may be partly attributable to the contribution of microbial nitrogen being ignored in measuring fecal nitrogen in the balance study.

The lysine requirements of starting emus has been found to be 0.86% of the diet [25]. This value is based on trials in which birds were fed diets with a lysine content ranging from 0.575% to 1.15%. Unsexed emus from 23 to 65 days of age were fed isocaloric (2,748 kcal, metabolizable energy/kg) but not isonitrogenous diets.

Nutrient Requirements of Ostriches

Information on the energy and amino acid requirements of ostriches is based on prediction equations from growth curves, carcass analysis, and feeding trials conducted in South Africa [11] (Table 2.12). Preliminary values are incorporated into recommendations used by members of the Klein Karoo Cooperative. Table 2.13 shows the seven dietary specifications used in the Cape [26].

Requirements for maintenance and efficiency of utilization were determined and validated using 7-month-old ostriches. The results were extrapolated to birds ranging in age from 1 to 80 weeks of age [11].

The energy and amino acid requirements for laying ostrich hens weighing from 220 to 245 lb are shown in Tables 2.14 and 2.15, respectively. These requirements are based on the studies conducted with Cape ostriches [27].

Macromineral trace element, and vitamin supplements suggested for ostriches are depicted in Tables 2.16 and 2.17 [26].

The adequacy of a nutrition program can be indirectly monitored in Cape ostriches by determining chest circumference (girth at the widest part of the body), which has been shown to be correlated with body weight. The relative values for chest circumference, weight and saleable hide area [26] are indicated in Table 2.18.

Nutrient Requirements of Rheas

There is no published scientific information relating to the nutrient requirements of rheas. Given the anatomical similarities of the digestive tracts of ostriches and rheas, it may be assumed that nutrient requirements are similar. There is no documentation to support this assumption at the present time.

TABLE 2.8. Nutrient Composition (Excluding Amino Acids) of Feedstuffs Commonly Used for Ratites (data on as-fed basis)

Ingredient	Dry Matter (%)	Mc$_n$ (kcal/kg) Poultry	Protein (%)	Ether Extract (%)	Linoleic Acid (%)	Crude Fiber (%)	Calcium (%)
Alfalfa, *Medicago sativa* meal dehydrated, 17% protein	92	1200	17.5	2.5	0.47	24.1	1.44
Barley, *Hordeum vulgare* grain	89	2640	11.0	1.8	0.83	5.5	0.03
Blood meal, spray dried	93	3420	88.9	1.0	0.10	0.6	0.41
Corn, Dent Yellow *Zea mays indentata* grain	89	3350	8.5	3.8	2.20	2.2	0.02
Cotton, *Gossypium* spp. Seeds, meal prepressed solvent extracted, 44% protein	91	1857	44.7	1.6		11.1	0.15
Fish, Menhaden *Brevoortia tyrannus* meal mechanically extracted	92	2820	60.05	9.4	0.12	0.7	5.11
Meat meal rendered	92	2195	54.4	7.1	0.28	2.7	8.27
Oats, *Avena sativa* grain	89	2550	11.4	4.2	1.47	10.8	0.06
Peanut, *Arachus hypogaea* kernels, meal solvent extracted (peanut meal)	92	2200	50.7	1.2	0.24	10.0	0.20
Poultry by-product, meal rendered (viscera with feet and heads)	93	2950	60.0	13.0	2.54	1.5	3.00
Poultry feathers, meal hydrolyzed	93	2360	81.0	7.0		1.0	0.33
Safflower, *Carthamus anctorius*, seeds, meal solvent extracted	92	1193	23.4	1.4		30.0	0.34
Sorghum, *Sorghum bicolor* grain, 8-10% protein	87	3288	8.8	2.9	1.13	2.3	0.04
Soybean, *Glycine max* seeds without hulls, meal solvent extracted	90	2440	48.5	1.0	0.40	3.9	0.27
Sunflower, common *Helianthus annuus* seeds, meal solvent extracted	90	1543	32.0	1.1	0.60	24.0	0.21
Wheat, *Triticum aestivum* grain, hard red winter	87	2900	14.1	2.5	0.59	3.0	0.05
Wheat, *Triticum aestivum* grain, soft white winter	89	3120	11.5	2.5		3.0	0.05

Adapted with permission from *Nutrient Requirements of Poultry*. copyright 1994 by the National Academy of Sciences. Courtesy of the National Academy Press, Washington, D.C.

TABLE 2.8. Nutrient Composition (Excluding Amino Acids) of Feedstuffs Commonly Used for Ratites (data on as-fed basis) (cont'd)

Ingredient	Biotin (mg/kg)	Choline (mg/kg)	Folacin (mg/kg)	Niacin (mg/kg)
Alfalfa, *Medicago sativa* meal dehydrated, 17% protein	0.30	1401	4.2	38
Barley, *Hordeum vulgare* grain	0.15	990	0.07	55
Blood meal, spray dried	0.20	280	0.4	24
Corn, Dent Yellow *Zea mays indentata* grain	0.06	620	0.4	13
Cotton, *Gossypium* spp. Seeds, meal prepressed solvent extracted, 44% protein		2685	0.9	46
Fish, Monhaden *Brevoortia tyrannus* meal mechanically extracted	0.20	3056	0.3	55
Meat meal rendered	0.17	2077	0.3	57
Oats, *Avena sativa* grain	0.27	946	0.3	12
Peanut, *Arachus hypogaea* kernels, meal solvent extracted (peanut meal)	0.39	2396	0.4	170
Poultry by-product, meal rendered (viscera with feet and heads)	0.30	5952	1.0	40
Poultry feathers, meal hydrolyzed	0.04	891	0.2	27
Safflower, *Carthamus anctorius* seeds, meal solvent extracted	1.43	820	0.5	11
Sorghum, *Sorghum bicolor* grain, 8-10% protein	0.26	668	0.2	41
Soybean, *Glycine max* seeds without hulls, meal solvent extracted	0.32	2731	1.3	22
Sunflower, common *Helianthus annuss* seeds, meal solvent extracted		3791		264
Wheat, *Triticum aestivum* grain, hard red winter	0.11	1090	0.4	48
Wheat, *Triticum aestivum* grain, soft white winter	0.11	1002	0.4	57

Adapted with permission from *Nutrient Requirements of Poultry*. 1994 NRC.

Total Phosphorus (%)	Non-phytate Phosphorous (%)	Iron (mg/kg)	Magnesium (%)	Manganese (mg/kg)	Sodium (%)	Copper (mg/kg)	Selenium (mg/kg)	Zinc (mg/kg)
0.22	0.22	480	0.36	30	0.09	10	0.34	24
0.36	0.17	78	0.14	18	0.04	10	0.10	30
0.30		3000	0.40	6	0.33	8		306
0.28	0.08	45	0.12	7	0.02	3	0.03	18
1.25	0.37							
2.88		440	0.16	33	0.65	11	2.10	147
4.10		440	0.38	10	1.15	10	0.42	103
0.27	0.05	85	0.16	43	0.08	8	0.30	38
0.63	0.13	142	0.04	29	0.07	15		20
1.70		440	0.22	11	0.40	14	0.75	120
0.55		76	0.20	10	0.69	7	0.84	54
0.75		495	0.35	18	0.05	10		41
0.30		45	0.15	15	0.01	10	0.20	15
0.62	0.22	170	0.30	43	0.02	15	0.10	55
0.93	0.14	140	0.68	34	0.2	35		100
0.37	0.13	60	0.17	32	0.04	6	0.20	34
0.31		40	0.10	24	0.06	7	0.06	28

Panto-thonic Acid (mg/kg)	Pyri-doxine (mg/kg)	Riboflavin (mg/kg)	Thiamin (mg/kg)	Vitamin B_{12} (mg/kg)	Vitamin E (mg/kg)
25.0	6.5	13.6	3.4	4	125
8.0	3.0	1.8	1.9		20
5.0	4.4	1.3	0.5	44	
4.0	7.0	1.0	3.5		22
14.5		4.7			
9.0	4.0	4.9	0.5	104	7
5.0	3.0	5.5	0.2	68	1
7.8	1.0	1.1	6.0		20
53.0	10.0	11.0	5.7		3
12.3	4.4	11.0	1.0	310	2
10.0	3.0	2.1	0.1	78	
33.9		2.3			1
12.4	5.2	1.3	3.0		7
15.0	5.0	2.9	3.2		3
29.9	11.1	3.0	3.0		
9.9	3.4	1.4	4.5		13
11.0	4.0	1.2	4.3		13

TABLE 2.9. Amino Acid Composition of Feedstuffs Commonly Used for Ratites (data on as-fed basis)

Ingredient	Dry Matter	Protein	Arginine	Histidine	Isoleucine
Alfalfa, *Medicago sativa* meal dehydrated, 17% protein	88.0	17.0	0.69	0.57	0.67
Barley, *Hordeum vulgare* grain	89.0	11.0	0.52	0.27	0.37
Blood meal, spray dried	93.0	88.9	3.62	5.33	0.98
Corn, Dent Yellow *Zea mays indentata* grain	88.0	8.5	0.38	0.23	0.29
Cotton, *Gossypium* spp. Seeds, meal prepressed solvent extracted, 44% protein	89.9	41.4	4.59	1.10	1.33
Fish, Menhaden *Brevoortia tyrannus* meal mechanically extracted	92.1	61.3	3.68	1.42	2.28
Meat meal rendered	92.0	54.4	3.73	1.30	1.60
Oats, *Avena sativa* grain	89.0	11.4	0.79	0.24	0.52
Peanut, *Arachus hypogaea* kernels, meal solvent extracted (peanut meal)	91.9	49.0	5.33	1.07	1.55
Poultry by-product, meal rendered (viscera with feet and heads)	94.2	59.5	3.94	1.07	2.16
Poultry feathers, meal hydrolyzed	91.0	82.9	5.57	0.95	3.91
Safflower, *Carthamus tinctorius* seeds, meal solvent extracted	92.0	27.0	2.21	0.61	1.02
Sorghum, *Sorghum bicolor* grain, 8–10% protein	87.5	9.1	0.35	0.22	0.35
Soybean, *Glycine max* seeds without hulls, meal solvent extracted	88.4	47.5	3.48	1.28	2.12
Sunflower, common *Helianthus annuus* seeds, meal solvent extracted	90.0	23.3	2.30	0.55	1.00
Wheat, *Triticum aestivum* grain, hard red winter	88.1	13.3	0.60	0.31	0.44
Wheat, *Triticum aestivum* grain, soft white winter	89.0	10.2	0.40	0.20	0.42

Adapted with permission from *Nutrient Requirements of Poultry*. 1994 NRC.

TABLE 2.10. Nutrient Composition of Forage Sources Commonly Used for Ratites (data on dry matter basis)

Ingredient	Dry Matter (%)	ME (Mcal kg)	Crude Protein (%)	Ether Extract (%)	Crude Fiber (%)	Neutral Detergent Fiber (%)	Acid Detergent Fiber (%)	Cellulose (%)
Alfalfa, *Medicago sativa* meal dehydrated, 17% protein	92	2.27	18.9	3.0	26.2	45	35	24
Bahiagrass, *Paspalum notatum* fresh	30	1.95	8.9	1.6	30.4	68	38	31
Barley, *Hordeum vulgare* hay, sun-cured	87	2.04	8.7	2.1	27.5			
Bemudagrass, Coastal, *Cynodom dactylon* hay, sun-cured, late vegetation	91	1.96	16.5	1.8	27.3	70	32	28
Bluegrass, Kentucky, *Poa pratensis*, fresh, mature	42	1.18	9.5	3.1	32.2	69	40	34
Brome, *Bromus* spp., fresh, mature	57	2.09	6.4	2.2	38.0	72	44	35
Clover, I adino, *Trifolium repens*, fresh, early vegetative	19	2.58	24.7	2.5	14.0			
Clover, Red, *Trifolium preatense*, fresh, full bloom	26	2.40	14.6	2.9	26.1	43	35	28
Corn, Dent Yellow, *Zea mays indentata*, cobs ground	90	1.78	3.2	0.7	36.2	89	35	28
Cowpea, Common, *Figna sinensis*, hay, sun-cured	90	2.18	19.4	3.1	26.7			
Fescue, Kentucky 31, *Festuca arundinacea*, fresh, vegetative	29	2.49	14.5	5.5	24.6			
Molasses and Syrup, sugarcane, molassos, more than 46% invert sugar, more than 79.5 degrees brix (Black strap)	75	2.76	5.8	0.1				
Oats, *Avena sativa*, grain	89	2.98	13.3	5.4	12.1	32	16	11
Oats, *Avena sativa*, hulls	92	1.11	3.9	1.8	33.4	78	42	30
Orchardgrass, *Dactylis glomerata*, fresh, early vegetative	23	2.76	18.4	4.9	24.7	55	31	25
Rice, *Oryza sativa*, hulls	92	0.08	3.3	0.8	42.9	82	72	33
Sorghum, Sudangrass, *Sorghum bicolor sudanese*, fresh, mid bloom	23	2.36	8.8	1.8	30.0	65	40	34
Soybean, *Glycine max*, hulls	91	2.98	12.1	2.1	40.1	67	50	46
Wheat, *Triticum aestivum*, flour by-product, less than 9.5% fiber (wheat middlings)	89	2.62	18.4	4.9	8.2	37	10	

Adapted with permission from *Nutrient Requirements of Dairy Cattle*, 1989 NRC.

Leucine	Lysine	Methionine	Cystine	Phenylalanin	Threonine	Tryptophan
1.19	0.73	0.24	0.19	0.81	0.69	0.23
0.76	0.40	0.18	0.24	0.56	0.37	0.14
11.32	7.88	1.09	1.03	5.85	3.92	1.35
1.00	0.26	0.18	0.18	0.38	0.29	0.06
2.43	1.71	0.52	0.62	2.22	1.32	0.47
4.16	4.51	1.63	0.57	2.21	2.46	0.49
3.32	3.00	0.75	0.66	1.70	1.74	0.36
0.89	0.50	0.18	0.22	0.59	0.43	0.16
2.97	1.54	0.54	0.64	2.41	1.24	0.48
3.99	3.10	0.99	0.98	2.29	2.17	0.37
6.94	2.28	0.57	4.34	3.94	3.81	0.55
1.74	0.90	0.42	0.45	1.10	0.85	0.37
1.14	0.21	0.16	0.17	0.47	0.29	0.08
3.74	2.96	0.67	0.72	2.34	1.87	0.74
1.60	1.00	0.50	0.50	1.15	1.05	0.45
0.89	0.37	0.21	0.30	0.60	0.39	0.16
0.59	0.31	0.13	0.22	0.45	0.32	0.12

Lignin (%)	Calcium (%)	Magnesium (%)	Phosphorus (%)	Sodium (%)	Cobalt (mg/kg)	Copper (mg/kg)	Iron (mg/kg)	Maganese (mg/kg)	Zinc (mg/kg)	Vitamin A (1000 IU/kg)
11	1.52	0.32	0.25	0.11	0.33	11	441	34	21	52
7	0.46	0.25	0.22							73
	0.23	0.18	0.26	0.14	0.07	24	101	27	48	21
4										
6	0.26	0.16	0.27	0.13		10	293	70		37
9	0.20	0.18	0.26	0.02	0.06	6	200	160	18	33
	1.35	0.48	0.31	0.13	0.16	10	413	95	17	141
7	1.31	0.51	0.27	0.20	0.12	10	173	47	16	83
7	0.12	0.07	0.04	0.47	0.13	7	230	6		
	1.40	0.45	0.35	0.27	0.07		300			14
	0.51		0.37							
	1.00	0.43	0.11	0.22	1.21	79	250	56	30	
3	0.07	0.14	0.38	0.08	0.06	7	85	42	41	
8	0.15	0.09	0.15	0.04		4	111	20		
3	0.58	0.31	3.58	0.04		7	169	96		193
16	0.10	0.83	0.57	0.12				334		
5	0.43	0.35	2.14	0.01			200			73
2	0.49		1.27	0.01	0.12	18	324	11	24	
	0.13	0.40	1.13	0.19	0.10	22	93	126	116	

TABLE 2.11. Element Composition of Mineral Sources Commonly Used for Ratites (data on as-fed basis)

Ingredient	Calcium (%)	Phosphorous (%)	Sodium (%)	Potassium (%)	Magnesium (%)	Iron (mg/kg)	Copper (mg/kg)	Manganese (mg/kg)	Zinc (mg/kg)
Bone meal	29.8	12.5	0.04	0.2	0.3	—	16	30	100
Calcium carbonate $CaCo_3$	38.0	0.0	0.02	0.06	0.05	300	24	300	2
Calcium phosphate, dibasic from defluorinated phosphoric acid	22.0	18.7	0.06	0.1	0.6	10,0000	10	300	100
Limestone, ground	38.0	—	0.05	0.1	2.1	2,000	—	—	—
Oyster, shells, ground	38.0	0.1	0.2	0.1	0.3	500	—	400	—
Phosphate, defluorinated	32.0	18.0	4.9	0.1	0.4	8,000	20	250	60

Adapted with permission from *Nutrient Requirements of Poultry*, 1994 NRC.
Note: Dashes indicate that no data were available.

PRACTICAL ASPECTS OF EMU AND OSTRICH NUTRITION

Actual nutrient requirements for Cape ostriches have been predicted from scientific studies. Additional practical information is available, based on ingredients commonly used in the United States. This information must be applied with care, as it is not derived by scientific research. Data and recommendations should be evaluated with regard to the specific conditions and limitations under which they were obtained. The basis of nutrient values published in lay periodicals is generally unknown, and information is difficult to interpret because of incomplete descriptions of diets, failure to analyze ingredients and rations, or to apply accepted experimental design with an appropriate number of subjects and controls.

Emu diets must be formulated with regard to the high variation in feed consumption, depending on season, climatic conditions, age, and breeding status. Adult birds will decrease feed consumption by 50 to 75% during the breeding cycle. The reduction in intake is more pronounced in the male than the female. As soon as the breeding season is over, emus tend to overconsume feed for 10 to 20 days before resuming customary intake of 0.9 to 1.8 kg/day (2 to 4 lbs/ day). Variation in feed consumed is also noted in growing birds which show a plateau during the late summer and fall and an increase during winter and spring [25]. Future trends in commercial production will influence dietary specifications.

Practical information on ostriches obtained from South Africa can be adapted to conditions in the United States. Under the management and feeding practices that prevail in the Klein Karoo slaughter, birds reach mature body size of 220–242 lb at approximately 14 months of age. These ostriches have traditionally been produced primarily for leather, with feathers a secondary source of revenue and meat are a

TABLE 2.12. Predicted Energy and Amino Acid Requirement of Ostriches (as is basis)[a]

Days of age	10	30	100
Intake, kg(lb)/day	0.11(.24)	0.29(.64)	1.2(2.64)
Metabolizable Energy, kcal/kg	3,080	2,630	2,390
Lysine, %	1.2	1.0	0.86
Methionine plus cystine, %	0.8	0.68	0.57

[a]du Preez et al. [27]. Based on "typical" Namibian males, with a mature body weight of 100 kg (220 lb).

TABLE 2.13. Nutrient Recommendations for South African Ostriches (as is basis)[a]

Diet	Age months	Body wt lbs	Energy kcal/kg	Crude protein %	Lys. %	TSAA %
Pre-Starter hatch	2	1.7–24	3,150	25.0	1.25	0.70
Starter	2–4	25–62	3,015	21.5	1.07	0.60
Grower	4–6	63–115	2,915	17.0	0.90	0.50
Finisher	6–10	116–200	2,600	13.5	0.84	0.46
Post-Finisher	10–20	201–235	1,950	8.5	0.63	0.35
Maintenance	mature	—	1,550	8.0	0.30	0.32
Breeder	laying	—	2,200	14.0	0.68	0.70

[a]duPreez [27].

TABLE 2.14. Estimated Recommendations for Energy Intake in Laying Ostriches[a]

	Energy requirements (Kcal ME) for maintenance and activity Body mass (lb)			Energy (Kcal ME) requirements for egg production Egg mass (lb)		
	220	230	240	2.6	3.0	3.5
Maintenance	3259	3375	3490			
Activity	327	337	349			
Egg lipid				550	640	734
Egg protein				855	999	1140
Shell (18% of egg mass)				62	72	84
Total	3586	3712	3839	1467	1711	1958

[a]du Preez [27].

TABLE 2.15. Estimated Amino Acid Requirements (g/day) for Mature Ostrich Hens[a]

	For maintenance of body mass (lb)			For egg production including shell (lb)		
	220	230	240	2.6	3.0	3.5
Protein, g	67	69	72	119	138	158
Amino acid, (g*)						
Arginine	5.70	5.87	6.12	3.56	4.15	4.74
Lysine	5.78	5.95	6.21	6.41	7.48	8.55
Methionine	1.86	1.90	2.00	2.67	3.10	3.56
Histidine	2.54	2.61	2.73	1.91	2.20	2.50
Threonine	3.54	3.64	3.80	6.85	8.00	9.13
Valine	4.32	4.46	4.65	5.50	6.40	7.30
Isoleucine	3.50	3.60	3.76	4.55	5.30	6.10
Leucine	6.90	7.14	7.45	9.00	10.50	12.00
Tyrosine	2.33	2.40	2.50	3.70	4.30	4.90
Phenylalanine	3.82	3.90	4.10	4.06	4.67	5.30
Cystine	0.89	0.92	0.96			
Tryptophan	0.73	0.75	0.78			

*Daily requirements of dietary amino acids per gram of egg produced. A female weighing 240 lb and consuming 2,000 g (4.4 lb) feed daily, producing a 2.6 pound egg at 2-day intervals, will require:

[a]duPreez [27].

$$\frac{(6.21 \text{ g} + 6.41 \text{ g})}{2000 \text{ g feed}} \times \frac{100}{1} = 0.63\% \text{ lysine in the diet.}$$

TABLE 2.16. Macromineral Recommendations for South African Ostriches (Starter, Finisher, Maintenance, and Layer Diets)[a]

	Total Calcium (%)	Available phosphorous (%)	Total sodium (%)
Pre-starter, starter and grower diets	1.2–1.5	0.4–0.45	0.20–0.25
Finisher and post-finisher diets	0.9–1.0	0.32–0.36	0.15–0.30
Maintenance diet	0.9–1.0	0.32–0.36	0.15–0.30
Layer diet	2.0–2.5	0.35–0.40	0.15–0.25

[a]Cillers and Van Schalkwyk [26].

TABLE 2.17. Trace Element and Vitamin Supplementation used in South African Ostrich Diets[a]

		Units or quantity per ton		
		Grower diets hatch-6 mo	Grower and finisher diets 6 mo-slaughter	Breeder diets
Vit A	IU	12,000,000	9,000,000	15,000,000
Vit D_3	IU	3,000,000	2,000,000	25,000,000
Vit E	IU	40,000	10,000	30,000
Vit K_3	g	3	2	3
Vit B_1	g	3	1	2
Vit B_2	g	8	5	8
Niacin	g	60	50	45
Calc,panth,A	g	14	8	18
Vit B_{12}	mg	100	10	100
Vit B_6	g	4	3	4
Choline chloride	g	500	150	500
Folic acid	g	2	1	1
Biotin	mg	200	10	100
Magnesium	g	50	—	40
Manganese	g	120	80	120
Zinc	g	80	50	90
Copper	g	15	15	15
Iodine	g	0.5	1	1
Cobalt	g	0.1	0.3	0.1
Iron	g	35	20	35
Selenium	g	0.3	0.15	0.3

[a]Cilliers and Van Schalkwyk [26].

by-product. The product mix in the United States will markedly influence the nutrient specifications and selection of ingredients.

Precise nutrient requirement or recommendation levels are not available for ratites as in commercial poultry. General guidelines are presented in Table 2.19. The information presented in this chapter should not be used as the only source of information for formulating diets. Nutritionists associated with the major feed manufacturers are aware of the quality and composition of ingredients and can draw from a wide range of available products to formulate balanced diets. Small scale producers should be dissuaded from mixing their own feeds. Purchase of commercial diets from a reputable supplier will generally provide an adequate intake of nutrients, approximating the requirements of the age and breeding status of the flock or individual ratite species.

TABLE 2.18. Growth Parameters of South African Ostrich[a]

Age months	Chest circum. inches	Live weight lbs	Saleable hide dm²	ft²
>1	7.1	1.8	—	—
	11.0	4.6	—	—
	14.2	8.0	—	—
1	17.3	12.5	—	–
2	24.4	29.9	—	—
3	28.3	46.6	—	—
4	32.3	70.2	—	—
5	34.6	88.8	100	10.76
6	35.8	103.4	102	10.97
7	37.0	111.1	105	11.30
8	38.5	126.2	112	12.05
9	39.3	137.7	115	12.37
10	40.5	147.4	120	12.91
11	41.0	157.5	122	13.13
12	42.5	179.1	128	13.77
13	43.3	190.3	131	15.06
14[b]	44.5	215.0	136	16.10
	48.0	261.0	150	1.61

[a]Cilliers and Van Schalkwyk [26]

[b]Range at slaughter: small frame, large frame.

TABLE 2.19. Nutrient Ranges in Practical Diets for Ratites[a]

Nutrient	Starter	Grower	Breeder	Maintenance
Ostrich				
Protein, %	17 to 28	17 to 24	15 to 24	10 to 20
Metabolizable energy, kcal/g	2.1 to 3.2	1.9 to 2.9	1.6 to 2.5	1.4 to 2.3
Lysine %	>.77	>.75	>.70	>.4
Methionine	>.38	>.36	>.34	>.28
Linoleic acid, %	>.75	>.75	>1.0	>.4
Fiber, %	>4.0	>4.0	>6.0	>8.0
Emu				
Protein, %	17 to 25	16.5 to 23	15 to 25	12 to 20
Metabolizable energy, kcal/g	2.0 to 2.8	1.9 to 2.7	2.0 to 2.8	1.7 to 2.6
Lysine, %	>.77	>.75	>.80	>.45
Methionine	>.38	>.36	>.40	>.28
Linoleic acid, %	>.75	>.75	>1.0	>.4
Fiber, %	>4.0	>4.0	>4.0	>4.0

[a]The information contained in this table is based on information on nutrient content available from scientific and nonscientific publications, and from nutrient levels commonly used in diets worldwide. Performance of birds on these diets is not documented.

REFERENCES

1. Herd RM, Dawson TJ. Fiber digestion in the emu, *Dromaius novaehollandiae*, a large bird with a simple gut and high rates of passage. *Physiol Zool* 1984; 57:70–84.
2. Swart D, Mackie RI, Hayes JP. For feathers and leathers. *Nuclear Active* 1987;36:2–9.
3. Fowler ME. Comparative clinical anatomy of ratites. *J Zoo and Wildl Med* 1991;22:204–227.
4. Cho P, Brown R, Anderson M. Comparative gross anatomy of ratites. *Zoo Biol* 1984;3:133–144.
5. Mackie RI. Microbial digestion of forages in herbivores. In: Hacher JB, Tenouth JH, eds. *2nd International Symposium on the Nutrition of Herbivores*. Academic Press, 1987;233–265.
6. Davies, SJJF. The food of emus. *Austral J Ecol*, 1978;3:411–422.
7. Kienholz EW. Why is water so important to animals? *Feedstuffs* 1978;50:17–18.
8. Levy A, Perelman B, Grevenbroek M, Creveld C, Agbaria R, Yagil R. Effect of water restriction on renal function in ostriches (*Struthio camelus*). *Avian Pathol* 1990;19:385–393.
9. Degen AA, Kam M, Rosenstrauch A, Plavnik I. Growth rate, total body water volume, dry matter intake and water consumption of domestic ostriches (*Struthio camelus*). *Anim Prod* 1991;52:225–232.
10. National Research Council. *Nutrient requirements of poultry*. 9th rev. ed. Washington, DC: National Academy Press, 1994.
11. Cilliers SC. *Feedstuffs evaluation in ostriches*. PhD thesis. South Africa: University of Stellenbosch, 1995.
12. Just A, Jorgensen H, Fernandez JA. Prediction of metabolizable energy for pigs on the basis of crude nutrients in feeds. *Livest Prod Sci* 1984;11:105–116.
13. Cilliers, SC, Hayes JP, Maritz JS, Chwalibog A, du Preez JJ. True and apparent metabolizable energy values of lucerne and yellow maize in adult roosters and mature ostriches (*Struthio camelus*). *Anim Prod* 1994;59.
14. Swart D. *Studies on the hatching, growth and energy metabolism of ostrich chicks (Struthio camelus var. domesticus)*. PhD thesis. South Africa: University of Stellenbosch, 1988.
15. Goering HK, Van Soest PJ. *Forage fiber analysis (apparatus, reagents, procedures and some applications)*. Agriculture Handbook No. 379. United States Department of Agriculture: Agriculture Research Service, 1970.
16. Angel CR. Research update: Age changes in digestibility of nutrients in ostriches and nutrient profile of eggs as an indicator of nutritional status of the hen and chick, in *Proceedings*. Assoc Avian Vet 1993;275–281.
17. Swart D, Mackie RI, Hayes JP. Fermentative digestion in the ostrich (*Struthio camelus* var. *Domesticus*), a large avian species that utilizes cellulose. *So Afr J Anim Sci* 1993;23:127–135.
18. National Research Council. *Mineral tolerance of domestic animals*. Washington, DC: National Academy Press, 1980.
19. Mellet FD. Ostrich production and products. In: Maree C, Casey NH, eds. *Livestock production systems: Principles and practice*. Pretoria, South Africa: Agri Development Foundation, 1993.
20. Scheideler SE, Wallner-Pendleton EA, Schneider N, Carlson M. Determination of baseline values for skeletal (leg bone) growth, calcification, and soft tissue (liver) mineral accretion, in *Proceedings*. Assoc Avian Vet 1994;111–120.
21. Rae M. Degenerative myopathy in ratites, in *Proceedings*. Assoc Avian Vet 1992;328–336.
22. Mertz W, ed. *Trace elements in human and animal nutrition*. Volumes 1 and 2. 5th ed. Orlando, FL: Academic Press, 1986.
23. McDowell LR. *Minerals in animal and human nutrition*. San Diego, CA: Academic Press, 1992.
24. Mannion PF, Kent PB, Barran KM, Trappet PC, Blight GW. Production and nutrition of emus, in *1995 Proceedings*. Austral Poult Sci Symp (in press) 1995.
25. O'Malley P. Feeding for the future. In: Smetana P, comp. *Emu farming*. Miscellaneous Publication No. 37/94. Western Australia: Department of Agriculture, 1993.
26. Cilliers SC, Van Schalkwyk SJ. *Volstruisproduksie Klein Karoo Landboukooperasie*. South Africa: Oudtshoorn, 1994.
27. duPreez JJ. Ostrich nutrition and management. In: Farrell DJ, ed. *Recent advances in animal nutrition in Australia*. Armidale, University of New England, 1991;278.
28. Dawson TJ. Digestion in the emu: Low energy and nitrogen requirements of this large ratite bird. *Comp Biochem Physiol* 1983;75A:41–45.

Chapter 3
Biosecurity and Control of Disease

Simon M. Shane and Lori Minteer

INTRODUCTION

The ratite industry is expanding rapidly in both the number of ostriches and emus and the size and geographic distribution of flocks within the United States. The need for biosecurity and comprehensive disease control programs is predicated on the following considerations:

- Suboptimal egg production or hatchability, or mortality in chicks and mature breeders represents a potential for significant financial losses.
- Recent experience has shown that ratites are susceptible to a wide range of primary disease-causing agents (pathogens) affecting domestic livestock and commercial poultry in addition to opportunistic organisms.
- It is necessary to establish a ratite industry based on foundation stock which is free of vertically transmitted diseases.
- The susceptibility of ratites to economically significant diseases of poultry will progressively result in intensification of disease control measures imposed by state and federal agencies dedicated to the preservation of the health and profitability of U.S. livestock. Currently, the ex-farm value of the poultry industry is approaching $20 billion annually.

The fact that similar disease-causing agents affect both ratites and commercial poultry suggests that preventive measures applied by chicken and turkey producers will be effective in ratites. With the emergence of the world's poultry industry from the 1930s onward, bacterial, protozoal, and viral infections successively limited expansion and performance. Introduction of programs to promote hygiene and to separate different types and ages of stock restored acceptable levels of production.

With the transition from multiplication of ratites to a mature industry with end-products, larger farms will be established with flocks at high density. The development of both horizontally integrated cooperatives and vertically integrated production units will impose special risks, relating to introduction and dissemination of disease. Significant losses have occurred in the Republic of South Africa and in Israel following exposure of large flocks on both single and multi-age units to bacterial and viral infections.

Ostrich and emu producers will have to adopt procedures as used in the poultry industry to prevent introduction and spread of disease. The importance and economic impact of infection increases exponentially with the number and the age spread of birds and the proximity of flocks in a limited geographic area.

The poultry industry has developed a series of procedures collectively termed "biosecurity," which limit the occurrence and impact of disease [1]. Biosecurity encompasses selection of farm sites, layout of buildings, drainage and facilities, and establishing barriers to the introduction of disease, including fencing and decontamination facilities for personnel. Operational practices in the poultry industry include closed flocks, all-in-all-out stocking programs, restriction of visitation or entry to farms, supplying bulk-delivered pelleted feed, rodent control, approved disposal of carcasses and litter, and the use of showers and protective clothing. Not all of these procedures are appropriate or practical for ratite production at the present time, but it will be necessary to select and adapt techniques from the highly successful poultry model.

Failure to establish sound practices to control infection at this early stage of development of the industry will limit expansion and disease may become a significant barrier to long-term efficiency. High productivity of breeding and growing stock will be necessary to maintain profitability in the future in a competitive end-products oriented industry.

METHODS OF DISEASE TRANSMISSION

A wide range of bacteria, viruses, protozoa, and fungi affect ratites. Pathogens can infect flocks by the following routes:

- **Vertical transmission.** Bacteria and viruses are spread from the parent female through the ovary or oviduct to the developing egg. Embryos which are contaminated either die early in incubation or infection may result in posthatch mortality or a weak, stunted chick. In the poultry industry, salmonellosis and mycoplasmosis, spread by the transovarial route, were formerly significant problems. There is evidence that ratites are susceptible to both *Salmonella* and *Mycoplasma* spp., although clearly defined disease entities will only be recognized as the industry increases in size. If a vertically transmitted agent becomes established in one of the segments of the ratite industry, appropriate control measures will be required. These will include serological identification of infected parents or flocks, which may have to be eliminated from breeding programs. Alternatively, their progeny would have to be assigned to specific growout farms to be slaughtered for meat, hides, and by-products. This will reduce the rate of expansion of the industry and may result in loss of valuable, genetic material. Sound biosecurity and prepurchase examination will limit the possibility of introducing pathogens to a flock through infected breeding stock or by contaminated personnel or equipment.

- **Mechanical transmission on the eggshell.** Bacteria, such as *Salmonella* and *E. coli* can infect the surface of the eggshell as it passes through the cloaca. Contamination of the shell also occurs frequently after oviposition due to improper nest construction, a delay in collecting eggs, or unfavorable hygiene or drainage in breeder pens. Pathogens can penetrate the eggshell and infect the embryo shortly after oviposition. The egg undergoes cooling to ambient temperature after lay, creating a negative pressure which favors the passage of microorganisms from the surface of the shell through the pores and the outer shell membrane. The moist film on the surface of the egg at the time of lay, or damp nest litter or saturated soil in the pen, will encourage shell penetration, particularly with motile organisms, such as *E. coli* and *Salmonella*. In addition to shell penetration, surface contamination can result in indirect infection of clean eggs during incubation. Although the inner shell membrane can inhibit penetration of fungal spores and many bacteria, the developing chick is infected when the beak penetrates the air cell at internal pipping or shortly before hatch. Bacteria, such as *Pseudomonas* sp. and *Staphylococcus* sp., may reduce hatchability and chick liveability, especially when infection persists in incubators and the hatchery environment.

- **Direct infection.** Mature breeders or young stock may become chronic carriers of disease-causing organisms, such as *Mycobacterium* sp. (tuberculosis) and *Salmonella* sp. (enteric and systemic infections). Excretion of pathogens in feces will contaminate pens, trailers, and infect contact birds. In many cases, ratites can be clinically normal carriers of infection, representing a danger to flocks. Often, the stress of movement, relocation, or a new diet may activate latent *Salmonella* infection, resulting in high rates of excretion of pathogens after transfer to a new facility. As with other avian species, ratites may excrete viral pathogens before onset of clinical signs, such as depression, diarrhea, respiratory distress, or dysfunction of the nervous system. In poultry flocks, the viruses causing Newcastle disease, laryngotracheitis, and influenza may be spread after a short incubation period before clinical stages of the disease are recognized.

- **Indirect infection.** Vehicles, trailers, footwear, clothing, equipment, and feed bags care commonly contaminated by pathogens. This results in the spread of viral and bacterial infection among farms. Ratite diseases known to be transmitted by this route include tuberculosis, mycoplasmosis, salmonellosis, and avian influenza. As endoparasites become more significant with intensification of the industry, indirect infection may result in spread of coccidiosis, giardiasis, and helminth ova.

- **Insect vectors.** Mosquitoes can transmit equine encephalitis virus to emus and pox virus, which affects most avian species including ostriches. Small flying insects, such as *Simulium* and *Culicoides*, transmit filariid parasites of the central nervous system (cerebral nematodes) from avian reservoir hosts to susceptible emu flocks. In Israel, bornavirus is transmitted among ostriches by blood-sucking insects. As the industry matures, and as greater numbers of birds are held in proximity, insects will become a more significant mechanism of disease transmission.

- **Vermin.** Rats and mice, which are ubiquitous in open pens and barns, serve as the reservoirs of diseases such as pasteurellosis and salmonellosis. These infections can be transmitted to ratites through droppings deposited near feeders or in nest areas. A number of avian viral and bacterial diseases known to be transmitted by rodents have not yet been diagnosed in ratites. The potential exists for adaptation of pathogens, such as *Leptospira* sp. and *Pasteurella* sp., to ratites in common with other intensively farmed

Biosecurity and Control of Disease

livestock. Raccoons may be infected with roundworms, such as *Baylisascaris*, which affect emus. Ingestion of eggs of the parasite results in infection and damage to the brain by the larva which develop in the accidental emu host.

PERPETUATION OF DISEASE IN THE U.S. RATITE INDUSTRY

The current state of development of the ratite industry favors the perpetuation and dissemination of disease. Common practices which should be avoided as the industry expands include the following:

- trading and speculation resulting in extensive and frequent movement of stock;
- sale of birds at auctions where large numbers of ratites are gathered from numerous farms without regard to control of direct and indirect infections;
- improper decontamination of trailers used to transport ratites;
- frequent visitation by owners to other units to examine stock or to assist in management;
- failure to apply precautions to prevent indirect infection when visiting farms or handling stock;
- indiscriminate purchase of eggs and uncontrolled contract hatching;
- misuse of antibiotics, including inappropriate and excessive medication which promotes the development of drug resistant pathogens;
- deficiencies in biosecurity, such as allowing unrestricted visitation or trespass, or the entry of free-roaming animals (Figure 3.1)
- spillage of feed or accumulation of foliage and debris in the vicinity of pens, which contributes to high populations of vermin;
- open feeders which attract wild birds into pens;
- location of farms or pens in low lying areas which may become waterlogged or poorly drained pens, resulting in large populations of insect vectors; and
- operation of multispecies units holding ostriches, emus and rheas or farms housing a wide variety of exotic and domestic birds and livestock species together with ratites.

PREVENTION OF DISEASE THROUGH BIOSECURITY

Ideal Location of the Ratite Industry

An organized, end-product based ratite industry will prosper in areas with low populations of commercial poultry. In the Republic of South Africa, the Klein Karoo Valley is the location of the world's largest and most successful commercial ostrich industry. The area is surrounded by desert and has no commercial chickens and only a small number of confined backyard poultry and few migratory or free-living wild species.

Ostriches introduced to Israel, and located on kibbutzim (communal multispecies farms) in proximity to poultry, developed Newcastle disease soon after initiation of farming activities.

Emus and rheas in many areas of the United States have been infected with avian influenza serotypes common to backyard poultry and shore-bird reservoirs.

The development of a viable, commercial ratite industry in the United States will require location in areas remote from chickens and turkeys. The poultry industry is heavily concentrated in the mid-Atlantic and southeastern states due to the availability of water, labor for processing plants, available contract growers, and access to feed grains and markets. High ambient temperature, dry climate, and distance from markets and infrastructure, which make commercial poultry farming uneconomical, favor ratite production. It is possible that ultimately large-scale ostrich and emu operations will locate primarily in the southwestern states based on climatic and health considerations.

Selection of Farms

Ratite farms should be located in well-drained areas remote from rivers, dams, swamps, or large bodies of water. Facilities should not be adjacent to a road carrying significant traffic, especially for livestock and

Figure 3.1 Effective biosecurity requires fences, secure gates and constant supervision. Courtesy: Pacesetter Ostrich Farm.

poultry. Pens should not be located near the boundary of a farm and should generally be visible from the homestead.

Separate facilities should be established for growout, breeding, and trading. Although these activities may be conducted on the same farm, the pens housing these three classes of stock should be separated by at least 1,000 ft. Operation of multispecies farms should be discouraged. In the poultry industry chickens serve as clinically unaffected reservoirs of diseases, such as histomoniasis, which affect turkeys and rheas. Some infections such as pasteurellosis are more severe in turkeys than chickens. A similar situation may arise with concurrent multiplication of emus, rheas, and ostriches. Because all three species are relatively recent introductions to the United States, it can be expected that many pathogens which currently do not cause specific problems in exotic and domestic poultry may emerge as disease-causing agents in ratites.

Breeding units should be operated on a closed basis. Frequent purchases or trading of stock will increase the probability of introducing disease to a farm. As the industry matures, minimum standards of operation and serologic tests for specific diseases will be developed. Farms achieving disease-free status will be able to sell certified "clean" stock, paralleling the situation in poultry, cattle, and swine.

If either immature or mature birds are acquired, the purchaser should obtain a certificate of health from a competent veterinarian before shipment. Newly introduced birds should be quarantined for at least 4 weeks before transfer to breeding or growout units. During this time, birds should be regularly examined to ensure freedom from obvious developmental or disease-related conditions, and laboratory tests should be performed to confirm absence of parasites and specific bacterial infections.

Aspirant and established ratite farmers should only purchase stock from reputable breeders with appropriate health certification. The status and soundness of birds acquired at public auctions is always in question, and there is considerable risk of introducing disease due to concentrating birds at a central point or following successive ownership by speculators.

OPERATIONAL PRECAUTIONS TO LIMIT INTRODUCTION AND SPREAD OF DISEASE

Farmers should limit contact with other ratites, exotic birds, and poultry. Occasional visits to recognized breeders to examine or purchase stock can be justified, but appropriate precautions should be taken to limit introduction or cross transmission of disease. Vehicles should be parked outside the perimeter of a farm. Visitors should wear outer protective clothing, such as coveralls, boots, and caps, supplied by the host farm. If protective clothing is unavailable on the farm to be visited, laundered coveralls and disinfected boots can be taken to the farm and worn during the visit. These should be decontaminated before transfer to the home farm or another facility. Extreme care should be taken to prevent contamination of both footwear and the interior of vehicles. Many pathogens can remain viable for extended periods in soil and on fecal material adherent to boots and clothing. Transport vehicles, trailers, and portable holding chutes should be thoroughly decontaminated at a remote site before return to the home farm.

Veterinarians should stringently observe biosecurity procedures consistent with the precautions applied by commercial poultry breeders. Current standards of hygiene and disinfection common to small-scale cattle and swine operations are inadequate for ratite units with high-value stock vulnerable to infection.

Water supplied to ratites should be free of pathogenic organisms and should be chlorinated to a level of 1–2 ppm. City water generally conforms to this specification. Water obtained from farm wells should be chlorinated and should be examined by a laboratory at 6-month intervals to confirm freedom from bacterial infection and mineral contaminants.

Ratite owners should be aware of current disease problems and follow the advice of their veterinarians and state, university and extension advisors. Membership in a local or national emu, ostrich, or rhea organization is advised. Attendance at technical seminars where speakers are selected on the basis of professional accomplishment and experience will improve knowledge of newly emerging diseases and their prevention.

Generally, ratites mask signs of infection and only display clinical abnormalities such as depression at a late stage in a disease. In addition, wasting is difficult to detect without regular handling or weighing of birds. For this reason, chronic disease, including parasitism, can remain undiagnosed for long periods and can severely impact livability and growth of immature birds and the reproductive efficiency of breeders. Owners should examine their flocks on a daily basis and request professional advice in the event of unusual behavior, a depression in egg production, or a reduction in feed intake.

With the high value of ratites, it is beneficial for owners to arrange for a visit by a veterinarian or to transport immature birds to a veterinary clinic or

Biosecurity and Control of Disease

teaching hospital for clinical examination in the event of a problem. Because diagnosis of ratite diseases depends on both clinical and laboratory examination, early intervention expedites diagnosis, facilitates treatment, and may limit the spread of disease within the flock or to adjoining farms. A routine annual visit by a veterinarian is recommended to update insurance health certificates, administer vaccines, and collect fecal specimens to monitor for parasites. Additional routine procedures, such as cloacal swabs or ultrasound examination of nonproducing breeders in season, can also be performed. A visit to the farm by an experienced ratite veterinarian will allow the advisor to become familiar with the farm layout and flock. This provides the health professional with a valuable background for subsequent investigation of disease outbreaks or deviations from normal performance.

DECONTAMINATION AND DISINFECTION

Decontamination involves the physical removal of biological and inorganic material from surfaces and equipment. Disinfection comprises the destruction of pathogenic organisms, including viruses, bacteria, fungi, and protozoa. Decontamination and disinfection are mutually complementary and should be performed sequentially for optimum destruction of disease-causing organisms.

Decontamination requires physical removal of fecal material and physical debris from equipment, feeding and nesting areas, transport trailers, and hatcheries. The process involves application of a detergent solution to solubilize organic material, preferably at a pressure exceeding 100 psi. After surfaces have been cleaned, a disinfectant with virucidal, bactericidal, and fungicidal activity should be applied. Disinfectants function most effectively on previously cleaned surfaces. The use of combined detergent-disinfectants is acceptable in relatively clean areas such as hatcheries, but destruction of pathogens by disinfectants is inhibited on potentially soiled surfaces in brooder sheds or pens.

All detergents and disinfectants used should be approved by the Environmental Protection Agency and be licensed by the United States Department of Agriculture. Label regulations should be strictly followed to prevent potential toxicity and environmental contamination and to promote efficacy.

The following disinfectants can be used by ratite producers:

- **Cresols**. Cresylic acid derivatives have broad virucidal and bactericidal action and are suitable for application to earth floors. Cresols are relatively inexpensive. Care should be taken to avoid application to drinkers or feeders. This group of compounds should not be used in hatcheries or brooder units.

- **Phenols**. These compounds are available commercially in formulations with detergents for dual purpose decontamination and disinfection. Phenols are effective virucides and have broad spectrum activity against bacteria, especially gram-positive organisms. Some commercial phenolic preparations have prolonged residual action and may include bisphenols, which are active against fungi.

- **Iodophors**. These compounds comprise iodine together with a solubilizing agent. When used at 100–500 ppm, iodophors are active against viruses and bacteria and have a pronounced residual effect. Use of iodophors is limited by cost, and some compounds may stain and corrode incubators.

- **Quaternary Ammonium Compounds**. The spectrum of QACs includes viruses and bacteria at use levels ranging from 100-500 ppm. These compounds can only be used in conjunction with nonionic detergents.

- **Chlorine Compounds**. Chlorine is an inexpensive and effective disinfectant with a broad spectrum of activity against viruses and bacteria. Chlorine compounds are inactivated by organic material and sunlight and do not function in high pH (alkaline) environments. Sodium hypochlorite, available as a 70% solution, can be diluted to an operating strength of 100–500 ppm.

- **Formalin**. Although this compound was previously used extensively in the poultry industry as a surface disinfectant and fumigant, formalin is not recommended for use by ratite owners. The compound is potentially carcinogenic and highly irritant, and strict regulations by the Occupational Safety and Health Administration and EPA limit availability and use.

Selection of disinfectants and detergents should be based on cost effectiveness. The cost per gallon of working solution should be calculated from the purchase price of the concentrate and the appropriate dilution rate. Generally, QACs range from 2–5¢/working gallon, phenolics 10–15¢/gallon, and iodophors 8–12¢/gallon.

DISINFECTION OF EGGS AND INCUBATORS

It is necessary to collect eggs and disinfect shells within 2 hours of lay to reduce the possibility of penetration by disease-causing agents. Eggs produced toward the end of the season, or from hens fed diets deficient in

Figure 3.2 Emu breeder in well-drained pen.

Figure 3.3 Ostrich eggs should be collected from a clean nest area without manual handling.

Figure 3.4 Impervious surfaces in stainless steel work areas facilitate disinfection of ratite eggs. Courtesy: Pacesetter Ostrich Farm.

calcium or phosphorous, or from hens with defective oviduct function will be more susceptible to infection.

Breeders should prevent contamination of eggs by providing clean, drained nesting areas (Figure 3.2). Eggs should be collected frequently, which in the case of emus requires observation during the night.

Eggs should be collected and handled with disposable gloves or a plastic bag (Figure 3.3). Eggs should be transferred from the nest area to a clean metal or plastic tray for subsequent disinfection and storage. In the processing area, eggs should be placed on a stainless steel rack to be evaluated for cleanliness. Eggs free of adherent organic matter can be lightly disinfected with a phenolic solution applied by spray. After air drying, eggs can be stored at 65°F and 70% relative humidity prior to setting. Disinfectants should be nontoxic to embryos at recommended use rates. Disinfectants should not produce any residue on the shell surface which may interfere with air exchange through pores during incubation.

Small areas of contamination can be removed using a dry, disposable paper towel to dislodge adherent soil or feces. The affected area can be disinfected by application of a phenolic or QAC by spray.

Severely contaminated eggs can be washed, but the procedure may reduce hatchability by encouraging movement of microorganisms from the surface into the interior of the egg. Eggs should be immersed for 30–45 seconds in a 100°F phenolic or quaternary ammonium solution. Eggs are then removed and rinsed in clean water and dried using a sterile towel. An additional disinfectant spray should be applied and allowed to air dry before eggs are placed in storage or an incubator. Disinfection and hygiene in hatcheries is promoted by using impervious wall and floor surfaces and stainless steel table tops and racks. Wood, fiber board, and other porous material is unsuitable in egg handling areas (Figure 3.4).

REFERENCES

1. Shane SM, Halvorson D, Hill D, Villegas P, Wages D., eds. *Biosecurity in the poultry industry*. 1st ed. Kennett Square, PA: American Association of Avian Pathologists, 1995.

Chapter 4

Restraint and Handling of the Ostrich

John R. Wade

INTRODUCTION

The general principles relating to handling and restraint of ostriches are consistent with the recommendations in the corresponding chapter on emus. Juvenile and mature ostriches are large and powerful birds that can injure handlers and bystanders by crushing and kicking. The relatively high center of gravity, weight, and muscular conformation renders these birds susceptible to injury when incorrectly restrained. During the breeding season, males protect their territory and hens and will aggressively attack anyone entering their enclosure.

Ostriches are vulnerable to specific injuries arising from handling. These include the following:

- fracture of the wings due to incorrect restraint,
- fracture of the tibiotarsal bones and luxation of the tarsometatarsal joint during capture,
- lacerations of the skin,
- dislocation of the cervical vertebrae,
- compression injury of the flexible trachea, and
- peroneal nerve damage if mature birds are restrained in lateral recumbency.

Capture and restraint of ostriches require a hood and experienced handlers, chutes, or restraining stocks. The use of chemical restraint should always be considered when it is necessary to perform a physical examination. Anesthesia is recommended for even superficial operative procedures of short duration. Reference to the section on anesthesia will provide details of handling and restraint during induction and recovery from anesthesia.

ACCOMMODATION FOR OSTRICHES

Chicks

Facilities for chicks should be operated according to an "all-in-all-out" placement system. Groups of chicks of similar age should be housed in a common pen provided with a shelter. Under ideal conditions, each unit should be separated from adjacent pens by double fencing, leaving at least 3 ft of neutral unused space. Chicks are normally held in the pen until 3–4 months of age. By keeping age units separate, horizontal transmission of disease by the aerosol or fecal route is limited. This system has the following advantages:

- Chicks remain in the same pen without social disruption arising from addition or removal of birds from the group.
- The all-in-all-out system allows complete depopulation and decontamination between successive batches. This is necessary to prevent the development of significant populations of pathogens.
- Batches of young ostriches of similar age develop uniform immunity when vaccinated or exposed to disease-causing agents.
- In the event of a disease outbreak, it is possible to confine infection to specific pens as contact between different age batches is limited.

Recommendations relating to selection of pen sites and management procedures to promote hygiene are provided in the chapter on biosecurity.

Chick shelters can be constructed of metal, concrete blocks, fiberglass, or translucent flexible plastic sheeting. The type of construction is determined by climatic conditions. In areas with high winds, buildings should be securely anchored to a foundation slab. Where heavy snowfall occurs, the roof should be sufficiently strong to withstand the weight of precipitation.

Stocking density within shelters and pens is dependent on climatic conditions. In the warm, dry climate of the southwest, chicks can be turned out to an enclosed pen early in the morning and returned to the shelter at nightfall. Time spent in the shelter will therefore be limited. Under adverse climatic conditions with persistent rainfall or extreme cold, chicks may have to be confined to the shelter for long periods of time. Under these conditions, substrate should be examined daily to ensure that it does not become damp or produce ammonia. If chicks are confined to a shelter exceeding 3 days at a time, an allowance of 15 ft^2 floor space/chick should be allowed.

Chick shelters should be adequately ventilated. Attempts to conserve heat by reducing ventilation will result in accumulation of ammonia. Ammonia is deleterious to young chicks abve 30 ppm, damaging the respiratory mucosa, conjunctiva, and cornea. The ventilation system will depend on prevailing climatic conditions. In areas with low rainfall and a warm climate, an open-sided unit with curtains is adequate. Convection ventilated shelters are similar to broiler houses used in the southeast tier of the United States.

Since young ostriches raised in areas with high rainfall or in cold climates spend a proportionately longer time in shelters, a forced-air ventilation system is required. Extraction ventilation is preferred, with air passing through a roof inlet, and entering the house through a baffle which prevents drafts. Ventilation rate should be based on an allowance of 0.01 ft^3/minute/degree F ambient temperature. For example, at 70°F the ventilation rate for 10 birds weighing 10 lbs each, will be $0.01 \times 100 \times 70 = 70$ ft^3/min. Allowing a space requirement of 5 ft^2/bird and a mean roof height of 8 ft, the volume of the shed will be 400 ft^3. With an air displacement of 70 ft^3/min, air in the shed will be replaced approximately 10 times per hour. A system incorporating variable speed fans controlled by a thermostat is recommended for chick shelters.

In many parts of the United States and Canada, it is necessary to provide supplementary heat for young chicks brooded in shelters. Various types of heating systems are available, including suspended radiant brooders, forced air units, and heated floors. Radiant heaters are convenient and economical to operate. It is important to remember that these heaters emit radiant energy which warms any object, including chicks entering the zone of heating. These units should be installed in accordance with the manufacturer's recommendations, specifically with regard to capacity and height above floor level. Radiant heaters do not affect the ambient temperature of the shelter but heat the floor, substrate, and chicks. Because the intensity of heating varies with distance, chicks have the opportunity to adjust their temperature by moving within the zone of heating. Unfortunately, thermo-regulation in young ostriches is poorly developed and severe dehydration can occur with overheating. Death can result from chilling and is a common occurrence in newly established units. When using forced air heaters, the ventilation rate should be adjusted to attain the required air exchange rate. This is necessary to replenish oxygen and remove carbon dioxide, ammonia, and water vapor.

Care must be taken not to overheat chicks and cause dehydration when using heated floor systems. Ideal floor temperature should range from 63°F to 68°F and should never exceed 72°F.

Flooring and Substrate for Chicks

Producers in South Africa now use electrical or hot-water pipe underfloor heating with a concrete floor overlaid with a nonslip rubber coating. This is cleaned daily to prevent accumulation of manure.

A wide range of substrates are successfully used in the United States to brood ostrich chicks. Rough-surface, sealed concrete is adequate, although requiring considerable labor to clean daily.

Coarse sand can be used as a substrate for newly hatched chicks. Urine is quickly absorbed, and solid waste can be removed by daily raking. Sand should be replaced between consecutive batches.

Organic substrates, such as straw and hay, can be used to overlay concrete. These materials may be contaminated with *Aspergillus* sp., other molds, and also bacteria. Hay promotes the multiplication of disease-causing organisms, especially if damp. Pathogens excreted by a few of the ostrich chicks in a batch may proliferate in litter, which serves as a reservoir of infection. Ingestion of straw or hay can lead to impaction.

Compacted dirt floors can be used only for limited periods because they cannot be decontaminated. Persistence of *Salmonella* spp. or *Clostridium* spp. will lead to infection of successive batches.

Fencing for Chick Pens

Chick pens should allow from 250 to 400 ft^2/chick. Long narrow pens provide greater opportunity for exercise compared to square pens of the same area.

The site and design of a chick pen is based on considerations, such as drainage, prevailing winds, proximity to the perimeter of the farm, access to food storage, and loading points for trailers.

Chain-link fencing with 2 in × 4 in twisted wire mesh or plastic is suitable for chick pens. Depending on the age at which chicks are released, fences should be 3–4 ft in height. It is important to install fencing that prevents chicks from becoming entangled in the mesh. There should not be any breaks in the fencing or gates which may injure or trap the feet or necks of chicks. In suitable climates, frequently mowed grass is an acceptable substrate. In desert areas, coarse sand allows drainage.

Juveniles and Adults

The design and layout of pens for juvenile and adult breeders should include considerations such as prevailing wind, drainage, accessibility to the homestead and trailer-loading point, convenience in moving birds among pens, location of feeders and waterers, and specific requirements imposed by the topography and management system of the farm.

Space requirements for adult breeders depend on the management system. Paired birds require one-third to one-half of an acre in a rectangular configuration. When using colony breeding, each bird requires one-third of an acre. It is beneficial to round the corners of pens to prevent injury to hens if they are cornered by an aggressive male. Double fencing is recommended to create a neutral area between adjacent pens. This prevents males in adjacent pens from injury by fighting.

It is recommended that natural cover should be provided in the breeder pen, or alternatively, a shelter should be erected. Birds may feel more secure if they have an area in which to hide from aggressive pen mates and to limit exposure to environmental stimuli such as strangers or vehicles adjacent to the pen. Depending on climatic conditions, three-sided lean-to structures or enclosed climate control shelters may be required in pens.

Fencing for juvenile and adult ostriches should be from 5–6 ft in height. Nonclimb twisted wire, chain-link and wild game fencing with a mesh of 2 in × 4 in is recommended. Single-strand barbed or plain wire causes injuries and should be avoided. If wire fencing is used, the bottom strand should be at least 1 ft from ground level to avoid injury to feet. Birds are vulnerable to predators and feral dogs with single strand or defective fencing. Depending on the risk of theft or predators, an outer perimeter fence with an inner electric barrier may be required.

Special precautions should be taken to eliminate gaps in fencing and to ensure that gates and other installations cannot trap the head, neck, or feet of birds. Protrusions and sharp edges which might cause harm to birds should be removed.

Care should be exercised to remove all nails, remnants of fencing and foreign objects from the pen, which will be ingested and lead to traumatic ventriculitis.

CAPTURE, HANDLING, AND RESTRAINT OF OSTRICHES

Chicks

Hatchlings and young birds up to 30 days of age can be grasped by the base of the neck and lifted while supporting the abdomen. Small chicks are restrained simply by holding them above the ground.

Growing birds ranging from 40–100 lbs in weight can be restrained by straddling with the tail cradled between the legs of the handler with the arms around the keel. Ostriches of this size are usually sufficiently small to be manhandled into a shipping crate or trailer.

Juveniles and Adults

Several safe and acceptable methods of capture and restraint of juvenile and adult birds are available to ostrich farmers. Selection of a specific method should be determined by the purpose of capture, such as transport, physical examination, or administration of drugs. Birds should be attracted to the most convenient site in a pen to carry out the procedure using food or relying on the innate curiosity of ostriches.

Capture and restraint of an adult require two to three experienced, physically fit handlers. Additional helpers may be required to herd birds within a large pen. Capture and restraint depend on the temperament of the bird, breeding status and environmental conditions.

If a bird is socially conditioned to respond to humans and is curious enough to approach the handler within arm's reach, it may be gently grasped by the neck. At this time, the handler holds the skin fold on the dorsum of the neck immediately behind the head, flexing the neck so that the head is approximately parallel with the ground. Simultaneously, the beak is held firmly with the other hand. As soon as the lead handler catches the bird, the second handler approaches the rear, grasps the tail, and steadies the body.

If the bird will not approach the handler, a shepherd's hook can be used for capture. Because this device may injure the bird, handlers should be trained by an experienced stockman. The bird should be grasped with the hook around the neck immediately below the head. The neck is then flexed so that it is approximately parallel with the ground. If the head is lowered too quickly or too far, the bird may kick, injuring the neck or head, resulting in serious injury or death. After immobilizing the bird, the handler approaches the bird maintaining tension on the hook. The nape of the neck is then grasped, as previously described, and the head is immobilized.

If a bird cannot be approached with a hook, it should be herded into a corner. It is then allowed sufficient space to attempt to break past the handler, who can then hook the bird around the neck just below the head. When immobilized, a second handler should grasp the tail from behind.

If the bird is to be moved a short distance and released into a new pen or transferred to a trailer, the handler at the rear of the bird pushes forward. Simultaneously, the lead handler applies gentle tension to the neck, steering the bird in the desired direction. Alternatively, the handler at the rear may pull the bird backwards by the tail. Hooding is appropriate when performing any procedure, such as physical examination, microchipping, or administration of medication. Before the bird is captured, the lead handler places the hood over the forearm, transferring it over the head when the neck is immobilized by hand or the shepherd's hook.

Simple procedures can be carried out on a hooded bird in the standing position. For longer periods or when carrying out potentially painful procedures, the hooded bird should be placed in a commercially available stock, chute, or in a trailer where firm restraint can be achieved. The bird should remain hooded until released.

Various commercial aids to restraint are available. A simple, permanent restraint stock can be constructed on a farm. This comprises a triangular structure with a single wooden post in front and two posts in the rear, topped by a single rounded rail. Covered horse trailers are also suitable for restraint. The hooded bird is placed into the corner of the trailer, with pressure applied by handlers against the rear and the side of the bird.

Using the wings to restrain birds during capture and examination should be discouraged, as fracture of the humerus can occur. Because this is a pneumatic bone, subcutaneous emphysema is a complication of this injury. A compound fracture may result in cellulitis or airsacculitis.

Chemical sedation or anesthesia is generally required for minor surgical procedures such as suturing lacerations. Restraining birds during induction and recovery from anesthesia are reviewed in Chapter 9.

Transport

Young chicks can be transported in portable pet carriers. It is necessary to provide adequate ventilation, and the container must be free of sharp edges or protrusions which may cause injury. Young ostriches may easily entrap or damage their heads, necks, or feet in slots or spaces in the container.

Juvenile and adult birds can be transported safely in an enclosed horse trailer. Open stock trailers should not be used.

Care should be exercised to prevent damage to birds when loading and unloading. Before loading, the trailer must be inspected for protruding objects, damage to door hinges, and panels which may result in injury. The floor of trailers should not be slippery to prevent potentially fatal or disabling damage to legs or handlers. A thin layer of clean hay or straw should be spread on the floor to absorb urine and droppings. Trailers should be insulated and ventilated.

When transporting ostriches during hot weather, birds should be adequately restrained and cooled. Ostriches should never be left in closed, stationary trailers, as they are vulnerable to fatal heat stroke. Transporting birds at night is recommended, as they are calmer and less subject to thermal stress.

Administration of Stresnil® at a dose rate of 0.4 mg/lb body weight by the oral or intramuscular route will calm birds without inducing ataxia or excitation.

Chapter 5
Restraint and Handling of the Emu

David Mouser

INTRODUCTION

The restraint and handling of emus requires a practical knowledge of both the bird and its behavioral characteristics. Normal and agitated responses and anatomical and physiological adaptation to stress should all be taken into consideration. Whether confinement consists of a physical enclosure, hands-on physical restraint, or administration of sedatives, basic understanding of emu behavior is important to achieve consistent results. Most of the problems in the restraint and handling of emus are due to underestimating the strength and activity of the bird or alternatively applying excessive force.

The powerful leg muscles of the emu allow it to cover ground quickly in search of food and to avoid predators. The emu has a sharp nail at the end of each of its toes which is capable of inflicting serious injury. The tarso-metatarsus (shank) is covered with a rough, scaly hide which can tear through thick clothing and cause severe abrasions to the skin of handlers. The wings of the emu are vestigial, but the small bony projections are often injured on fences. The skin of the emu is thin and easily lacerated. The head and long neck are particularly vulnerable to injury. There is little muscle or supportive tissue over the cervical vertebra to cushion trauma or to provide protection during vigorous handling. The trachea of the emu can be easily collapsed.

By nature, the emu is inquisitive and curious of anything introduced into its environment. Often, birds may be examined or even captured by standing in the pen and rattling a keychain until the bird approaches to investigate. Most emus are fairly calm and many will not resist mild restraint. The female emu seems to be more nervous and harder to control, although some males may be extremely difficult to handle safely. Bonded pairs tend to defend each other and, although rare, unprovoked attacks by both male and female adults on handlers can occur during the breeding season.

The emu has limited ability to tolerate stress and violent or prolonged agitation. Respiratory and circulatory collapse, exertional myopathy, self-trauma, hyperthermia, and rupture of the great vessels have all been reported in stressed emus. It is wise to evaluate individual birds before, during, and after restraint to prevent injury or death.

FENCING AND PENS

A variety of types of fencing are currently used to enclose emus. Desirable characteristics of fencing material for the species include reasonable cost, high-breaking strength, elimination of the potential for climbing, and small openings to prevent head, neck, or foot entrapment or injuries. Fencing should also exclude predators and reduce trauma to the bird from "fence rubbing." Chain link fencing appears to fulfill most requirements for emu pens (Figure 5.1). Chain link has a certain amount of give which cushions impact. When mounted on polyvinyl chloride or fiberglass, the possibility of injury is further reduced due to the flexibility of the frame. Chain link fencing should be strung only moderately tight.

Shade cloth over fencing provides protection from wing rubbing and also prevents fighting between birds in adjoining pens. Shade cloth also provides protection from wind, blowing precipitation, and sun. The visual stimulus of other birds often distracts breeding pairs early in the laying season. Breeding hens often become aggressive toward their mates when exposed to younger birds in adjacent pens. Shade cloth strung over fencing prevents this problem even though the hens can still hear chicks.

Most enclosures should include a shelter for the birds. Feed should be supplied in this area and a dry,

Figure 5.1. Six-foot chain link breeder pens on large farm. Shade cloth between pens prevents fighting between breeding pairs. Scrub brush on fence is for cleaning and disinfecting footwear before entering pen.

Figure 5.2 The circular "carousel" breeder pen arrangement allows for eocnomical use of fencing and allows for easy access to feeders and rapid egg collection.

raised nest should also be provided for laying pairs. Some large emu farms use a multiple breeding pair enclosure system incorporating natural vegetation for shelter from weather and other birds. Feed is provided in feeders with covers activated by weather vanes.

Ideal pen size for emus had not been determined. Although many producers hold breeder pairs in small enclosures without obvious problems, some units report injuries and behavioral problems in small pens. Emus in small enclosures seem more likely to develop vices such as geophagia and pecking at fencing and shade cloth. When confined to pens, birds will remain within 4 feet of the fence most of the day, beating a path around the perimeter. Emus seldom move across the center of the pen.

If space is limited, pens should be long and narrow rather than square. This allows the bird more room to run or walk before encountering a corner. The minimum size for a breeder pen should be 30 × 100 feet. Pens should be equipped with portable dividing panels in the event of overly aggressive courting behavior. Long pens (150–200') are desirable for juveniles because they promote exercise essential for robust development of chicks. Exercise may reduce the prevalence of gastrointestinal impaction and angular and rotational limb deformities.

The carousel or round pen system (Figure 5.2) is becoming popular, as the areas for shelter, feeding, storage, and maintenance for all pens is centrally located. Because pens radiate from a central hub, labor for egg gathering and feeding is minimized. The carousel offers the advantage of security, limiting exposure to dogs or other predators along the perimeter of a conventional pen, and facilitates transfer of birds among pens. Cost of shelter and fencing is also reduced.

Care should be taken to prevent "Y-shaped" spaces around gates and corners which might entrap the neck over or under openings. Damage to the cervical spinal cord, leading to paralysis or death, or even decapitation, may occur when an emu attempts to free a trapped head. Posts should be on the outside of the pen to prevent injury to running birds. Double fencing of adjacent pens or padding the posts with "emu-proof" cushions may be considered when pens have a common border. All fencing should be inspected to ensure absence of nails, loose wires, or ties which might injure birds. Remnants of discarded fencing material which are easily ingested should be removed from pens before introducing birds. Even with all these precautions, unforeseen problems can occur. The author once removed a large bone from the intestine of an adult emu. The pen where the bird was housed was built over a cow cemetery and the bone apparently was exposed at the surface and was ingested.

TRANSPORT

Emus can be successfully transported in containers ranging from pet carriers to sophisticated, environmentally controlled trailers. In practice emus can be moved safely in any trailer or enclosure which allows sufficient room for the bird to stand or lie without being trampled. There should not be any openings in the trailer large enough for a head or foot to protrude. Emus should be transported during moderate climatic conditions. The transporter should be insulated and adequately ventilated. Most birds tolerate travel in semidarkness. Emus of different ages and other ratites should not be transported in the same compartment. The flooring of trailers should be solid, and seamless to avoid entrapment of toes or nails. Since most flooring soon becomes wet from droppings, hay, or some other suitable bedding should be placed on the floor to absorb moisture and provide a secure footing.

HANDLING OF THE EMU

Due to their small size and the relative ease in capture, chicks up to 3 months of age present no problems in restraint. It is important to remember that rough handling may lead to serious injury or death.

Juvenile emus from 3 to 8 months of age can be restrained easily for most examinations and treatment procedures. A young emu can normally be caught by approaching from behind and scooping the bird up with one arm between the legs with the hand on the sternum. The other hand is used to restrain and immobilize the head and neck. It is important to remember that emus may briefly struggle, sometimes violently, when caught. Care should be taken to keep the neck and head of the bird away from the thrashing feet. The same is true for the face and arms of handlers or bystanders who should be aware of the possibility of injury. Because emus very often defecate when captured, the vent should be directed away from other handlers. Most treatments and sample collections on young birds may be accomplished with light but firm manual restraint. Vent sexing and cloacal swabs may be obtained from young emus using two-handed restraint. Collection of blood from the medial tarsal vein can be accomplished safely by an assistant with an appreciation of the force and range of motion of the leg.

It is often advantageous to gain more extensive exposure of the vent area for sexing. The bird should be restrained with both legs held in one hand and the back cradled in the handler's other arm. Young birds seldom resist this form of restraint, and vent sexing is facilitated with the bird in this position. The examiner and handlers should be aware of the probability of projectile defecation.

Older juvenile and adult emus present more difficulty in handling. These birds have well developed legs, considerable body mass, and some individuals may be extremely fractious or aggressive. The safety of handlers, patients, and spectators should be considered

Figure 5.3 Restraint technique which allows acceptable control of bird. After the bird settles down, the hand that controls the neck can be used to examine and palpate the patient.

when attempting capture and restraint of juvenile and mature birds.

Many adult emus are tame and may be caught by simply approaching the bird slowly in a nonthreatening manner. Natural curiosity inhibits emus from running away. The bird can be grasped by one or both wings from behind and the body is then pulled slowly towards the handler. Some birds, especially those which have experienced human contact will not resist restraint and will tolerate handling. When restraining by grasping the wings, it is important to remember that injury and fractures will occur if the bird jumps or struggles when it becomes agitated. For this reason, it is usually better not to restrain any but the most gentle birds in this manner.

Generally, emus should be caught and restrained by placing one arm around the junction of the neck with the sternum. The body of the bird is pulled back against the handler and the bird is tilted slightly more upright than normal (Figure 5.3). The handler's other hand can be used to help restrain the bird or to administer oral or injectable medication. Blood may be drawn from the

Figure 5.4 "Flipping" an emu to allow vent sexing of young adult birds and examination of the genitalia and cloaca in adult birds. Subject positioned between two handlers and lower legs are simutaneously grasped.

Figure 5.6 The subject is gently lowered to the ground. A moderate upward lift on the legs will prevent struggling by the bird. Care should be taken when the bird is released to prevent injury to both handlers and the patient.

Figure 5.5 The bird is immediately picked up and rolled up on its back which is cradled by the handler's free arm.

jugular vein, and the neck, abdomen, and legs may be palpated while the bird is restrained in this manner. It is important for the handler to position legs and feet wide enough apart to provide good balance but the knees should be kept close together to position the bird in front of the handler.

Capture of nervous or untamed birds may be accomplished by other methods. Often, frightened birds will run parallel to the fence from corner to corner of the pen. If the handler is positioned 4 to 6 feet from the fence, the bird will often run up to the handler, hesitate slightly, before continuing, or will turn to run in the opposite direction. It is relatively simple to catch these emus as they pause or slow down. Some birds may not slow when passing the restrainer. The handler may be able to step into the path of the bird and "hook" the sternal area with one forearm, swinging the emu away from the fence. The back of the bird should then be firmly positioned against the upper legs and trunk of the handler. Emus caught in this way often "buck" briefly but can soon be restrained. It should be noted that some individuals may try and climb the fence or jump and strike at the handler as they pause in running. These emus may present a real danger to the restraining crew and caution should be exercised whenever this method is used.

Very nervous or aggressive birds are best captured in small enclosures or in the corners of the pens. Several people are required to corner these birds and satisfactory restraint is sometimes impossible. Chemical sedation should be considered when handling fractious emus for other than quick procedures. The use of xylazine, diazepam, and ketamine is relatively effective. Postsedation and recovery restraint is essential and the Ratite Wrap® (Jorgensen Labs) has been developed for this purpose.

Restraint for visualization or exploration of the cloaca of the adult emu is best done with the bird in dorsal recumbency. Two handlers are required for this procedure. They position themselves on either side of the bird facing forward. Both handlers simultaneously grasp the distal lower leg of the bird on their respective sides with their "outside" hand (Figure 5.4) and lift the bird off the ground (Figure 5.5). The "inside" arms cradle the bird's back and neck as it is inverted and gently lowered to the ground. A moderate upward lift on the legs of the bird seems to limit struggling and leg movement (Figure 5.6). Care should be exercised to avoid injury to the head and neck which may easily be stepped on while restraining the legs. Dorsal recumbency allows a veterinarian to perform necessary procedures.

A modification of the juvenile restraint procedure can be applied to adult birds to allow visualization of the cloaca or to insert a cervical microchip. The bird is approached from the rear. One arm is placed under the body between the legs as far forward as possible. The other arm is passed around the anterior sternum. The bird is then lifted and balanced on the hip of the handler. Few birds resist this technique and struggling is rare. While restrained in this manner, the adult emu has full range of motion of its feet. Care should be taken to avoid injury to the handler or other crew members. Severe self-injury to the head and neck by flailing legs is possible. This method of restraint requires a handler with above average upper body strength and experience.

Restraint devices such as walking halters and bird stocks may lead to a false sense of security and actually increase the possibility of injury to the bird.

Restraint of emus requires experience, patience and a knowledge of the individual habits of the bird and realization of the handler's limitations and the bird's physical ability.

Chapter 6

Reproduction

Karen D. Hicks-Alldredge

INTRODUCTION

Expansion of commercial ostrich and emu production during the past decade has resulted in a concerted effort to understand more about the reproductive biology of these species. There has been limited research on ratite reproduction and incubation in the United States. Much of the understanding of the basic principles of reproductive physiology and incubation technology has been extrapolated from poultry and other avian species [1].

ANATOMY

Female

Hens have a single left ovary and oviduct. The ovary lies dorsal to the abdominal air sacs and ventral to the cephalic pole of the left kidney. The size and appearance of the ovary varies considerably, depending on the physiological state of the bird (Figure 6.1). The inactive ovary is a small pink organ, the cortex consisting of numerous small white developing follicles and a medulla of highly vascular connective tissue. The follicle consists of granulosa cells, the oocyte, and varying amounts of yolk.

The yolk components are manufactured in the liver and transported via the blood stream to the ovary where deposit occurs in concentric rings of white and yellow yolk. The composition of each concentric ring is biochemically different allowing absorption of nutrients specific to demands of the stage of the development of the embryo. The yolk is surrounded by the vitelline membrane. The ovum consists in turn of the yolk, germinal disc, and vitelline membrane.

In poultry, yolk deposition occurs over a 7–11 day period, while the emu deposits yolk for 25–26 days. Yolk deposition studies have not been performed on the

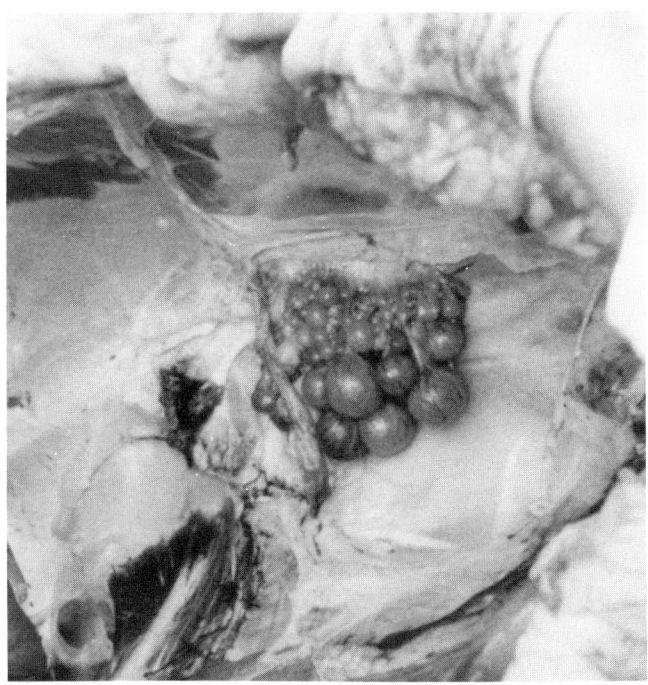

Figure 6.1 The single (left) ovary of a 2-year-old ostrich hen prior to onset of production.

ostrich. The time between onset of yolk deposition in the emu and oviposition (egg laying) varies from 31–39 days. This interval should be somewhat shorter in ostrich hens.

The functional oviduct lies on the left side of the abdominal cavity. The oviduct is divided into six anatomically distinct component parts:

- the infundibulum, where fertilization occurs within the first 15 minutes after ovulation;
- the magnum, where the dense albumin is deposited around the yolk;
- the isthmus, where the inner and outer shell membranes are formed sequentially;

- the uterus or shell gland, where water is absorbed into the albumin, the shell is secreted around the developing egg, and the mucin coat is secreted; and

- the uterovaginal sphincter, which separates the uterus from the vagina.

The vagina opens into the urodeum at the 10 o'clock position. Its size is subject to considerable variation, depending on the physiological state of the bird. The cells lining the oviduct actively synthesize the organic proteinaceous material which form the egg during passage down the oviduct. The histology and ultrastructure of the ostrich oviduct closely resemble that of the domestic hen and appear to have no special features relating to egg size [2]. The oviduct of the immature hen may be difficult to locate. The oviduct of a hen which has laid eggs retains considerable size and may occupy much of the abdominal cavity.

The infundibulum is transparent, funnel shaped and is located close to the ovary. It receives the extruded ova and is the site of fertilization. The presence of sperm host glands has not been documented; however, it is known that spermatozoa can survive for extended periods of time in the oviduct.

The longest portion of the oviduct is the magnum, lined with a highly glandular mucosa. Grossly, the mucosa has longitudinal folds that are tan to white in the normal oviduct. The short isthmus has more mucosal folding than the magnum and connects the magnum with the uterus.

The uterus is the most dilated portion of the oviduct and has a glandular mucosa which is pink to red in color, depending on the state of production. The vagina is the terminal portion of the oviduct and is responsible for forcefully expelling the egg into the cloaca. The muscles of the vagina are responsive to calcium and oxytocin. The role of prostaglandins in oviposition in the ostrich has not been determined.

The oviduct is suspended from the abdominal wall by cranial and caudal vascular ligaments.

Males

Paired testes are located dorsal to the abdominal air sacs and ventral to the cephalic pole of the kidneys, deep in the body cavity (Figure 6.2). The testes consist of seminiferous tubules, rete tubules, and vasa deferens. The simple epididymis lies on the dorsomedial surface of the testes. The ductus deferens leaves the epididymis as a fairly straight tube parallel to the ureter near the midline [3]. The ductus deferens has a saclike ampulla proximal to the ejaculatory ducts, which project into the dorsal part

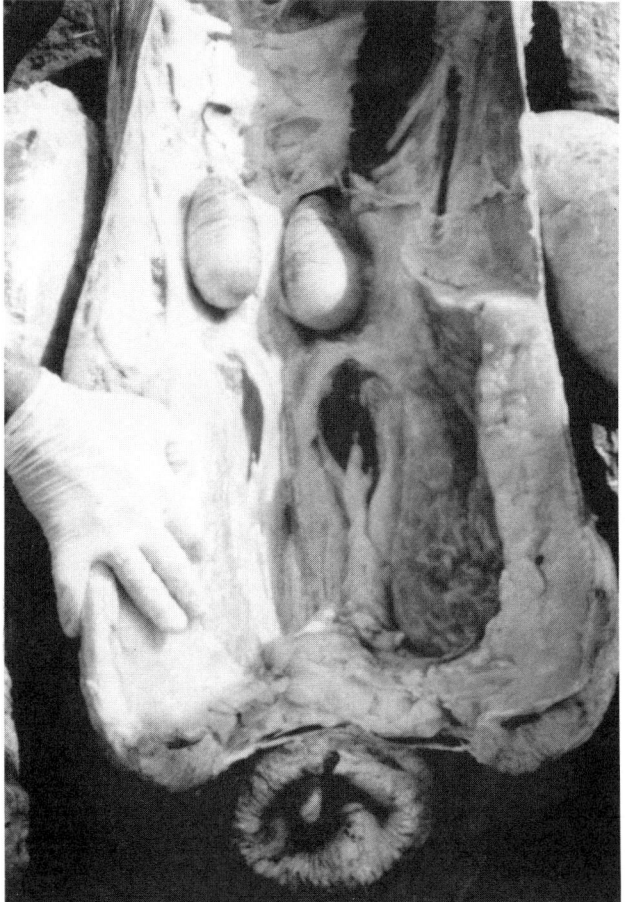

Figure 6.2 Paired testes of mature male ostrich located equidistantly from the midline in the body cavity.

Figure 6.3 Phallus of a mature ostrich, protruding from cloaca.

Reproduction

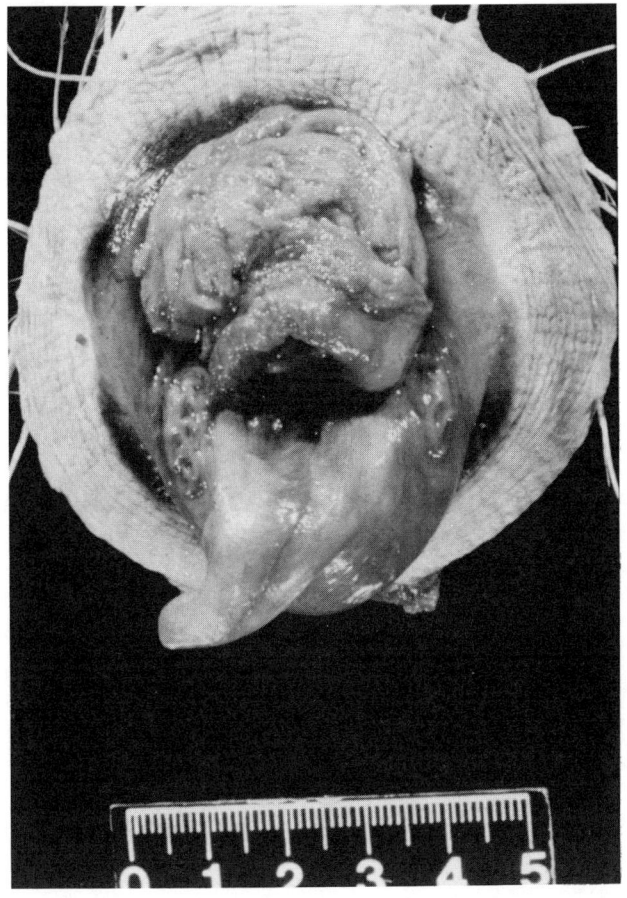

Figure 6.4 Phallus of a mature emu, everted from cloaca.

of the urodeum. The terminal proctodeum contains the phallus.

The phallus in the mature male ostrich is approximately 20–30 cm in diameter at the proximal end and 34–40 cm in length (Figure 6.3). The phallus of the emu has more of a spiral shape than the ostrich and is approximately 5–10 cm in length (Figure 6.4). There is a prominent dorsal groove with erectile tissue on either side in both species. When the phallus is erect and engorged with lymph, the groove functions as a tube to transport ejaculated semen from the urodeum to the cloaca of the hen. The phallus doubles in size during the breeding season.

The male ostrich chick has a phallus that is conical in cross section, contains a palpable core of fibroelastic tissue, and is characterized by the presence of a prominent dorsal seminal groove. The clitoris in the hen is laterally compressed, soft and lacks the seminal groove. Emu and rhea males are usually distinguished from the females by the spiral conformation of the phallus.

DNA sexing is an expensive technique that in the author's opinion is not accurate enough at this time to warrant the expense. Evolution to a commercial market will make DNA sexing prohibitive to most producers.

DNA fingerprinting as well as sexing is available from several commercial sources. The technique has been used in population dynamic studies in some avian species to establish genetic information in large aviaries and zoos. The technique has potential for helping establish registries for valuable ratite breeding stock.

PHYSIOLOGY AND BEHAVIOR

The ostrich becomes sexually mature at 2 to 4 years of age. Factors affecting the onset of maturity include:

- Subspecies. The smaller south African subspecies matures earlier than the larger north African subspecies.
- Season. Birds hatching during a period of increasing day length mature faster than those hatched during a period of decreasing day length.
- The nutritional plane of the bird.
- Environmental and housing conditions.

Ostriches are photoperiod dependent, coming into season during periods of increasing daylight. In the United States, the onset of the breeding season varies according to latitude. Birds in the northern United States have a defined laying season from May to September. Birds in the southern United States may produce all year. The effect of temperature on season is unknown, but climatic extremes inhibit production.

The rhea is similar to the ostrich and is a long-day breeder.

Emus reproduce during from November to March in the United States in response to a decreasing photoperiod. In the southern hemisphere, all three birds reproduce during the corresponding opposite season.

Males

The testes of the ostrich and, presumably, other ratites may quadruple in size during the breeding season. Cocks do not produce spermatozoa during the non-breeding season. In the ostrich and rhea, testosterone production increases with day length. Secondary sexual characteristics, such as reddening of beak and legs, vocalization, and territorial displays, appear in the male ostrich. Production of spermatozoa, which is controlled by follicle-stimulating hormone (FSH), commences at this time. High exogenous testosterone levels exert a negative feedback on gonadotropin (FSH) production.

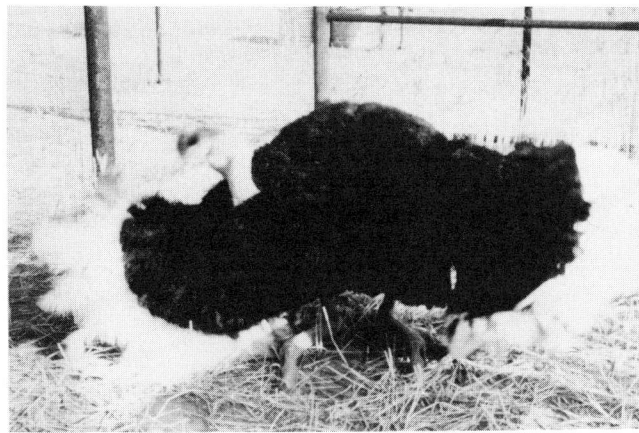

Figure 6.5 Male ostrich in typical courtship position (kanteling).

Indiscriminate administration of parenteral testosterone to males is generally counterproductive and will not enhance fertility.

The ostrich cock displays to the hen during the breeding season by dropping to his hocks, fanning his wings, and striking the back of his head on either side of his back, which is called kanteling (Figure 6.5). This display and the booming vocalization characteristic of males in season is also demonstrated when a visitor approaches the enclosure. The ostrich hen flutters her wings, drops her head and makes a snapping motion with her beak. A receptive female will drop to the ground with her head extended. The male mounts from the left placing his right foot on the back of the hen. He drops to his hocks and intromission occurs. During copulation, the male strikes the back of his head along his back on either side of midline. When ejaculation occurs the male extends his head forward making a guttural sound.

The male emu makes a growling sound rather than booming. During copulation, the female sits on the ground with her cloaca extended. The male drops to his hocks behind her and intromission occurs. The male pecks the back of the hen during mating and makes guttural noises during ejaculation.

Emu hens are dominant and vocalize by making a drumming sound during the breeding season. Male emus gather the eggs and are responsible for incubation in the wild. Ostrich share incubation; the male gathers the eggs and sits on the nest at night, the female incubates during the day. Ostrich also share in brooding the chicks (Figure 6.6).

The male rhea incubates eggs and broods chicks, aggressively displacing the female. All three orders of birds are gregarious by nature.

Avian semen can be collected by electroejaculation, forced massage or voluntary ejaculation. Electroejaculation is used in zoos and requires the bird to be anesthetized. Forced massage is used in the poultry industry and requires physical restraint of the bird.

Ostrich semen has been collected by a combination of forced massage and voluntary response. Collections are heavily contaminated with urine, making assessments of sperm concentrations, volume, and pH unreliable. Administration of 10 IU of oxytocin intravenously before collection may facilitate sampling in most birds.

A male emu that is imprinted sexually on a human can be induced to ejaculate. The ventral neck area should be lightly rubbed (Figure 6.7), resulting in the bird dropping to his hocks with an erect phallus. Using a gloved hand or a bull semen collection cone (Figure 6.8), pressure is applied over the dorsal phallus (Figure 6.9). As the cock ejaculates, he will peck the collector, who should be provided with padded clothing. In Australia, semen has been stored for 15–18 hours. Commercial chicken semen extenders have been used successfully as well as fresh unextended semen for insemination (Figure 6.10).

Females

In the ostrich, follicular maturation is controlled by FSH, which is released from the pituitary gland under stimulation of gonadotropin hormone during increasing photoperiod. The ovary contains approximately 200,000 immature follicles at hatch. Gonadotropins stimulate the ovary at sexual maturity, and a hierarchy of follicular development is established. The follicles undergo a period of slow development over months to years, followed by a rapid growth phase lasting 7 to 11 days before ovulation. During this period, concentric layers of yolk are deposited in the follicle. Each maturing follicle is steroidogenically active, as is the postovulatory follicle. This hierarchy of follicles, F1, F2, F3, and postovulatory follicles, must be maintained for ovulation to occur. In most avian species, the administration of exogenous gonadotropins results in a preponderance of F2 follicles blocking ovulation [5].

Ovulation is controlled by luteinizing hormone (LH) as in mammals. Avian species have three LH peaks during a 24-hour period rather than the two in a 21-day mammalian cycle. Ostrich hens are indeterminate layers; they will continue to lay as long as eggs are removed from the nest. In the wild, an average clutch for ostrich comprises 20 to 22 eggs. In the United States, ostrich hens average 45 eggs/year, laying every

Figure 6.6 Young ostriches held in pen with mature male "caretaker."

48 hours. In Australia, emu hens average 21–25 eggs in captivity and normally lay at 72-hour intervals.

EGG FORMATION

The infundibulum engulfs the follicle prior to ovulation and propels the ovum down the oviduct. Fertilization occurs in the infundibulum. Occasionally, due to injury, disease, obstruction or neural dysfunction or malformation of the infundibulum, the ovum will not enter the oviduct and is released into the abdominal cavity (see Figure 6.11a).

The magnum secretes the dense albumin protein around the ovum, in response to pressure on the secretory glands. The ovum is propelled into the isthmus where the two shell membranes are secreted loosely around the albumin. While in the uterus, water, vitamins and mineral salts pass through the shell membranes into the egg and migrate toward the yolk in response to an osmotic gradient. This results in a thin layer of albumin followed by a thick layer of albumin next to the yolk. The egg shell is deposited and finally a coat of mucin is applied before oviposition. While in the uterus the egg is constantly rotated as new components are added. Understanding normal oviductal function is important for practitioners performing and interpreting ultrasonography.

The air cell is formed between the two shell membranes at the most porous end of the shell as the contents cool. As water evaporates across the shell the air cell increases in size. Eggs of all avian species lose approximately 15% of their initial weight through evaporative water loss during incubation. The majority of embryo respiration takes place across the air cell. The mucin coat, which is water soluble, dries and forms a barrier for the first 7–10 days of incubation. It forms the first line of defense of the embryo against bacterial invasion. Other defense mechanisms include the inner shell membrane. Lysozymes in the albumen destroy invading bacteria and avidin binds with B vitamins which then are unavailable for bacterial growth.

In most avian species, the ovum remains about 15 minutes in the infundibulum, 3 hours in the magnum, 1.5 hours in the isthmus and 20–21 hours in the uterus before passing through the vagina. In the emu, rapid yolk deposition occurs for 26 days, but eggs are not laid until 10 days after yolk deposition is completed. Whether the delay in egg formation is due to holding the mature ovum within the follicle thus delaying ovulation, or is associated with slow passage through the oviduct is unknown. Ostrich eggs can be visualized in the uterus, using ultrasonography for periods exceeding 24 hours before oviposition, supporting the theory of delayed passage rate.

There are no anatomic structures to prevent the egg from undergoing reverse peristalsis. This can result in egg peritonitis as well as intraabdominal deposition of partially formed eggs.

BREEDER MANAGEMENT

Breeder management is an area dominated by anecdotal information and observations, and opinions are often advanced without scientific verification.

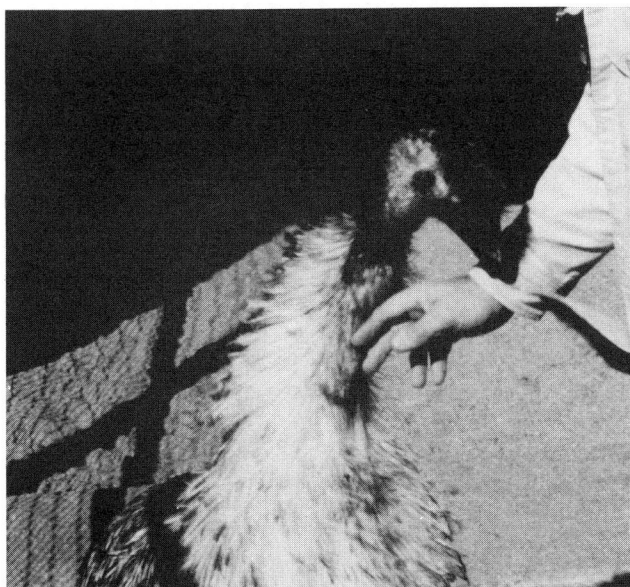

Figure 6.7 Stimulating the neck of a male emu to initiate release of semen.

Figure 6.8 Cone used to collect semen.

Ostrich, emu, and rhea are all gregarious by nature. Group breeding situations seem to be best in relation to economics, fertility, and ease of management. The standard in the United States is to keep birds in pairs or trios. This system not only costs more from the standpoint of feeding extra males and added fencing and labor costs, but fertility is lower with this system compared to group-bred flocks. If breeders are to be maintained in groups, ostrich and rhea should have 20–30% males while the emu should have 60% males. Reports from South Africa claim that up to one-third of ostrich males are infertile. Although there have been no controlled studies in the United States, practitioners recognize male infertility as a significant cause of economic loss to multipliers.

Separating males and females during the nonbreeding season is an important management procedure that is often neglected. Separation allows the male and female birds to be fed separately so that body weight can be controlled and a 1% calcium diet can be fed during the off-season. The results of reuniting males and females appears to have a positive effect on inducing lay, but it is unknown how this is brought about.

The nutritional requirements of ostrich, emu, and rhea vary considerably. A balanced breeding ration supplied by a reputable feed company is the best feeding system at present. Excessive feeding leading to obesity is more of a problem than specific nutritional deficiencies at this time.

REPRODUCTIVE FAILURE

On average, 30–40% of ostrich eggs are infertile in the United States. This represents a significant economic loss at current market values. A diagnosis of infertility is established by examining ostrich eggs that are designated as "clear" by candling at 7–10 days of incubation. Egg breakout differentiates between infertility and early embryonic death. The embryonic disc floats up on breaking the egg shell at the air-sac end and can be examined for possible development (Figure 6.11b). This procedure is best performed during the 5- to 10-day period of incubation, because early-dead embryos cannot be visualized at hatching time. Early embryonic death has many causes, including nutritional deficiency, toxins, improper egg storage, and inappropriate incubation.

Failure to copulate, due to behavioral problems, is the most obvious cause of infertility. Although ostriches are gregarious in nature, with one male breeding several hens, they display individual preferences. Incompatibility is common when birds are not allowed to select their mates. Environmental conditions may also result in behavioral infertility. Proximity to high-voltage power lines, machinery or oil field equipment, presence of predators, and transfer of ostriches among pens interferes with normal breeding behavior.

Investigations of the effect of flock size on fertility have yielded conflicting results. Only small numbers of birds have been evaluated and their age distributions were skewed. This inconsistency suggests that further studies are necessary to determine if pairs or trios have fertility rates different from larger groups. Managing breeders in groups rather than pairs or trios should increase fertility.

Reproduction

Figure 6.9 Pressure on phallus with cone releases semen for collection.

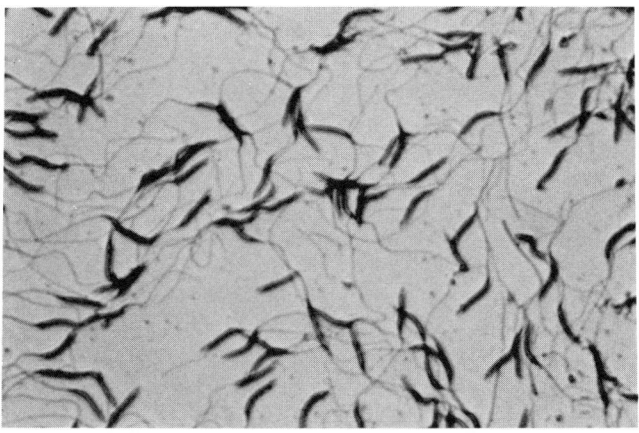

Figure 6.10 Micrograph of emu semen showing spermatozoa.

Seasonal infertility is common early in the breeding season when the hen comes into production and before the cock produces mature spermatozoa. Infertility also increases toward the end of the season or under temperature extremes.

Nutritional causes of infertility have not been documented.

Anatomical causes of infertility include anomalies in the phallus that prevent copulation, such as deviation or lack of a seminal groove. Hens may ovulate internally when the oviduct is deformed or if conformation displaces the infundibulum from the ovary. Hens that ovulate internally develop a pot-bellied appearance.

Intersexes occur in ostriches. The black pigment of male plumage results from low levels of estrogen. A mature black bird judged to be a hen by cloacal examination will not reproduce and may have inactive ovaries, or testes, or ovotestes. Some young hens may be very dark brown or even have black feathers. These hens should not be confused with mature "black hens" because many young 16 to 24-month-old hens will turn brown as they mature and will subsequently reproduce.

REPRODUCTIVE DISEASES

Many diseases may result in reproductive dysfunction, either suppressing formation of ova, ovulatory failure or production of abnormal or contaminated eggs.

Bacterial

Oviduct infection (salpingitis) occurs in the ostrich and emu. Etiological agents vary, as does the severity of the infection. In mild cases, only the uterus of the shell gland is affected (metritis), and clinical signs range from abnormal appearance of the shell to failure to produce eggs. There may be an associated salpingitis or peritonitis, depending on the duration or source of the disease process. Infection may result from muscular fatigue of the oviduct during egg production, allowing an ascending bacterial infection from the cloaca. Salpingitis may extend from airsacculitis or follow a perforation of the abdominal cavity or intestinal tract by a foreign body.

Common isolates cultured from the reproductive tract include *Escherichia coli*, *Pseudomonas*, *Acinetobacter*, and other gram-negative bacteria. *Mycoplasma* spp. which have been isolated are ostrich-specific, distinct from poultry pathogens, and their role in disease has not been determined.

Affected hens generally present with a history of erratic egg production, malformed or malodorous eggs, or a sudden cessation in production. On physical examination, there are no consistent abnormalities in temperature and respiration. Accumulation of a fluid discharge below the cloaca may be noted, resulting in hens having a peculiar odor. Affected hens often have leucocyte counts ranging from 20,000 to more than 100,000 cm^3. The differential ranges from a pronounced heterophilia in acute cases to lymphocytosis in cases of longer duration. Serum chemistries are frequently unremarkable. Ultrasonography and radiology are useful in assessing the amount and consistency of exudate in the oviduct. Paracentesis is indicated to evaluate the nature of abdominal fluid (Figure 6.12).

FERTILE YOLK

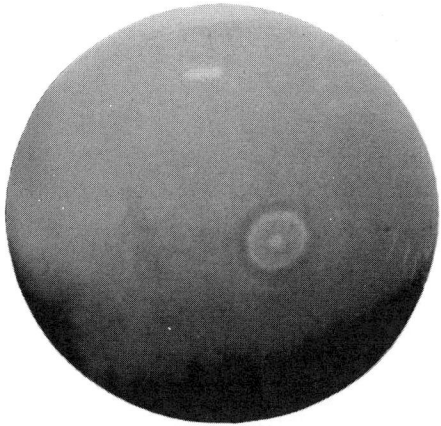

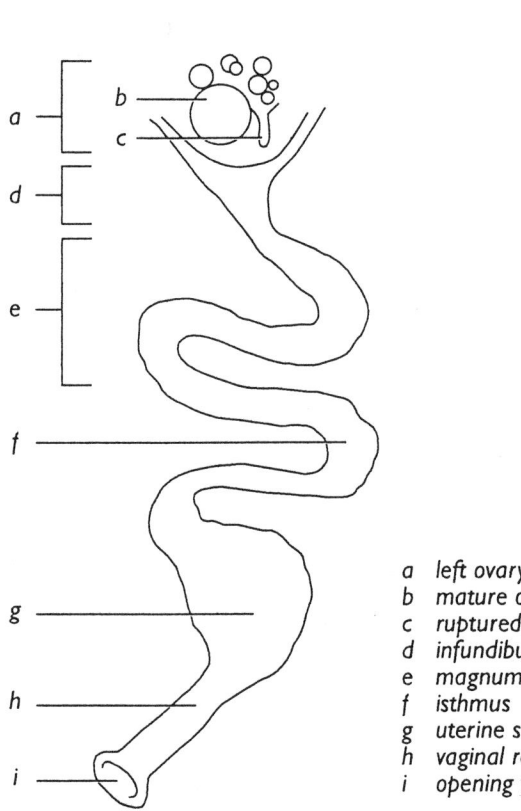

Figure 6.11b Diagram of avian egg showing principal structures including the shell, membranes, albumen, and yolk. Courtesy: Inner-Vision Bio-Technologies Corp.

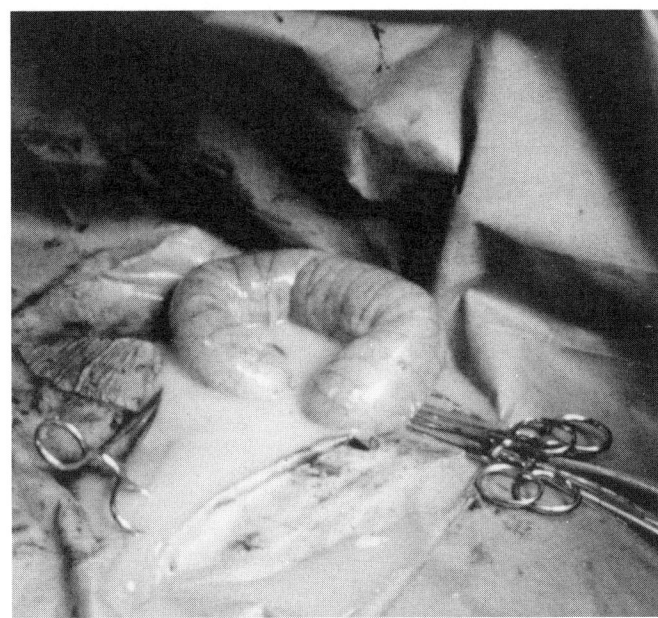

Figure 6.12 Presence of egg yolk material in the body cavity of an ostrich due to internal laying is removed at surgery.

a left ovary
b mature ovum
c ruptured follicle
d infundibulum
e magnum
f isthmus
g uterine shell gland
h vaginal region
i opening to urodeum

Figure 6.11a Diagram of female reproductive tract.

Treatment should be based on culture and sensitivity of the organism isolated. In cases of oviduct exudate, lavaging of the tract may be indicated. This procedure requires surgical placement of a Foley catheter in the magnum or isthmus with normograde flushing of debris from the oviduct. This is the method of choice when large amounts of exudate are present. A Foley catheter (size varies according to the bird) may also be placed in the vagina or uterus and retrograde lavage performed; using a 1% povidone-iodine (Betadine) solution; or buffered solution of gentamicin (Gentocin). Retrograde flushing does not allow visualization of the oviduct and can result in introduction of exudate into the abdominal cavity.

The hen should also be treated with systemic antibiotics as determined by sensitivity. Ciprofloxacin hydrochloride at 20 mg/kg twice a day or doxycycline at 10 mg/kg twice a day for 14 days or cetifour sodium at 20mg/kg once a day have all been used without adverse effects.

Peritonitis and death following a reproductive tract infection is a significant sequel of salpingitis in mature hens.

Salmonella spp. can frequently be isolated from cloacal and vaginal cultures. This is significant not only in regard to vertical transmission but also has implications for food safety. *Salmonella* sp. are frequently isolated from emu and rhea, and infection is attributed to contaminated feed and rodent infestation. *Salmonella enteritidis* was identified in emus imported into the United States from Europe. Although not commonly associated with metritis, *Salmonella* spp. may be transmitted mechanically or vertically through the egg.

Cocks may develop ascending infections of the seminiferous tubules. *Enterobacteriacea* are generally isolated and treatment should be based on isolation and sensitivity of the organism cultured. The prognosis is guarded and the role of the male in transmitting salpingitis or metritis from hen to hen should be considered.

Egg Peritonitis

Peritonitis due to passage of yolk material into the peritoneal cavity is a common condition in avian species. Peritonitis may occur from either the failure of ova to enter the infundibulum of the oviduct or from rupture of the oviduct, or from infection of surrounding tissues and weakening of the ovarian wall or from injury. The condition may also occur as a result of obstruction of the oviduct or reverse peristalsis.

The peritoneal cavity of affected birds contains varying amounts of yolk material which may be sterile or contaminated by bacteria. Generalized peritonitis with adhesions among various organs is common.

Diagnosis of egg peritonitis is based on clinical signs and hematology but may be confirmed by ultrasonography and abdominocentesis. A severe inflammatory response is reflected in leukocytosis. Abdominocentesis to obtain a culture and sensitivity is recommended. Ultrasonography may help to determine if surgery is required. Most birds recover following appropriate antibiotic therapy, but if large amounts of yolk are present, surgical intervention may be necessary.

Viral Infection

Papillomavirus has been isolated from the reproductive tracts of both male and female ostriches, although pathogenicity has not been determined. Most avian species are affected by viral infections of the reproductive tract. Research is needed to determine the possible role of adenovirus, coronavirus or paramyxovirus in reproductive dysfunction.

Egg Bound Hens

Hens may become egg bound. It is accepted that this tendency may be partly genetic but is complicated by malnutrition, cold weather, or lack of exercise. Although some hens may present with a history of straining or with a vaginal prolapse, the majority exhibit no clinical signs. Although an egg may be palpated in the caudal abdomen in the emu, it is difficult to locate in the ostrich. Radiology or ultrasonography may be required to diagnose the condition (Figure 6.13).

Ultrasonography is a valuable technique to assess the status of the reproductive tract in ostrich hens. The ovary and oviduct can only be visualized during the reproductive phase of the cycle or if pathology exists. A 3.5 mhz probe is required for most hens. The focal distance of the 5.0 mhz probe is too short for large hens but is adequate for thin ostrich hens or emus. Cystic structures within the ovary or oviduct can be visualized as well as eggs or exudate within the oviduct or abdominal cavity. Assessing the normal nonproductive tract is of no value.

Medical treatment of egg binding involves increasing the ambient temperature of the bird and administration of calcium and oxytocin. An attempt to break the egg in the uterus is contraindicated in both the ostrich and emu because of the thickness of the eggshell. Surgical intervention is, however, frequently required. If the vagina has prolapsed, the egg may have to be broken to allow repositioning of the vagina before extraction of the egg remnants.

Cystic Ovaries

Prolonged egg laying is a problem in companion and exotic avian species as well as in ostrich. Once the number of eggs for a clutch size is attained, hormone levels change as the female exhibits broodiness. Continually removing eggs from the hen creates a

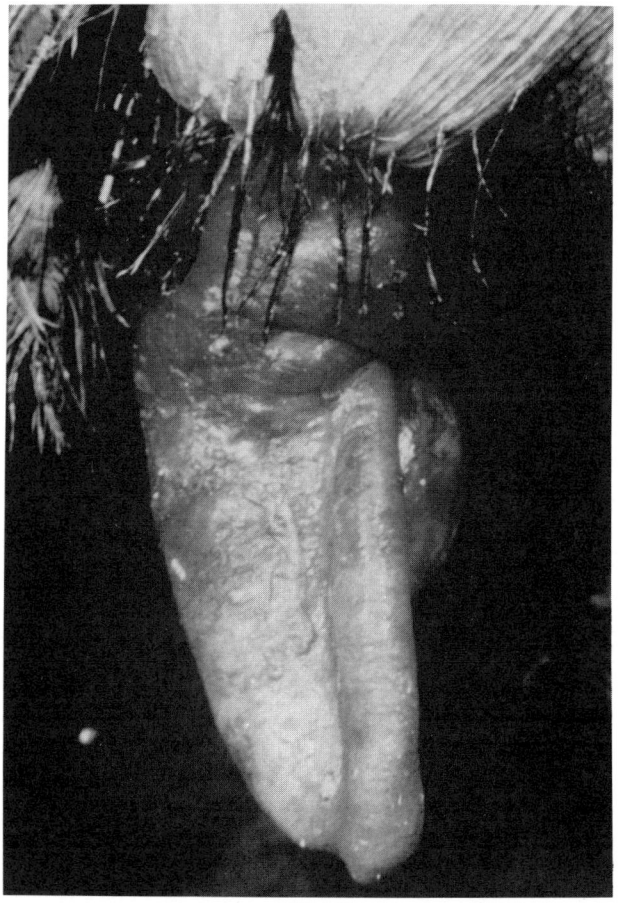

Figure 6.13 Prolapse of the phallus in an ostrich.

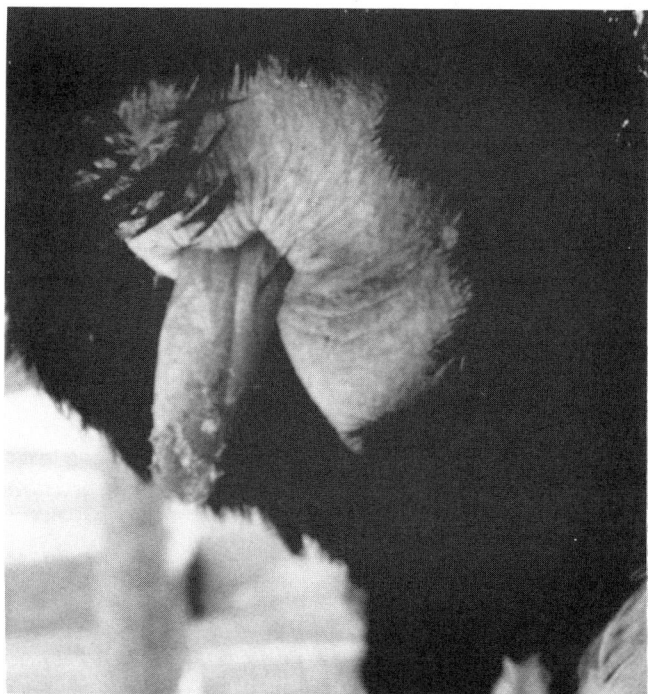

Figure 6.14 Partial prolapse of the phallus of an ostrich.

condition in which the ostrich hen does not cease production and can result in a cystic ovary and cessation of egg production. Diagnosis is based on ultrasonography and abdominocentesis or folliculocentesis. Treatment with progesterone or gonadotropins is generally unsuccessful.

Tumors

Ovarian neoplasms include granulosa cell tumors and adenocarcinomas. A recent case of lymphoid leukemia affecting the reproductive tract, and abdominal viscera has been recorded in an ostrich. Lipomas involving the abdomen and subcutaneous tissue are seen, frequently associated with low egg production and infertility. Tumors of the male reproductive tract have not been reported.

Prolapse

Prolapse of the phallus is common in birds that are debilitated from disease, or are at the end of the breeding season, or have been subjected to climatic extremes (Figure 6.14). The specific cause of prolapse is unknown. Frostbite and necrotizing dermatitis are frequent sequelae to prolapse. The prognosis is good if prompt therapy is instituted to avert damage.

Treatment consists of repositioning the phallus after obtaining cultures and thoroughly cleansing the organ. This should be repeated 3–4 times daily, together with administration of parenteral antibiotic and application of a topical antibiotic selected on the basis of sensitivity. Affected males should be separated from females. If the prolapse does not respond to repositioning and antibiotic therapy, a purse string suture should be inserted into the cloaca to retain the phallus in the cloaca. There is a fossa (pocket-like structure) in the dorsal aspect of the cloaca that contains the distal tip of the phallus when it is retracted into the cloaca. Sutures should not be placed through this structure. Some male ostrich do not have this structure, and when they become sexually mature they continually display 5 cm of phallus (Figure 6.15). There is no treatment for this abnormality and these males should not be bred.

Prolapse of the vagina and oviduct may occur without egg laying and has been seen in 12-month-old hens. Prolapse is associated with cold weather stress. Replacement of structures and retention with a purse string suture is indicated. The long-term prognosis is unknown. In Australia, young ostrich hens coming into

Reproduction

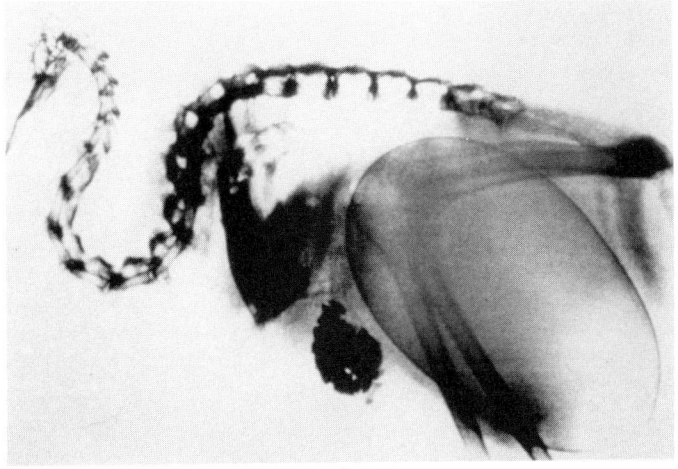

Figure 6.15 Radiograph of a kiwi showing egg retention. The kiwi produces the largest egg in relation to body mass of any avian species.

production show prolapse with the presence of a fluid-filled structure which requires lancing and repositioning of the tract.

Some ostrich hens coming into production may also have what appears to be a prolapsed cloaca which, when palpated, reveals a pedunculated mass originating from the dorsal surface. Microscopic examination of these masses reveals connective tissue interspersed with large round cells. The etiology and diagnosis are unknown at this time, but surgical removal restores reproductive performance. Tissues should be screened for *Mycobactorium* spp. (Avian tuberculosis).

Peritoneal hernias may occur in the caudal abdominal cavity, allowing the intestine and uterus to prolapse into the pericloacal region. The hen appears to have a large swelling around the vent. Ultrasound examination is diagnostic, and prompt surgical intervention is required.

Abnormal Eggs

Eggs with rough-textured surfaces, ridges, a lack of mucin coat, or soft shells may be an indication of metritis (Figure 6.4). Soft shells may also occur as a result of calcium deficiency or any factor that interferes with neurogenic control of oviposition or retention of the egg in the uterine shell gland. Some medication can cause dysfunction of the oviduct as well as interfering with shell production. Drugs such as sulfonamides inhibit carbonic anhydrase and result in thinning of the egg shell. Yolkless eggs may be caused by metritis, deposition of yolk into the peritoneal cavity, or abnormalities of the ovary. Double yolks reflect abnormal egg passage, overfeeding, and, in the emu, immaturity at onset of production. A detailed discussion of drug and nutritional interactions on egg shell quality is included in *Pathogenesis of the Avian Embryo* by Romanoff [4].

If metritis is suspected, appropriate antimicrobial therapy should be instituted. Obvious dietary deficiencies should be corrected by supplying a balanced commercial diet. Breeding diets should be stored under temperate conditions for periods not exceeding 6 weeks from the time of manufacture.

REFERENCES

1. Romanoff AL. *Pathogenesis of the avian embryo*. New York: Wiley Interscience, 1972.
2. Mawazi RT, et al. The oviduct of the ostrich *Struthio camelus massaicus*. *J Orn* 1982;123:425–433.
3. Bezuidenhout AJ. The topography of the thoraco-abdominal viscera in the ostrich (*Struthio camelus*). *Ondersteport J Vet Res* 1986;53:111.
4. Hicks KD. *Zoo and wild animal medicine*. Philadelphia: Saunders WB, 1993.
5. Sturkie PD. *Avian physiology*. Ithaca, NY: Springer-Verlag, 1986.

Chapter 7

Hatchery Management in Ostrich Production*

James S. Stewart

INTRODUCTION

Artificial incubation and hatching of ostrich eggs for commercial production has been an ongoing practice since Arthur Douglass of Albany, South Africa, patented the "Eclipse" brand incubator for ostrich eggs in 1867. The techniques that he developed were followed with little change for over a century. Concurrent with the international expansion of commercial ostrich production in the 1980s, more modern methods have been introduced. Although the basic principles of commercial poultry technology form the foundation of ostrich incubation, there are important differences in the specific parameters that apply to ostriches.

Hatchery management is evaluated by the percentage of fertile eggs incubated that successfully produce viable chicks. A hatchability rate of over 80% can be attained in well-managed commercial ostrich operations. It is important to examine nonviable eggs in order to determine their fertility and to ascertain the developmental stage at which death occurred to determine the etiology of embryonic losses (Appendix 7.1). This chapter reviews factors that affect hatchability and provides recommendations to resolve incubation problems encountered in commercial ostrich production.

HATCHING EGG QUALITY

Genetics

Egg quality has a significant genetic component. In commercial poultry production, laying hens have been intensively selected so that the eggs produced are uniform in conformation and content. Egg uniformity has not yet been attained in commercial ostrich production. The relatively small number of birds in the gene pool and the long generation time make selection a difficult undertaking. Ostrich breeders in South Africa have traditionally emphasized commercial traits such as quality of feathers and, to a lesser extent, quantity of eggs. Egg uniformity, shell quality and hatchability have been ignored in the development of the U.S. industry until the present time resulting in problems of variability in these parameters. Incubation practices with regard to individual hens are of necessity a compromise and may result in a wide seasonal or annual variation in hatchability in a flock. This genetic diversity is illustrated by the average performance parameters and ranges recorded by commercial producers in Oudtshoorn, South Africa (Table 7.1). The repeatability of various reproduction parameters in consecutive annual seasons, expressed as a percentage, was determined for a large population of ostriches in Oudtshoorn, South Africa (Table 7.2). Repeatability can be regarded as a measure of the heritability value for a specific characteristic. In reviewing production records, South African investigators documented the following observations:

- There was no significant difference in body mass between high- and low-producing hens.

- High-producing hens had a relatively small chest circumference, longer legs and visibly lighter thighs than low-producing hens.

- Extremely large and small eggs yielded lower hatchability than eggs of average weight.

- Male body weight was negatively correlated to fertility.

Studies in Israel and South Africa have confirmed the value of performance selection on improving reproduc-

*From *Zoo and Wild Animal Medicine*, edited by Murray E. Fowler. Copyright © 1993. Reprinted with permission of W. B. Saunders Co., publisher.

TABLE 7.1. Egg Production and Hatchability of South African Ostriches, 1994

Parameter	Average	Range
Duration of laying season (days)	120	20
Egg production per mature hen (eggs)	50	20
Clutch size (natural incubation) (eggs)	20	10
Fertility %	80	20
Hatch of fertile %	80	20
Overall hatchability of eggs laid	50	30

TABLE 7.2. Repeatability of Reproductive Parameters in South African Ostriches

Parameter	% repeatability
Seasonal egg production	20
Egg weight	74
Infertility	27
Embryonic mortality	10
Overall hatchability	14

tive potential. The transition of the ostrich industry to end-products and away from speculative multiplication will stimulate advances in egg production, fertility, and hatchability through genetic selection.

Egg Size

Ostrich eggs may vary from 350 g to over 2200 g. Average egg size varies with the subspecies and management but would typically fall between 1300 g and 1700 g. Oversized eggs have a proportionately small surface area for exchange of gas and water vapor and are likely to produce weak, edematous chicks. Conversely, undersized eggs may lose excessive water during incubation, which results in weak, dehydrated chicks. Small eggs usually produced early in the season, or by young birds, are frequently infertile. Very small eggs may have very thick shells and often contain little yolk.

Shell Quality

The shell acts as a selective barrier to allow adequate exchange of oxygen, carbon dioxide, and water vapor but inhibits the entry of infectious agents. Shell thickness is variable in ostrich eggs, extending from 1.4 mm to 2.4 mm with an average ranging between 1.7 mm and 2.0 mm. Thick shells inhibit water loss, resulting in weak, edematous chicks. Thin shells allow excessive water loss, and are significantly more subject to bacterial and fungal penetration. Shells with cracks, chips, ridges, pimpling, chalkiness, or other irregular textures are deficient in structural integrity making them more prone to invasion by pathogens.

Breeder Nutrition

An egg is a preformed package that contains all the ingredients necessary to produce a healthy chick. If a hen is marginally deficient in protein, carbohydrate, or fat, fewer eggs will be produced or they will be smaller in size than normal for the hen. Deficiencies in specific vitamins or minerals seldom result in a nutrient-deficient egg. The effects of single nutrient deficiencies on embryonic morphology and mortality have been well described for poultry [1] and must at present serve as the basis for analyzing nutrition-related hatchability problems. A high energy grain-based diet may lead to mid-embryonic mortality characterized by edematous chicks with subcutaneous hemorrhages. This suggests pantothenic acid deficiency but correction with multivitamin supplementation will obscure a specific diagnosis. Clinical evidence has related skeletal anomalies of the head and legs with diets deficient in manganese, since the problem has responded to appropriate dietary supplementation.

Breeder nutrition also has an effect on shell quality. Dietary calcium, phosphorus and vitamin D_3 must be balanced. In current small-scale operations, oversupplementation of limestone or shell grit is a more frequent problem than calcium deficiency. An excess of chloride, copper, or other minerals is detrimental to shell formation in poultry and is assumed to have a similar effect in ostriches.

Disease

Infectious agents within the reproductive tract may be transmitted to the egg during formation of either the yolk or albumen, resulting in embryonic mortality, reduced hatch, and nonviable chicks (see Chapter 13). Active infections of the reproductive tract or the anatomical and physiological abnormalities resulting from previous disease may affect the formation and integrity of eggs. Defects in egg contents, the membranes, or shell occur commonly in ostriches. Gastrointestinal conditions may reduce vitamin or mineral absorption and result in nutrient deficiency eggs even from hens fed a balanced ration.

HANDLING EGGS BEFORE INCUBATION

Collection and Transportation

Ostriches generally lay eggs in the late afternoon and evening. Eggs should be collected several times per day with the last collection at sundown (see Chapter 3). Eggs left in the nest are frequently rolled about and damaged or even eaten by adults. Eggs left overnight in the nest will cool, drawing bacteria into the egg through the pores. Immersion sanitization of the shell of internally contaminated eggs is of little value and may even lower hatchability. Eggs left in the nest for several days are subject to extremes of environmental temperature, particularly when exposed to direct sunlight, resulting in a high incidence of early embryonic mortality due to preincubation.

Ostriches are bred in outdoor enclosures and collection of eggs by hand is necessary. Large numbers of eggs are usually placed in padded crates to minimize shaking during prompt transport to the hatchery. Eggs collected from extensive management systems or in game parks have usually been subjected to preincubation and particular care must be taken to prevent shaking and cooling of the developing embryo which results in a substantial decrease in hatchability.

Eggs should be identified at the time of collection. The pen or hen number, or other means of identifying parentage, is necessary to evaluate productivity. Numbers may be safely written on the egg shell with a pencil. All production and identification data should be collected and logged in incubation records before the eggs are sanitized and stored.

Sanitization

An egg possesses several natural defenses against penetration and growth of infectious agents. The cuticle comprises a layer of mucus that dries onto the shell surface shortly after the egg is laid. This initially seals the pores of the shell and acts as the principal physical barrier to infectious agents. The pores of the shell are not small enough to prevent bacterial penetration. The outer shell membrane serves as a third but relatively inefficient physical barrier to infection. The inner shell membrane provides greater protection against penetration by bacteria and fungal spores, and lysozymes in the albumen are bacteriocidal. Several vitamins and minerals are bound to proteins within the albumen and are unavailable to bacteria, thereby inhibiting growth. The developing embryo acquires some passive immunity from immunoglobulins present within the yolk which may provide limited protection.

Production of a clean egg is the most important consideration in reducing contamination (Figure 7.1). Ostriches lay eggs in nests formed from shallow depressions in the ground. It is essential that nests be as clean and dry as possible and that eggs be collected promptly. Sand placed in the selected nest site provides a reasonably clean substrate with good drainage and is usually acceptable to adult birds. Disinfection of the nest site is advocated by some breeders, but the efficacy of this procedure has not been proven. Because ostrich eggs are laid on the ground, there is always some dust and dirt on the shell surface.

Methods of sanitizing eggs remain controversial even within the commercial poultry industry. Modern mechanical egg washers are not used in ostrich hatcheries because of the limited scale of production. Egg sanitation is therefore performed manually. A common practice is to remove adherent dirt with sand paper or a plastic scourer. Eggs may be rinsed, sprayed, or immersed in warm solutions, including sodium hypochlorite, chlorhexidine, quaternary ammonium compounds, or phenolics. These methods damage the cuticle and lower resistance to subsequent bacterial penetration. Practical experience has demonstrated that these procedures generally increase the occurrence of infection in clean eggs to levels above that obtained without immersion or spray disinfection.

Current practice for the hand cleaning of ostrich eggs involves dry removal of dust and dirt followed by the application of a disinfectant mist. A dry egg should be lightly buffed with a soft-bristle brush only on specific areas to remove adherent dirt without destroying the cuticle. If the egg is wet or contaminated with moist dirt, the surface is dried with a blow drier and then buffed. The egg surface is then coated with a fine layer of commercial disinfectant solution containing either a quaternary ammonium or phenolic compound formulated and registered for chicken eggs. Disinfectants are applied either by a hand spray, aerosol, or by fogging. The egg is allowed to air dry and is then placed in storage. Formaldehyde gas, although effective as a fumigant, requires a specially constructed chamber and is extremely irritating to humans and chicks. Formalin is subject to EPA and OSHA regulations as an environmental hazard. Because safe and effective alternatives are readily available, the use of formaldehyde gas fumigation is not recommended.

Storage

Storage of eggs before incubation significantly increases efficiency of hatchery operations by allowing batch processing of eggs. When properly controlled, storage is not detrimental to hatchability. The threshold temperature for avian embryonic growth is 85°F. Ostrich eggs held at or above this temperature show a marked increase in early embryonic mortality after the second day of incubation. Eggs maintained between 55°F and 65°F may be safely stored for 7 days, but hatchability will be significantly reduced after 10 days. Relative humidity near 75% is recommended to prevent excessive evaporative water loss from eggs during storage.

Eggs should be placed on clean wire racks within the storage cabinet or room. If eggs are positioned with the air cell upwards in the incubator tray, the entire tray can be transferred without handling, reducing contamination of individual eggs. The air cell can be identified shortly after the egg is laid by illuminating the end of the egg with a high intensity light source in a completely darkened room. A tight seal between the lamp and the egg surface facilitates candling. The margin of the air cell should be marked with a pencil line on the shell. Eggs do not require turning if stored for one week or less. For longer periods, eggs should be turned once daily but prolonged storage is not recommended because of the reduction in hatchability.

INCUBATION

Temperature

Ostrich eggs can be successfully incubated from 95–98.5°F, but most producers operate at a setting between 96.8°F and 97.5°F. Constant temperature is critical because temperature fluctuation results in lowered hatchability and reduced viability in chicks. The optimum incubation temperature for ostriches has not yet been determined, but should be expected to vary slightly with egg quality, ambient temperature and humidity, and design of the incubator. Results of a South African trial to relate hatchability and temperature showed that increasing temperature from 96.8°F to 99°F reduced hatchability from 73% to 44%.

Increased temperature shortens the incubation period for the ostrich by approximately one day for each 1°F rise in temperature. Incubation attains an average of 45 days at 95.0°F and 42 days at 98.0°F. Excessively high temperature will increase the rate of embryonic mortality at any stage of development. High temperature contributes to malformed eyes and limbs of chicks and results in premature pipping and hatch.

Low incubation temperature also increases embryonic mortality and results in soft, weak chicks that hatch late. Embryos have a limited ability to withstand transient periods of low temperature resulting from electrical failure or incubator malfunction. Chicken embryos can tolerate a decrease in temperature to 65°F for a few hours during early and mid-embryonic development with little effect on hatchability. In contrast, cooling is disastrous during the last few days of incubation. This is also consistent with clinical observations in ostriches. Power failure lasting several hours in ostrich hatcheries results in mortality at 35 or more days of incubation but has little effect on younger embryos. It is recommended that a standby generator be installed in ostrich hatcheries.

Humidity

The level of humidity in the incubator regulates evaporative loss of water from eggs determined by the size of the chick and yolk sac and the volume of the air cell. Suboptimal water loss results in a weak, edematous chick and a small air cell. High levels of water loss cause in weak, dehydrated embryos and eggs with large air cells. Embryos also adhere to membranes after pipping, resulting in low hatchability and poor viability, despite intervention by the incubationist. Water loss is monitored by measuring the decrease in egg weight during incubation. The desirable weight loss for ostrich eggs from set to internal pipping is 12 to 15% of initial weight. Ostrich eggs with a weight loss below 12% show poor hatchability. Below 10% weight loss, chicks rarely hatch without considerable assistance. Ostrich chicks from eggs with excessive weight loss have good hatchability, but hatch early and chicks show reduced viability.

Humidity settings to incubate ostrich eggs currently range between 25% and 40% RH. Humidity depends on selected temperature, egg quality, and air circulation in the setter. Higher incubation temperatures require lower humidity to maintain hatchability. Higher temperatures increase the metabolic rate of the embryo and hence the amount of metabolic water produced. This reduces the incubation period with proportionately less potential for evaporation. The reverse holds true for lower temperatures. On average, eggs of high weight or with thick shells lose less weight than small or thin-shelled eggs. As indicated previously, wide variation in egg characteristics in ostrich eggs results in a wide variation in hatchability. In a group of 179 ostrich eggs (weight range 1319 to 1925 g, mean 1633 g) incubated at 96.5°F and 33% relative humidity in a single incubator, weight loss varied from 8.5% to 23.6%, with a mean of 13.2%. Ideally, eggs should be classified by size and shell quality and incubated under different conditions to compensate for variability. Most commonly, the actual humidity used to incubate ostrich eggs is determined by the acceptable average weight loss for all eggs derived from the flock.

Air Circulation and Ventilation

Ventilation describes the exchange of air between the incubator and the surrounding environment. Circulation refers to the movement of air within the incubator. Air circulation is critical to uniform temperature, humidity, oxygen, and carbon dioxide within all areas of the setter. Circulation is provided by paddles or fans that move air in a specific pattern characteristic of incubator design. In general, high rates of air flow create uniform mixture of air and ensure removal of moisture and carbon dioxide (Figure 7.2).

Hatchery Management in Ostrich Production

Figure 7.2 Modern incubator with adequate air exchange capacity and installations to maintain temperature and humidity. Courtesy: Natureform Incubator Co.

The velocity of airflow has little apparent effect on hatchability in poultry, but ostrich eggs incubated in cabinets with low internal air velocity routinely have low total water loss, even at very low humidity settings. This may be explained by the boundary layer effect of a very large egg that produces a zone of still air near the surface. Because the dissipation of both water and heat generated by the embryo are reduced, the egg is housed in a microenvironment characterized by higher humidity and temperature than indicated by the machine settings. Consequently, evaporative water loss is reduced and metabolic water production is increased, resulting in weak, edematous chicks.

Ventilation is required to supply an adequate level of oxygen to the incubator environment and to exhaust excess carbon dioxide. Ambient air should contain 20.95% oxygen and 0.03% carbon dioxide at sea level. Recommendations for commercial poultry include an oxygen concentration above 20.5% and carbon dioxide below 0.5% [2]. Hatchability is reduced by about 5% for each 1% decrease in oxygen. Carbon dioxide concentrations above 1.5% depress hatchability severely. Oxygen consumption of ostrich eggs has been measured [3], and it can be extrapolated that at peak metabolism, 200 ostrich eggs represent the metabolic equivalent of 1,000 chicken eggs. Ventilation recommendations for commercial chicken hatcheries are generally 90 to 120 ft^3 of fresh air/1,000 eggs/hour. This corresponds to the equivalent ventilation rate for 200 ostrich eggs.

Ventilation regulates humidity in most setters (refer to Appendix 8.1). Cool room air raised to incubation temperature has the potential to hold more moisture, thereby lowering relative humidity. Increased ventilation draws more cool air into the incubator and lowers humidity. Conversely, closure of the ventilation ports increases humidity. To attain low humidity settings required for the incubation of ostrich eggs, it is necessary to cool and dehumidify the incubator room in many areas of the United States. Air temperatures in the range of 72°F to 75°F are recommended. The introduction of cool air directly into the setter can create low-temperature pockets near the inlet ports. Commercial ostrich incubators should incorporate a premixing chamber to warm fresh air by recirculation.

Position

The embryo normally is positioned with the legs bent on either side of the abdomen with the feet ventral to the shoulders. The neck is bent ventrally, and the head is rotated to the right, with the beak situated adjacent to the right foot and shoulder. The spine follows the long axis of the egg, and the head is at the air cell pole. This allows the chick to penetrate the inner shell membrane at pipping and commence pulmonary respiration within the air cell space (Figure 7.3, upper).

Embryonic malposition II, commonly referred to as the "backward chick," is frequently observed in ostrich eggs (Figure 7.3, lower). The embryo is rotated in the egg with the head located at the opposite end from the air cell. These embryos are unable to pip into the air cell before hatching and have a high mortality rate. In poultry, the prevalence of this malposition is 2% in eggs incubated with the air cell upward and 4% in eggs incubated horizontally. In the ostrich, an acceptable level of 3% malposition is obtained from eggs incubated with the air cell upward. This value increases to 20% when the eggs are incubated in the horizontal position.

Orientation of the ostrich embryo is determined between the 7th and 10th day of incubation when the yolk and embryo cease to float freely within the

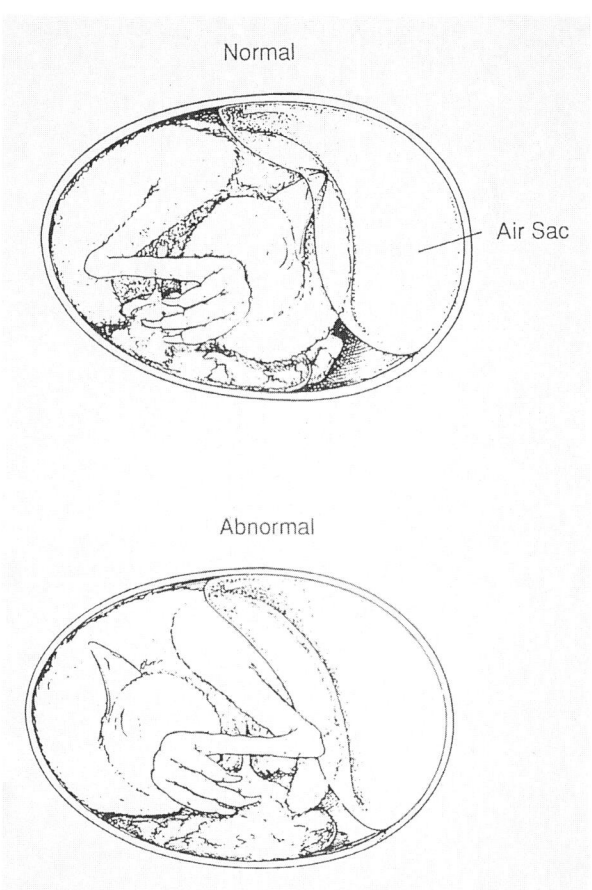
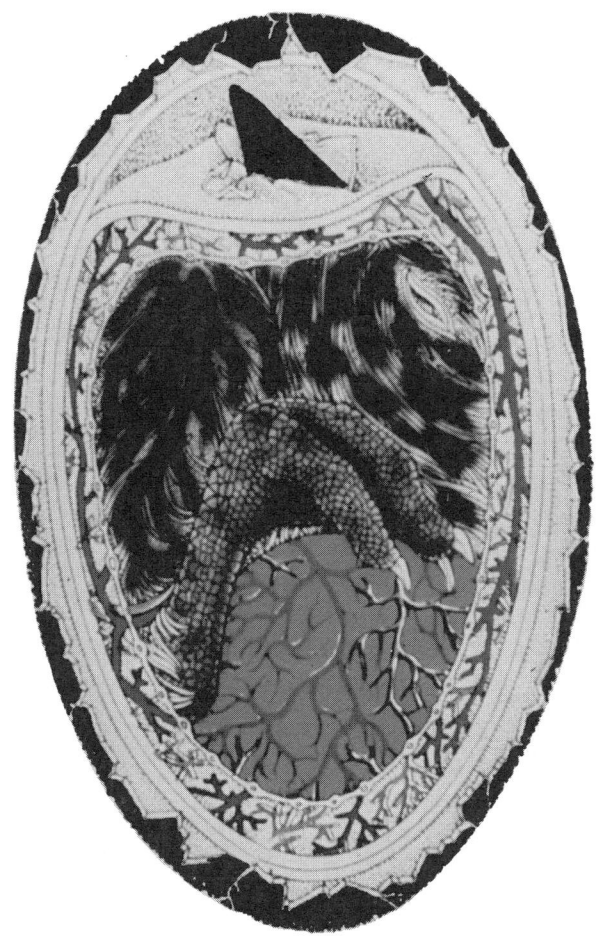

Figure 7.3 Normal position of the ratite embryo at the time of pipping (upper) and malposition Type II (lower). Courtesy: Inner-Vision Bio-Technologies Corp.

albumen. A common procedure for ostrich eggs has been to incubate horizontally for 10 days. At this time, the eggs are candled for fertility and then repositioned with the air cell upward. This procedures produces a rate of malpositions similar to horizontal incubation. In experiments performed in South Africa, optimal hatchability (85%) was obtained when eggs were rotated at 12 days. It is imperative that eggs should be positioned with the air cell upward with vertical orientation. The air cell is readily identified using a candler or high intensity light source.

Turning

Avian eggs should be rotated periodically during incubation to agitate the layers of nutrients and waste products around the developing embryo. Eggs turned infrequently show low hatchability and a high prevalence of embryonic malpositions. It has been demonstrated in poultry that rotating eggs 45° from vertical at least six times per day results in optimum hatchability. Until research has demonstrated otherwise, it should be assumed that the same applies to the ostrich. Most commercial ostrich incubators automatically turn eggs through 45° once or twice per hour. The traditional practice of hand turning eggs three times per day should be regarded as inadequate and contributing to increased bacterial contamination.

Candling

Candling identifies infertile and nonviable eggs, which should be removed from the setter because they waste space, are a source of excess evaporative water vapor, and are a potential source of contamination. Commercial egg candling devices consist of a bright light encased in a box or tube with an opening in the top. Eggs may be individually placed over the opening, and in a dark room the contents are illuminated. An alternative is to illuminate eggs while still in the incubator trays from below using a high-intensity flashlight. This procedure is more efficient and eliminates unnecessary handling and contamination of viable eggs.

The shadow created by the yolk, embryo, and extraembryonic membranes changes during the course of incubation and confirms continued embryo viability. The common practice of candling weekly is educational to the novice but is of no practical value in commercial ostrich production. Candling is recommended only at 14 days of incubation and again at transfer to the hatcher. Egg viability is readily determined at approximately 2 weeks of development. With this schedule, egg loading and candling are performed on the same day of the week, reducing contamination within the setter.

HATCHING

Temperature and Humidity

Recommended parameters for hatcher temperature and humidity are controversial. It has been suggested that the temperature should be decreased in the hatcher to compensate for the increase in metabolic heat generated at hatch. Conversely, it has been suggested that hatcher temperature be increased to compensate for the decrease in heat produced by the smaller number of eggs in the hatcher compared with those in the setter. Some producers recommend an increase in hatcher humidity once most eggs have pipped to prevent membranes from drying and sticking to the chicks. Others advocate a decrease in humidity in order to increase the partial pressure of oxygen at the time of greatest demand.

It is recommended that the temperature and humidity in the hatcher be the same as in the setter. The hatching of ostrich eggs, even under the best incubating conditions, routinely extends over 3 to 6 days. The use of different temperature and humidity levels in the setter and hatcher requires that the eggs be individually monitored in the setter before transfer. This is not only inefficient but requires frequent opening of the setter with increased risk of infection. If setter and hatcher conditions are the same, transfer time is independent of the specific developmental stage, and eggs can be moved as a batch. The recommended time of transfer is 3 days before average hatch time which ranges from 39 to 40 days in routine ostrich production.

Position and Turning

Eggs should not be turned while in the hatcher. During the last few days of development the embryos rotate into position for hatching. Turning eggs will reduce hatchability. It is least disruptive to the embryos if the eggs continue to be positioned with the air cell upward in the hatcher. If the eggs are placed horizontally, the upper side should be marked with a pencil and the eggs returned to the same position after handling.

Assisted Hatch

The sequence for the normal hatching process is for the embryo to break into the air cell on day 40, to pip the shell during day 41 (Figure 7.4), and to complete hatching by day 42. This sequence is subject to biological variation. In addition, weak embryos may fail to hatch because of defective incubation, infection, nutritional deficiencies, developmental abnormalities, or other causes of weakness. Normal chicks may also fail to hatch because of improper incubation and hatching conditions, or simply as a result of a lack of external stimuli. The social facilitation of hatching is well developed in the ostrich, and the absence of the stimulus from parents and other chicks to promote the hatching process may be a significant factor in reduced hatchability with artificial incubation.

The hatchability of ostrich eggs can be significantly increased if each egg receives individual attention after being placed in the hatcher. The weighing of the egg at collection and again at transfer is useful to identify inadequate or excessive evaporative water loss that may require special procedures.

Commercial production guidelines for an efficient and effective routine must also be established to accommodate to variation in hatch time. Assisted hatching on a specified day is too early for some chicks and too late for others. Development of the embryo and the progress of hatching can be monitored by observing the air cell daily at candling. Eggs in which the embryos have penetrated the air cell are examined twice daily and identified with a pencil to become part of a hatching group. The majority of these embryos should pip the shell within 36 hours. If pipping is delayed, a 2-cm opening in the shell over the air cell can be made. By the following day, most chicks in the group should have completed the hatching process. Those that remain may be assisted by the gradual removal of shell to free the chick. Assisting malpositioned or other abnormal chicks is a skill that is developed with experience. The viability of assisted chicks is considerably lower than with unaided hatching.

Posthatch

The hatcher should provide a clean environment in which the new chick dries and begins to develop strength to walk. The umbilicus of all chicks should routinely be treated with antiseptic ointment or spray. If the umbilicus is open, it should be protected with a clean dressing to prevent infection. The hatcher surface should not be injurious to the soft skin of the newly hatched chick but must provide adequate traction to prevent spraddled legs. Hatching ostrich in individual 10-in^2 cubicles reduces the potential for spraddled legs and provides support for the chicks in their initial attempts to stand. The holding time for most chicks after hatch should not exceed 24 hours, at which time they should be transferred to a brooding unit.

REFERENCES

1. Beer AE. A review of the effects of nutritional deficiencies on hatchability. In: Carter TC, Freeman BM, eds. *The Fertility and Hatchability of the Hen's Egg.* Edinburgh: Oliver and Boyd, 1969;93.
2. North MO. *Commercial Chicken Production Manual.* Westport: AVI Publishing Co. Inc., 1984.
3. Hoyt DF, Vleck D, Vleck CM. Metabolism of avian embryos: Ontogeny and temperature effects in the ostrich. *Condor* 1978;80:265.

Appendix 7.1

Identifying Incubation Problems in Ostrich

Observation	Probable causes
Infertility	Incompatible breeders
	Sterile or obese male
	Distractions around breeder pens
	Breeders too young or too old
	Deficiencies or imbalances in breeder diet
	Inclement weather
	Handling or managemental stress
	Early or late in lay cycle
	Systemic or reproductive disease
Early embryonic mortality	Delayed egg collection
	Storage temperature too high
	Storage time prolonged
	Infected eggs
	Formaldehyde fumigation during 24 to 96 hours of incubation
	Deficiencies or imbalances in breeder diet
Midembryonic mortality	Deficiencies or imbalances in breeder diet
	Infected eggs
	Inadequate egg turning
	Rough handling of eggs
Late embryonic mortality	Deficiencies or imbalances in breeder diet
	Inadequate ventilation
	Infected eggs
	Fluctuating incubation temperature
Air cell pip, no hatch	Hatcher temperature too high or too low
	Hatcher humidity too high
	Inadequate ventilation rate
	Lack of social facilitation
Early hatch	Setter temperature too high
	Setter humidity too low
	Undersized eggs
Late hatch	Setter temperature too low
	Oversized eggs
Malpositions	Horizontal egg position after first week
	Inadequate turning
	Rough egg handling around 35 days
	Deficiencies or imbalances in breeder diet
Malformed chicks	Setter temperature too high
	Deficiencies or imbalances in breeder diet
	Rough egg handling
	Malpositioned chicks
	Genetic factors, especially with related parents
	Teratogens
Edematous chicks	Inadequate circulation in setter
	Setter humidity too high
	Oversize eggs
	Thick shells
	Deficiencies or imbalances in breeder diet
Small chicks	Setter humidity too low
	Undersized eggs
	Thin, porous shells
Sticky chicks	Setter or hatcher humidity too low
External yolk sacs	Premature intervention to assist chick
	Infected eggs
	Edematous chicks
	Setter temperature too high
	Fluctuating setter temperature
Yolk sac infections	Premature intervention to assist chick
	Inadequate umbilical disinfection
	Contaminated hatcher
	Infected eggs
Spraddled legs	Smooth hatcher surface
	Edematous chicks

Chapter 8

Hatchery Management in Emu Production

Simon M. Shane

INTRODUCTION

All commercial emu eggs are hatched artificially, adapting techniques from the poultry industry, although the general principles of hatchery management and hygiene, as outlined in Chapter 7 on ostrich production, are relevant to emus.

There is considerable variation in the quality and characteristics of eggs produced by hens. Differences in shell thickness and egg weight influence hatchability. In addition, sperm production by the male, willingness and ability of the pair to mate, and environmental conditions influence fertility.

Emus are produced throughout the United States and the southern tier of Canada, often under conditions of high humidity which require special installations and management to attain acceptable performance. Annual surveys conducted in Louisiana and Arkansas during the mid 1980s confirmed that hatchability was influenced by experience of the incubationist and flock size. Large units regularly attained 80% hatchability of all eggs set in incubators, but producers with only a single pair of young birds experienced extremely low hatchability in the first year of operation. Emu producers lack the breadth of information concerning production characteristics which is available to ostrich producers because of the relatively recent emergence of emus as a propogated species.

HANDLING EGGS BEFORE INCUBATION

Emus produce eggs during the winter breeding season at 3-day intervals. Since eggs are generally laid after sundown, it may be necessary to collect eggs during the early evening or night to prevent incubation by the male (Figure 8.1). For optimal hygiene and hatchability, eggs should be collected as soon as possible after laying. If eggs are incubated for longer than 8 hours, embryonic development is initiated. If preincubated eggs are collected and cooled for storage, the embryo will die. On subsequent breakout examination at 40 days, preincubated eggs will resemble infertiles.

Figure 8.1 Typical emu nest in pen.

Eggs should be identified at the time of collection using a white grease pencil to record date and hen sequence number. Production data should be recorded manually or in a computerized system.

It is advisable to wear disposable plastic gloves when handling eggs, which should be transferred from the nest or the area that the hen selected, to a clean metal or plastic tray. Eggs should be transferred to an egg room and evaluated for cleanliness. Small areas of dirt adherent to the shell can be removed with a plastic scourer. Some producers routinely spray eggs with a phenolic or quaternary ammonium disinfectant at a concentration recommended by the manufacturer. Eggs should be allowed to air dry on a stainless steel rack prior to storage.

Severely contaminated eggs can be washed, but this procedure will reduce hatchability due to removal of the outer protective cuticle. Immersion in a disinfectant solution may also facilitate penetration of bacteria through the pores of the shell. Eggs can be dipped for 30–45 seconds in a freshly prepared solution of an approved phenolic or quaternary ammonium disinfectant warmed to 100°F. The surface is then wiped clean with a paper towel, rinsed with clean water, and dried by gentle patting using a freshly laundered towel. A disinfectant spray can then be applied to the shell surface and the egg allowed to dry before storage.

Eggs should not be immersed in an antibiotic solution as a routine, as this practice may lead to development of drug resistant strains of pathogens. The frequent occurrence of bacterial or fungal contamination should be investigated, and appropriate remedial action should be taken. This may include microbiological sampling of the reproductive tracts of both the male and the female. The drainage of pens should be evaluated, clean, dry nesting material provided, and decontamination and hygiene in the egg storage area and incubators should be upgraded.

All work surfaces and equipment used in handling eggs should be cleaned and disinfected daily. Eggs should contact only stainless steel or impervious plastic. Wood, fiberboard, and other porous materials are unsuitable, as they cannot be cleaned, leading to accumulation of potentially pathogenic microorganisms. Storage of emu eggs for up to 6 days is acceptable and will not depress hatchability. Specific recommendations are detailed in Chapter 7 dealing with incubation and ostrich eggs.

Incubation

Large forced-air commercial incubators should be used to achieve optimal hatchability. Acceptable results are obtained with a dry bulb temperature ranging from 97°F to 97.5°F and a relative humidity of 25–30%. This corresponds to a wet bulb temperature of 70°F to 72°F. These parameters usually result in an incubation period of 51 days.

Control of humidity is critical to attaining adequate and consistent loss in weight through the incubation period. Generally, emu eggs should lose 15% of the original setting weight by the time of pipping (Figure 8.2). The ideal loss varies according to the mass of the egg and shell thickness. Generally, eggs produced by individual mature hens are consistent in quality, but considerable variation exists among hens. Accordingly, selection of incubation parameters represents a compromise, and producers should modify suggested temperature and humidity values to optimize hatchability. The application of the psychrometric chart in the operation of setters and hatchers is critical to achieving acceptable performance. Adjusting heating, ventilation, humidification, and cooling systems depends on the physical principles depicted by the psychrometric chart (Appendix 8.1). Limiting the amount of evaporative water loss from the egg will result in an edematous chick with reduced viability. In most cases, "waterlogged" chicks fail to hatch. Egg weight obtained using an accurate electronic scale should be recorded at 4–6 day intervals during incubation to confirm a linear pattern of weight loss.

Eggs should be positioned vertically in the incubator with the air cell uppermost. Eggs should be turned through 180° at 4–8 hour intervals during the setting phase, which is usually 48 to 49 days in duration.

To achieve the appropriate range of air temperature (70°F to 75°F) and relative humidity of 25–40% in the incubator room, it may be necessary to install either a heater, air conditioner, or dehumidifier, depending on environmental conditions. To attain a level of 21% oxygen and 0.05% carbon dioxide in ambient air, a ventilation exchange rate of 25 ft^3/min/40 eggs should be maintained. A small incubation room should be operated with at least 2 air exchanges per hour.

Air circulation within small wooden incubators is relatively inefficient, allowing abnormal air distribution or stagnation, especially when units are operated with very few eggs or at full capacity. This results in variability in temperature and humidity within the setting compartment. Suboptimal circulation of air or impeding air intake will result in variability in the time of internal pipping. In the event of oxygen deprivation, late embryonic mortality will occur or chicks will show poor posthatch viability and activity (Figure 8.3).

Embryo development can be monitored in emus using infrared candling (Figure 8.4). A solid state source passes an infrared beam through the egg, and an image of the embryo and contents is displayed on a video screen. Experienced operators can distinguish between fertile and infertile eggs and monitor the development of embryos by observing vascularity of the chorioallantoic membrane. Embryonic malposition can be diagnosed at the stage of internal pipping, allowing for appropriate intervention.

EGG CHART

LAYING SEASON ________

SPECIES ____________________ INCUBATION PERIOD ________

PARENT ID:

FEMALE ____________ MALE ____________ PEN # ______

DATE LAYED ________________

INITIAL WEIGHT(FRESH): ____________

x 15%= WT. LOSS ____________

÷ INC. PERIOD ____________

=WT. LOSS PER DAY ____________

WT. LOSS PER 2 DAYS ____________

EGG # ____________

INCUBATOR # ____________

DATE TO INCUBATOR ____________

DATE TO HATCHER ____________

Day #	2	4	6	8	10	12	14	16	18	20	22	24	26	28	30	32	34	36	38	40	42	44	46	48	50
Date																									
Target Wt.																									
Actual Wt.																									
Room RH																									
Inc. RH																									
NOTES																									
% Vein Growth																									

Desired Incubator Temp:

Desired Relative Humidity:

Figure 8.2 Chart to document moisture loss during incubation.

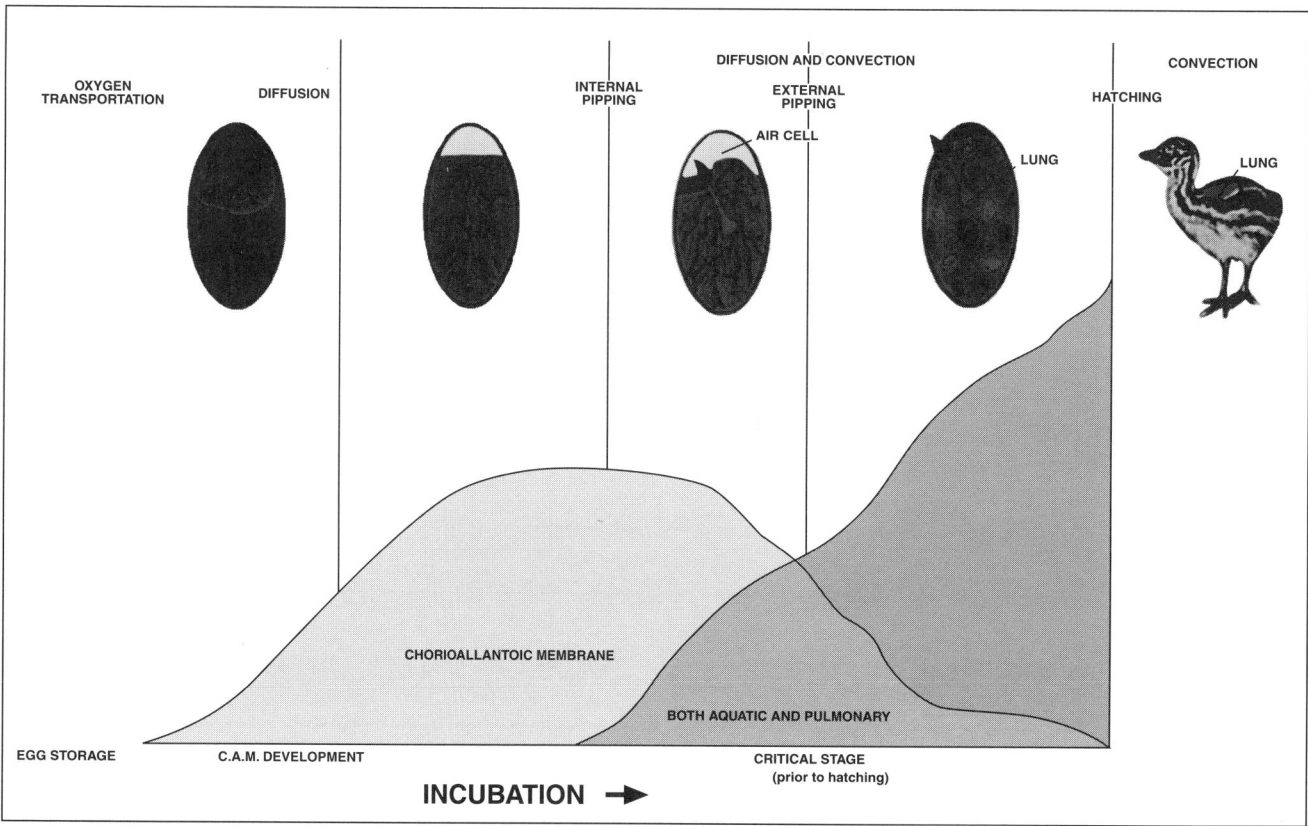

Figure 8.3 Oxygen requirements of the developing embryo. Courtesy: Inner-Vision Bio-Technologies Corp.

Figure 8.4 Infrared imaging system to determine size of air cell and viability of embryo. Courtesy: Inner-Vision Bio-Technologies Corp.

Eggs are transferred to the hatcher compartment 1–3 days before anticipated hatch. The optimal time corresponds to internal pipping, which is denoted by a hollow sound when the egg is tapped. Hatchers should be operated at 90°F. Relative humidity should range from 25% to 35%, corresponding to a wet bulb temperature of 70°F to 74°F.

During the early 1980s, the high value of emu chicks together with the relative inexperience of owners resulted in premature and aggressive intervention to assist hatching and to attempt to reposition defective embryos. Generally, efforts were unrewarding because neonatal mortality was high due to omphalitis, septicemia, and devitalization caused by anoxia and edema.

A high rate of malpositioned embryos in eggs derived from a specific mating should be investigated. If other pairs on the same farm, subjected to similar management, egg handling and incubation demonstrate normal hatchability, the problem may be attributed to genetics. With the advent of a market-oriented industry, breeders will be able to cull birds and eventually eliminate inherited defects.

The normal practice of allowing emerged chicks to remain in the hatcher compartment for up to 36 hours should be discontinued. When the shell is broken, pathogens including bacteria and fungi will spread within the hatcher and may infect other

Hatchery Management in Emu Production

pipped and emerging embryos. Shell material should be promptly removed from the hatcher, and chicks should be transferred to a heated and ventilated prebrooding cabinet within 4 hours. To prevent omphalitis, the navel should be treated with a suitable antiseptic. An open umbilicus should be protected with a clean dressing.

Problems relating to infertility and embryonic mortality in emus conform to the information provided in Chapter 7 on ostrich incubation.

Appendix 8.1

The Application of the Psychrometric Chart in Incubation

Storage and incubation of ratite eggs require precise control of temperature and humidity. An understanding of how these parameters are related is critical to controlling setter and hatcher compartments to achieve optimal hatchability. The psychrometric chart relates dry and wet bulb temperature, relative humidity, and changes in the energy content of air. Although at first glance the chart (Figure 8.5) appears complicated, a step-wise approach provides an understanding of the variables which are critical to successful incubation. Dry bulb temperature in degrees Fahrenheit (°F) is shown on the bottom axis and ranges from 20°F to 110°F. Dry bulb temperature is the ambient temperature within a room or an incubator recorded by a suspended thermometer or electronic instrument. Dry bulb isotherms (lines indicating the same temperature) are depicted vertically on the psychrometric chart.

Wet bulb temperature in degrees Fahrenheit is shown on the curved axis of the left upper quadrant of the chart and ranges from 25°F to 80°F. Wet bulb temperature is the value recorded when the bulb of a conventional mercury thermometer is surrounded by a saturated wick. As moisture evaporates from the wick, a depression in temperature occurs due to the change in state of water from liquid to vapor. Evaporation of water requires a change in latent heat content producing the cooling effect responsible for the difference between dry and wet bulb temperature readings. Under conditions of low humidity, the rate of evaporation increases and there is a larger differential between dry bulb and wet bulb values. On the psychrometric chart wet bulb isotherms run at 45° to the vertical.

Relative humidity curves radiate outward from the left side of the psychrometric chart increasing from 10% to 90% with specific values depicted for each of the curves. Relative humidity (RH) expressed as a percentage is a measure of the degree of saturation of air.

For the purpose of understanding the psychrometric chart and operating an incubator, the relative humidity expresses as a percentage, the quantity of moisture in vapor form in a given quantity of air, compared to the total amount of vapor the air could hold at a given temperature.

The following values indicated in the text should be identified to gain familiarity with the psychrometric chart.

- Select a dry bulb reading of 70°F on the bottom of the chart. The 70°F dry bulb temperature isotherm successively intersects the 10%, 20%, 30% and 40% relative humidity curves. At 30% relative humidity, a 70°F dry bulb temperature will correspond to a wet bulb reading of 53°F as indicated by the 45°-angled 53°F isotherm. At 30% RH, air at 70°F dry bulb temperature contains only 30% of the moisture that would be present at saturation, which is theoretically 100% RH. The psychrometric chart shows that the higher the relative humidity (or proportion of air saturation) the smaller the differential between the dry bulb and wet bulb reading. As an exercise, the reader should calculate the wet bulb readings in °F corresponding to 10%, 20%, 30%, and 50% RH.

- The quantity of moisture in a given mass of dry air is indicated on the vertical scale on the right of the chart. Values range from 0 to 170 grains of moisture per pound of dry air. A grain is a unit of mass with 6000 grains to each pound. Note that the lines depicting the quantity of moisture in air run horizontally across the psychrometric chart parallel to the dry bulb axis. On the psychrometric chart, the 80°F dry bulb isotherm intersects with the 56°F wet bulb isotherm on the 20% RH curve. Under stated conditions, one pound of air will contain 30 grains of moisture in vapor form.

- Dew point values, expressed in °F, correspond to the wet bulb readings on the axis comprising the left upper quadrant of the psychrometric chart. Dew point corresponds to the temperature at which moisture in air will precipitate as condensation. The appearance of a film of water on the surface of a glass containing ice cold liquid illustrates the lowering of air temperature adjacent to the glass to a level

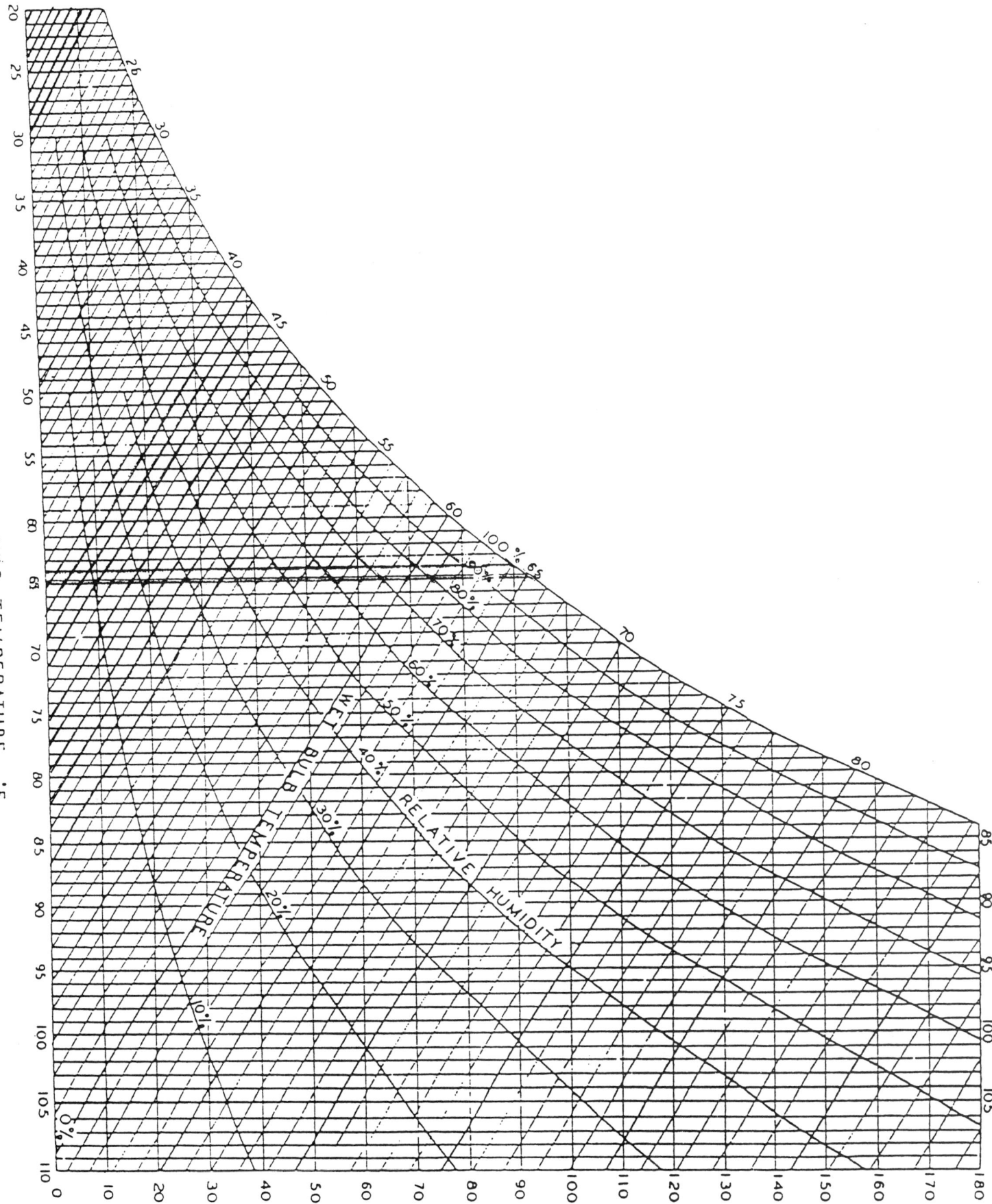

Figure 8.5 Simple psychometric chart showing relationship of temperature and humidity.

below the dew point, resulting in condensation. In the given example a dew point value of 35°F would correspond to a dry bulb reading of 80°F at 20% RH. The higher the relative humidity value, the greater will be the quantity of moisture contained in a given volume of air. Warm air contains more moisture in vapor form than an equivalent volume of cold air. Dew point follows a horizontal line parallel to the dry bulb axis. Note that dew point lines do not follow the wet bulb isotherms which are angled at 45° to the vertical.

More comprehensive psychrometric charts (Figure 8.6) include curves to show the amount of energy present in a given quantity of air (enthalpy) and the volumes occupied by a pound of air at a given temperature. For operating an incubator, this additional information is unnecessary and a simplified chart has been used to illustrate general principles.

To demonstrate the application of the psychrometric chart, assume that air in an incubator room in an arid region of the United States has a dry bulb temperature of 80°F and a relative humidity of 40%. Ambient air, when drawn into the incubator, will be heated to 98° and will have a resultant relative humidity of only 10%. To achieve the desired internal incubator RH, it will be necessary to add water to the atmosphere by spray or evaporation from a pan. At a dry bulb temperature of 90°F and with 10% RH, the moisture content of air is approximately 28 grains/pound. Reference to the psychrometric chart shows that the 98°F dry bulb isotherm intersects the hypothetical 25% RH curve (midway between the 20% and 30% curves) at a point corresponding to 70 grains moisture. This means that every pound of dry air in the incubator must receive additional (70—28 grains) or 42 grains of moisture. Assuming that the dimensions of the incubator are 6 ft × 4 ft × 4 ft (96 ft^3) and that an air exchange rate of 20 changes per hour is maintained, the total volume of air which must be humidified is 1,920 ft^3. This volume of air occupies 14 ft^3 per pound at approximately 100°F dry bulb temperature (from tabular values). The mass of air circulated through the incubator each hour corresponds to 137 pounds, each of which requires the addition of 42 grains of moisture. The total quantity of moisture which should be added each hour is 5754 grains (137 pounds × 42 grains/pound), corresponding to almost 1 pound or 15 fluid ounces of water. The evaporation rate from an exposed pan can be calibrated by weighing at approximately 1 hour intervals and calculating the loss. Alternatively, the output of an incubator water spray system controlled by the humistat can be measured. This situation would be typical for an arid area of the United States where air entering the incubator is cooled from an ambient temperature of 98°F.

Along the Gulf Coast, the problem faced by producers is to attain a sufficiently low relative humidity in the incubator to achieve a required 13% to 17% reduction in egg mass during incubation. Although eggs from different hens with characteristic egg-shell thickness and porosity require different relative humidity settings, most incubators should operate in the range of 22% to 28% RH (wet bulb readings of 69°F to 73°F, respectively) at 98°F dry bulb temperature. If incoming air at 90°F and 60% RH is supplied to the incubator, internal humidity will attain 40% RH corresponding to a moisture content of 110 grains/pound of air. The target of 25% would correspond to approximately 70 grains of moisture/pound of air at 97°F. This means that incoming air must be dehumidified to achieve optimal hatchability. Reference to the psychrometric chart shows that cooling the incoming air to 85°F and reducing humidity to 50% RH will provide the desired level of humidity in the incubator. Various other permutations of dry bulb and RH values will also result in the desired conditions, providing that dehumidification and cooling are achieved.

Reference to a more sophisticated psychrometric chart (Figure 8.4) shows an enthalpy value of 34 BTU per pound dry air at 90°F dry bulb temperature and 60% RH. At the desired values of 85°F dry bulb and 50% RH enthalpy will be 26 BTU per pound of dry air. A hatchery room of 10 ft × 20 ft × 9 ft (1,800 ft^3) would normally require an air exchange rate of eight cycles per hour to provide sufficient oxygen for the incubator. This means that 14,400 ft^3 air must be cooled each hour. The mass of this volume of air is 1,000 pounds, assuming a density of 14 ft^3/pound of air at approximately 95°F. Accepting an 8 BTU/pound differential, an 8,000 BTU/hour air conditioner would be required. The quantity of moisture to be removed from the air in the room would be (127 grains—90 grains) or 37 grains/pound of air. This corresponds to 6 pounds, or 1 gallon of water. The air conditioner mounted in the window of the room and the dehumidifier adjacent to the incubator inlet should have a combined total dewatering capacity of 1 gallon/hour to achieve optimal conditions in the incubator.

The psychrometric chart can be used to understand the phenomenon of "sweating" on the shell surface after removal of an egg from refrigerated storage. Usually eggs are held at 60°F dry bulb and a relative humidity of 70%. Under these conditions, wet bulb and dew point temperatures will be 54°F and 50°F, respectively, and atmospheric moisture content will attain 53 grains/pound of air. If eggs are transferred from the cooler to a handling room or transport vehicle, condensation will occur on the surface at any combination of dry bulb temperature and humidity values correspond-

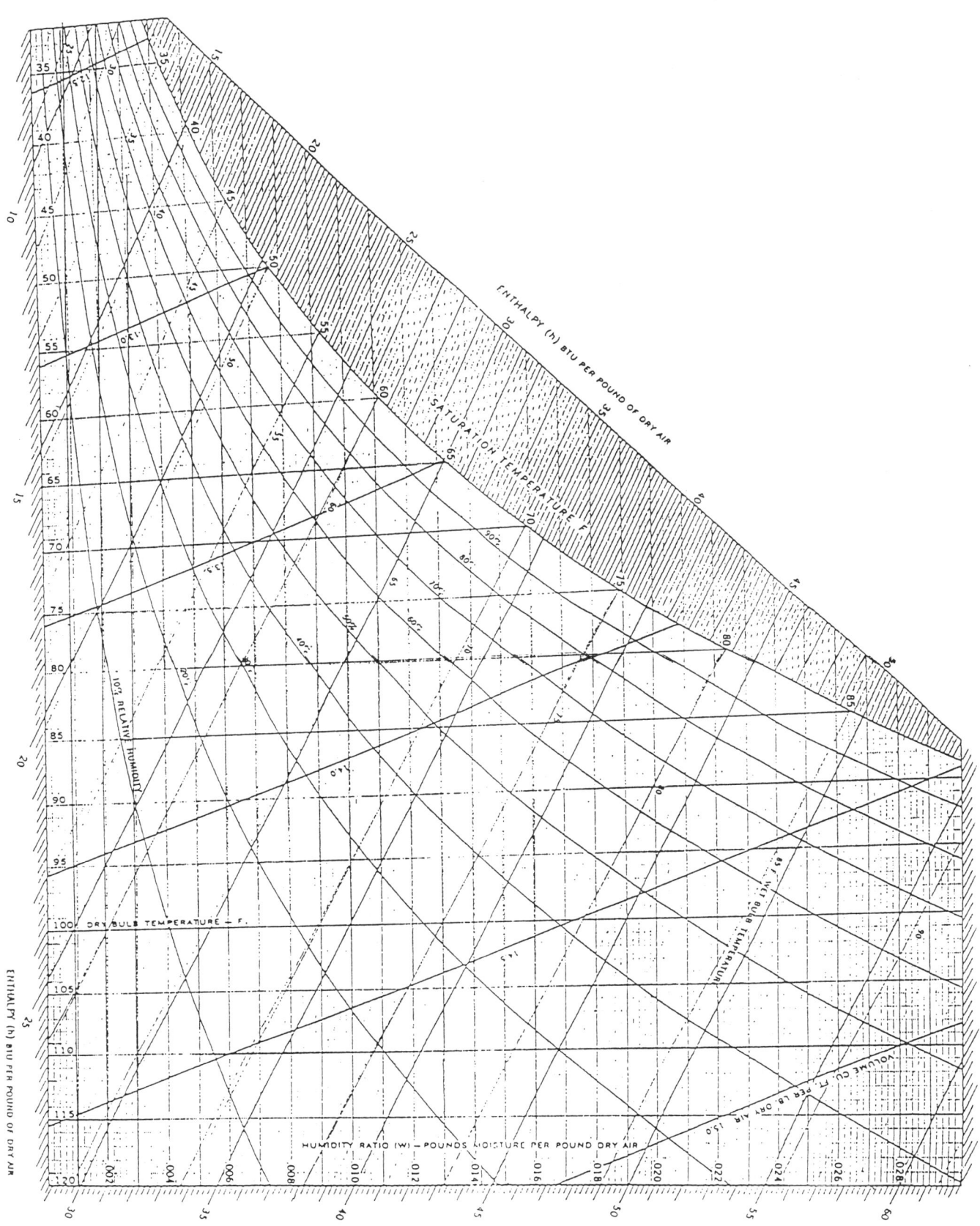

Figure 8.6 Sophisticated psychometric chart, including parameters such as enthalpy.

ing to a dew point of 60°F. From the psychrometric chart it may be determined that at a room temperature of 80°F dry bulb relative humidity exceeding 30% will result in condensation. At a room temperature of 90°F dry bulb, condensation will occur at a relative humidity of approximately 14%. Condensation will continue until the surface temperature of the egg rises to a value above the dew point. In the case of ostrich and emu eggs this will take many hours since the interior of the egg will be close to the storage temperature of 60°F after 2 days due to the high specific heat capacity of the egg contents. The film of moisture which condenses on the shell, facilitates penetration of motile bacteria and spores of fungi from the surface through the pores to the air cell. Contamination will be reflected in rotten eggs, embryo mortality, decreased chick viability, and omphalitis (navel infection). Eggs should be transferred rapidly to incubators after storage, and should not be subjected to periodic or cyclic changes in temperature and humidity at any time between collection and incubation.

The application of the psychrometric chart to the operation of setters and hatchers is critical to achieving optimal hatchability. Normal embryonic development depends on a narrow range of temperature and humidity during the setting and hatching phases of incubation. Adjusting heaters, selecting ventilation rates, and operation of humidification and cooling systems depends on the physical principles illustrated by the psychrometric chart. Emu breeders are subjected to a greater challenge than commercial poultry producers due to the relative diversity of characteristics of eggs from specific hens. The factors which influence hatchability include shell thickness, porosity, egg weight, and genetics.

Chapter 9

Anesthesiology of Ratites

Janyce L. Cornick-Seahorn

INTRODUCTION

Large ratites, including ostriches, emus, rheas, and cassowaries, present a unique challenge to veterinarians in handling and restraint. Historically, treatment of these species was largely confined to zoo veterinarians. Growth of ostrich and emu production in the United States since the early 1980s has created a demand for veterinary services in a field where very few advances have been made. A major problem facing veterinarians is the paucity of information on suitable chemical restraint agents and anesthetic protocols. These are required to facilitate physical examination, sample collection, and induction and maintenance of surgical anesthesia.

Adequate techniques for physical restraint are essential to safely manage ratites. Since adult male ostriches may attain a weight of 175 kg, their size, speed, and leg strength all contribute to significant danger to both handlers and the patient undergoing restraint.

Anesthesia may be induced and maintained with parenteral or inhalation agents. Monitoring techniques used for mammalian and other avian species are applicable to ratites and are important to a successful outcome. As with other species, complications are most likely to occur during induction and recovery. Careful preoperative examination and planning will minimize complications during anesthesia.

PREANESTHETIC PREPARATION

Restraint

Adequate restraint is essential to safely handle ratites. While it is impossible to manually restrain free-ranging animals, most hand-raised ratites are docile and inquisitive. Unless handled in a deliberate and calming manner, the response to environmental and physical stress is often unpredictable and violent. All ratites can deliver a powerful forward kick, and handlers must never stand directly in front of adult birds. Ostriches are the most dangerous of all ratites due to their powerful legs, but the emu and cassowary have prominent nails on the medial digit which can inflict severe damage [1]. Emus can also kick sideways while the smaller rhea may kick in any direction [2].

Young chicks are easily immobilized by wrapping them loosely in a towel and holding them next to the body of the restrainer. Juveniles of up to 6 kg body weight may be held by flexing the legs and holding them firmly in an upright position [1]. Birds between 6 and 15 kg may be restrained by holding around the abdomen, leaving the legs free [3]. Adult emus and rheas may be adequately restrained by placing one forearm between the bird's legs and lifting the bird rapidly so that the body is perpendicular to the ground with the dorsum resting on the abdomen of the restrainer [1]. This technique will provide 3–5 minutes of restraint, depending on the strength of the handler and the size of the bird.

Adult ostriches require up to four experienced handlers to provide a short period of restraint. Ostriches are not generally aggressive toward humans by nature except for males during the breeding season. They hiss, fan their wings, and thrust their chests against gates when approached by humans [1]. When this behavior is displayed, the bird should not be approached without protection such as plywood or plexiglass shields or a forked pole. When escape from an attacking ostrich is not possible, lying flat on the ground is best because the forward thrust of the leg usually strikes 50–80 cm above the ground [1,2].

Darkness is one of the best aids in handling all ages of ratites and has a calming effect which facilitates restraint, examination, and induction of anesthesia. Hooding with light opaque fabric, such as surgical

Figure 9.1 Hood used on juvenile and adult ostriches for restraint.

TABLE 9.1. Normal Physiologic Values for Ratites

Parameter	Ostriches	Emus
Body Temperature (°C)	37.2–40.0[3]	†37.8–39.8
Heart Rate (/minute)	*60–72[20]	†42–76
	*57–103	*152–192
Respiratory Rate (/minute)	6–12[3]	†4–17
	*2–20[20]	*13–21
	*9–22	
Mean Blood Pressure (mm Hg)	*118–171[14]	†60–109
Systolic Blood Pressure (mm Hg)	*145–206[14]	†99–146
Diastolic Blood Pressure (mm Hg)	*90–147[14]	†43–66
Tidal Volume (ml/kg)	13–18[35]	

*Values collected during isoflurane anesthesia.
†Values collected during xylazine, ketamine, and guaifenesin anesthesia.

Note: The nonreferenced value ranges and those for reference 14 represent the mean minimum and maximum values from a series of clinical cases or from a controlled study.

stockinette, calms ostriches and can be applied for handling, restraint, and transport. Hoods should cover the eyes without occluding the nostrils and be long enough to extend at least one-third down the the neck [1] (Figure 9.1). Hooding rarely elicits anxiety in ostriches but hoods should be removed immediately if frantic behavior follows placement. Other ratite species are much less tolerant of hooding and the effect is unpredictable [1,4].

An additional restraint technique useful for adult ostriches is to gently grasp the head with both hands and abruptly lower it to the ground. Ostriches are reluctant to kick and are relatively helpless in this position. They may hop backward a few steps and then assume sternal recumbency. Birds may be held for several minutes in this position especially if external stimuli are minimized [1]. It is imperative that the head is held close to the ground to dissuade the bird from standing. The same effect may be achieved using a shepherd's hook to restrain the head with the neck parallel to the ground.

To move an ostrich, a handler should grasp the tail root with the lead handler at the head. The bird must be moved carefully, avoiding excessive pressure on the wings, which may fracture or dislocate if grasped distally.

Stanchions ("plucking boxes") have been used for several decades for restraint [1]. These are 115 cm in both height and length, and 50 cm and 67 cm wide in the front and back, respectively.

Physical Examination

A physical examination should be performed to determine the risk status of each patient. The thoroughness of the examination is influenced by adequate restraint. Fortunately, higher risk patients are easier to restrain and examination and collection of necessary preoperative samples is facilitated. Care must be taken when handling sick birds because they are more susceptible to physical stress. Examination should cover assessment of body condition, hydration, and vital signs, including pulse quality. Normal vital signs are listed in Table 9.1.

Normal hematology and chemistry values are described in the chapter on clinical pathology. A minimum database for clinically healthy patients comprises a hematocrit and total protein while a complete blood count and chemistry panel (including electrolytes) is indicated for sick birds. Abnormalities, such as dehydration or acid-base and electrolyte imbalances, should be corrected prior to induction of anesthesia. The abdominal skin cranial to the thigh is an acceptable site for evaluation of hydration [5]. Some emaciation indicates debilitation; consequently, selection of the anesthetic protocol and monitoring are both critically important in these animals. Very young and very thin birds are especially susceptible to hypothermia during anesthesia. A distended proventriculus indicates the need for rapid induction and intubation to prevent regurgitation and fatal aspiration. Birds that are too debilitated to stand should not be anesthetized. If anesthesia is necessary, however, mask induction and maintenance with isoflurane probably provides the safest method, regardless of the size and age of the bird.

Fasting

Adult ratites should be made to fast for 12 to 24 hours prior to induction of anesthesia to minimize the risk of regurgitation and aspiration [6]. Fasting for periods longer than 24 hours may complicate the anesthetic period due to the high metabolic rate of ratites [5]. Fasting is not recommended for ratite chicks because they become hypoglycemic during the perianesthetic period due to limited glycogen storage and high metabolic rate.

PREANESTHETIC AGENTS

Intramuscular Injection

Preanesthetic agents, such as tranquilizers, sedatives and opioids, are usually administered intramuscularly (IM). The purpose of these agents is to calm the ratite patient prior to anesthetic induction. Information regarding the effectiveness of these agents is limited. Suitable sites for IM injection in the ostrich include the cranial and caudal thigh muscles and the lumbosacral area [3,6–9]. The pectoral muscles are not developed in ratites and are useless as an injection site.

Although ratites have a renal portal system similar to other birds, the precise physiologic function is unknown [10,11]. The impact of the renal portal system on drugs injected into these muscle masses has not been studied. The high dose requirement and short duration of effect of drugs such as etorphine in ostriches [6,7] suggest that the renal portal system and the high metabolic rate of ratites may have a significant effect on drug metabolism and excretion [5,6].

Anticholinergic Agents

Anticholinergic agents are not used routinely in avian species because they increase the viscosity of respiratory secretions which can lead to airway occlusion [12]. Bradycardia is usually not a problem in avian species. An increase in heart rate has been reported in ostriches during isoflurane anesthesia [3]. Atropine has been used in conjunction with xylazine-ketamine administration in a juvenile emu [13] and glycopyrrolate has been administered as specific therapy for bradycardia during isoflurane anesthesia in ostriches [14]. Anticholinergic agents should be reserved to treat bradycardia in ratites. A heart rate of less than 80 per minute has been reported to be abnormal for adult ostriches immobilized in the field [6]. Based on clinical findings in isoflurane-anesthetized mature ostriches, anticholinergic therapy is indicated when the heart rate drops below 30–35 beats per minute or if a precipitous decrease from baseline rate occurs [14]. The decision to administer an anticholinergic agent should be based not only on the heart rate but also on pulse quality and mucous membrane color. Table 9.2 lists suggested dosages.

Tranquilizers and Sedatives

Dosages for acepromazine, azaperone, diazepam, midazolam, xylazine, detomidine, and medetomidine have been reported for ratites. These agents have a calming effect which may facilitate blood collection and induction of anesthesia in individual ratites. None of the agents produce a significant tranquilizing effect except in the debilitated patient. Acepromazine is reported to produce satisfactory muscle relaxation [3] but should only be used in healthy animals due to its side effects. Azaperone produces minimal muscle relaxation but provides safe tranquilization for up to 24 hours which may be useful for shipping or adaptation to a new environment [15]. Diazepam and midazolam have minimal cardiorespiratory effects and are safe to use in both debilitated and healthy birds [14]. Midazolam has the advantage of being more rapidly absorbed IM, but its use in adult ostriches may be cost prohibitive.

Xylazine, the most commonly used alpha-2 agonist, is most often used in combination with ketamine to produce immobilization but has been used alone for both sedation and immobilization in ratites [5,7,14,15]. Xylazine has a significant cardiorespiratory depressant effect and should not be used in debilitated birds. Xylazine should not be administered as a pre-medication for inhalation anesthesia because it may greatly exacerbate the cardiodepressant effects of the inhalant agents [14]. A dosage for detomidine has been reported with the major benefit being good gastrointestinal analgesia [14]. The sedative effects of detomidine appeared negligible based on limited experience in emus. Medetomidine[a] administered IM to ostriches produced mild drowsiness and ataxia, drooping of the wings and head, and a decrease in heart rate [16]. No clinical studies on its usefulness as a preanesthetic agent in ratites have been reported.

Opioids and Neuroleptanalgesic Combinations.

Use of opioids in ratites has been limited to the potent but potentially hazardous agents, carfentanil and etorphine, alone or in combination, for the immobilization of free-ranging ratites [6–9]. Intramuscular administration of carfentanil in combination with xylazine to adult ostriches produced recumbency within 5 minutes of injection but was preceded by a short period of excitement [8]. Carfentanil alone was associated with an evident excitable phase characterized by uncontrolled running and circling [8]. Two birds that demonstrated a prolonged period of frenzied running following carfentanil administration developed exertional rhabdomyolysis [8]. Carfentanil as a premedication with isoflurane anesthesia was also reported to produce a short period of frenzied running prior to the onset of sedation [14]. Etorphine with acepromazine provided

[a] Domitor, Farmos Group Ltd, Finland. Not approved for use in the United States.

TABLE 9.2. Dosages of Anesthetic Agents Used in Ratites

Drug	Dosage (mg/kg)	Reference
Anticholinergics		
Atropine	0.035 IM	13
Glycopyrrolate	0.011 IV	14
Tranquilizers/Sedatives		
Acepromazine	0.1–0.2 IV	5
	0.25–0.5 IM	5
Azaperone	0.5–2.0 IM	15
Diazepam	0.1–0.3 IV	17
	0.22–0.44 IM	15
	0.5–1.0 IM	
Midazolam	0.15 mg/kg IM	14
Xylazine	0.2–1.0 IM (sedation)	5
	1.0–2.2 IM (immobilization)	5
	0.4–0.9 IM	14
	1500 mg*	7
Detomidine	1.5 IM	15
Medetomidine	0.1 IM	16
Opioids/Neuroleptanalgesic Combinations		
Butorphanol	0.05–0.5 IV, IM	
Carfentanil	0.01–0.02 IM	18
	3.3 mg IM*	8
Etorphine	0.02 IM	18
	3–10 mg IM†	9
Carfentanil/Xylazine	0.03 IV / 0.5 IM	14
	3 mg IM / 150 mg IM*	8
Etorphine/Acepromazine	3.6 mg IM / 15 mg IM‡	7
	2.5–3.0 mg IM / 50 mg IM‡	9
Etorphine/Acepromazine/Xylazine	6.0 mg / 25 mg / 200 mg IM*	7
Etorphine/Medetomidine	8–9 mg IM / 4–8 mg IM*	6
Fentanyl/Droperidol (Innovar)	1 ml / 9 kg	18
Dissociative Agent Combinations		
Ketamine	20–25 IM (rheas, emus)	18, 20
	10–15 IM (ostriches)	18
	1000 mg IM*	4
Ketamine/Xylazine	5 IM / 1 IM	18
	2.2 IV / 0.25 IV	18
	2.2–3.3 IV / 2.2 IM (IV ketamine given 15 minutes after IM xylazine)	5
	3000 mg IM / 400 mg IM*	7
	15 IM / 3 IM (emus)	
Ketamine/Diazepam	5 IV / 0.25 IV (mixed)	14
	2.2–3.3 IV / 0.22–0.5 IM (IV ketamine given 15–30 minutes after IM diazepam)	15
Ketamine/Etorphine	150–200 IM / 4–6 IM*	9
	120–180 IM / 5–9 IM*	6
	100–300 IM / 7–12 IM†	9
Tiletamine/Zolazepam	2–10 IM	18
	1–3 IV (high end of dosage range needed for emus and rheas)	18
Other Combinations		
Ketamine/Xylazine/Alphaxalone-alphadolone	5 IV / 1 IV / 12–18 IV	19
Metomidate	2000 mg IM*	6
	15–20 IM	16
Metomidate/Azaperone	10–20 IM / 3.3–6.6 IM	16
Ketamine/Xylazine induction followed by 0.1% ketamine/ 0.05% xylazine/ 5% guaifenesin infusion	15 IM / 3 IM KXG infusion: 2–4 ml/kg/hr IV	
Antagonists		
Diprenorphine	4.5 mg IV*	7
	12–30 mg IV*	6
Naltrexone	300 mg IV*	8
Naloxone	2 mg IV*	14
	0.044 IV	18
Yohimbine	12.5 mg IV*	8
Atipamezole	5–20 mg IV*	6

*Total dose administered to adult ostriches.
†Total dose administered to adult cassowaries.
‡Total dose administered to juvenile ostriches (10–12 months of age).

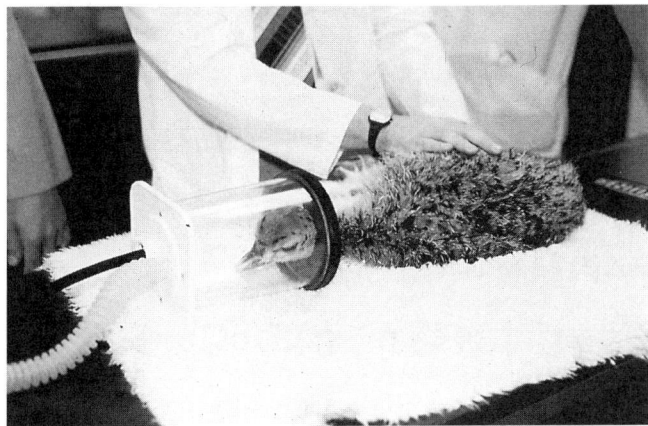

Figure 9.2 Juvenile ostrich restrained on a nonslip surface for induction of anesthesia via mask delivery.

sedation following a short period of excitement but muscle relaxation was poor [9]. The level of sedation was adequate for minor procedures and for transport but did not produce complete immobilization [6,8]. One death due to regurgitation and aspiration was observed with this combination [7]. Etorphine and acepromazine in combination with xylazine in adult ostriches provided sedation, muscle relaxation, and immobilization for 30 minutes without death [7]. Etorphine in combination with medetomidine has been used to capture free-ranging adult ostriches [6]. This combination produced complete immobilization and good muscle relaxation within 6–10 minutes of injection. Of the 8 birds receiving etorphine-medetomidine, one developed myopathy and eventually died and a second bird suffered fatal respiratory arrest.

A dosage for fentanyl-droperidol (Innovar) in ratites is available but its effects have not been reported (Table 9.2). A dosage for butorphanol has not been published but it has been used as an adjunct to isoflurane anesthesia as an analgesic.

INDUCTION

Induction of anesthesia may be accomplished by IM or intravenous (IV) injection of parenteral agents or by delivery of inhalation agents via a mask (Figure 9.2). Induction should be performed in a small, dark enclosure with seamless walls and a nonslip floor surface, which is free of sharp edges and gaps adjacent to the door which may entrap feet or the head [2]. Intramuscular injection is indicated when the birds are too large or intractable to restrain adequately for IV injection. Intramuscular injection of ketamine alone and of tiletamine-zolazepam combination can produce an excitable and dangerous induction [5,17]. Intravenous injection provides the quickest and least traumatic method of induction in healthy adult birds [18]. If ratites are difficult to restrain, IM injection of a tranquilizer, such as azaperone or diazepam, followed by an intravenous agent or combination may provide the most effective method of induction. Several drug combinations have been reported in the literature (Table 9.2), but it is emphasized that the limited number of patients and details of the response, effectiveness, and safety prevents generalization to all cases.

Intravenous Access

In the ostrich and rhea, the right jugular, brachial, and medial metatarsal veins may be used for placement of catheters, collection of blood, and administration of drugs. In the emu, the right jugular and medial metatarsal veins are accessible but, due to vestigial wing development, the brachial vein is very small. As with other avian species, the left jugular vein is relatively small in ratites [10]. Although veins other than the right jugular should be used for catheterization and injection [3,10], this vein may have to be used when other veins are inaccessible or for long-term maintenance of a catheter. If a ratite cannot be restrained adequately for catheterization or for IV injection with a needle and syringe, a butterfly catheter placed in a superficial vein will facilitate IV injection of induction drugs [19].

Dissociative Combinations

Although dosages are presented in the literature, administration of ketamine alone either by the IM or IV route is an unacceptable method to induce anesthesia in ratites. Ketamine has been associated with excitable induction and recovery and convulsive behavior during recovery [3,17]. Xylazine IM or IV with ketamine IM or IV has been used to induce anesthesia [14,15,18]. Xylazine-ketamine prior to isoflurane maintenance in one adult ostrich resulted in a fair induction, bradycardia and apnea during the anesthetic period and a poor recovery period [14]. Xylazine-ketamine is not recommended as an induction combination for inhalation anesthesia and is contraindicated in debilitated patients because of detrimental cardiorespiratory effects. Diazepam and ketamine administered simultaneously IV provide excellent induction of anesthesia and is recommended for debilitated patients due to minimal cardiorespiratory effects [14]. It is imperative that the full calculated dosage should be used in healthy patients (Table 9.2). Excitation during induction is likely and depth of anesthesia may be inadequate for intubation if suboptional doses are used. Diazepam (0.3–0.5 mg/kg IM) may provide adequate sedation for subsequent induction with IV diazepam-ketamine in patients that are difficult to restrain. In debilitated patients, 50% of the dose should be administered and the remainder titrated to effect intubation to facilitate a smooth intubation.

A range of dosages for IM and IV tiletamine-zolazepam (Telazol) have been reported. Tiletamine-zolazepam is associated with rough induction and recovery when administered IM [5,16,20]. Unfavorable inductions are more likely with a low dose. The quantity administered should be metabolically scaled and should also take into account the status of the patient. As a general rule, smaller ratites (emus and rheas) should receive a dosage from the high end of the recommended range. In a clinical study, IV tiletamine-zolazepam provided smooth and rapid induction of anesthesia, although apnea occurred and recovery was very rough [14]. Administration of diazepam before and/or after tiletamine-zolazepam administration may help to smooth both induction and recovery.

Antagonists to Anesthetic Agents

Antagonists are available to reverse opioids, alpha-2 agonists, and benzodiazepines. Table 9.2 provides dosages for the opioid antagonists diprenorphine, naloxone, and naltrexone in ratites [7,8,14,18]. These agents have been used to reverse etorphine, carfentanil, and fentanyl (Innovar-Vet). Yohimbine has been used in adult ostriches IV to reverse the effects of xylazine IM [7] and should also be effective for the reversal of detomidine and medetomidine. Atipamezole has been administered IV to adult ostriches to reverse the effects of IM medetomidine [6]. Flumazenil, a benzodiazepine antagonist, will effectively reverse diazepam, midazolam, and zolazepam. Due to the cost of the drug and the minimal side effects of benzodiazepines in ratites, flumazenil would rarely be indicated.

Inhalation Agents

Isoflurane and halothane may be used to induce anesthesia in ratites if the bird can be safely and adequately restrained. Methoxyflurane is unsatisfactory for this purpose due to its high solubility and thus slow onset of anesthesia. Induction using inhalation agents is recommended for juveniles and debilitated adults. Two approaches to induction with an inhalation anesthetic in birds have been documented [12]. One method describes a slow increase in concentration to 2–3% maximum. The alternative suggests a setting of 5% initially followed by a decrease to maintenance levels to facilitate a smooth and rapid induction. This latter approach is recommended for ratites. Possible exceptions may include patients at high risk for which high gas concentrations should be avoided. Induction of anesthesia with inhalation agents in healthy adult ratites is possible if adequate preanesthetic sedation and restraint is applied, but this method is often ineffective and stressful, being prolonged and dangerous to both handler and patient. Additional details are provided in the following section relating to maintenance with inhalation agents.

MAINTENANCE

Dissociative Combinations

While many of the injectable protocols reported for ratites provide immobilization, none have been documented as consistently effective in maintaining adequate anesthesia for invasive surgical procedures. Dosages for IM ketamine alone as an immobilization agent has been reported for ostriches, emus, and rheas [4,20]. Ketamine administered to an adult ostrich (11 mg/kg IM) failed to produce complete immobilization or adequate muscle relaxation and resulted in a rough induction [4]. High doses of ketamine and xylazine (Table 9.2) may provide adequate immobilization for field procedures [7]. Ketamine (5 mg/kg IV) at 10–15 minute intervals may be used to prolong and maintain immobilization following induction [20].

In a study on healthy young emus (mean weight of 14.5 kg), ketamine and xylazine IM was used for induction of anesthesia. This was followed by an IV infusion of 0.1% ketamine and 0.05% xylazine in 5% guaifenesin, administered at an approximate rate of 2–4 ml/kg/hour (Table 9.2). This infusion maintained immobilization for 60 minutes. Although heart and respiratory rates and arterial blood pressure decreased during the infusion, arterial blood gases remained near normal and all patients survived. The pedal reflex in response to a standardized toe pinch was absent in some but not all birds.

In a study on healthy ostrich chicks, anesthesia could be induced with IV ketamine and xylazine and maintained with the steroid anesthetic mixture, alphaxalone/alphadolone[b] [19]. This combination was safe and provided good muscle relaxation, loss of the pedal reflex, and smooth induction and recovery.

In a group of healthy ostrich chicks, three anesthetic protocols were studied: tiletamine/zolazepam, metomidate[c] (an imidazole anesthetic agent), and metomidate with azaperone [16]. Pedal reflexes remained intact with all three protocols in some or all of the birds. Inductions were often rough, and several birds moved

---

[b] Saffan, GlaxoVet Ltd, United Kingdom. Not approved for use in the United States.
[c] Hypnodil, Janssen Pharmaceutica, Belgium. Not approved for use in the United States.

when handled. Four of 40 birds salivated or regurgitated, and one subject became apneic on azaperone/metomidate. All birds made a smooth recovery.

Etorphine with ketamine provided excellent immobilization within 15 minutes of injection but caused varying degrees of excitement in adult ostriches and cassowaries [6,9]. Physical restraint was not required to maintain immobilization and muscle relaxation was acceptable [6,9]. Diprenorphine was administered to reverse etorphine in both studies. Recoveries were reported to be smooth though somewhat prolonged in the cassowary [9]. In the second study, recovery was rough and required assistance [6].

Other Agents

Barbiturates, including pentobarbital and thiamylal, have been used in avian species [21]. One report described the use of pentobarbital IV to induce and maintain anesthesia in an adult ostrich which resulted in intermittent regurgitation and a prolonged recovery [22]. This class of anesthetic agents is not recommended for induction or maintenance of anesthesia in ratites. Propofol,[d] a short-acting IV anesthetic agent with noncumulative effects, has been used in buzzards, owls, and pigeons with the most significant side effect being transient apnea [23,24]. Although there are no reports on propofol in ratites, this agent may be useful.

Inhalation Agents

Detailed information on injectable protocols to maintain anesthesia in adult birds is unavailable. When a surgical plane of anesthesia is required, inhalation agents are the most reliable and safest [18,25]. Isoflurane, halothane, methoxyflurane and nitrous oxide have all been used to maintain anesthesia in ratites [13,14,20,26]. Isoflurane is the agent of choice due to rapid induction and recovery and relative safety [18]. Recommended concentrations for inhalation agents are listed in Table 9.3.

A nonrebreathing system, such as Bain's or Ayre's T piece, should be used for birds less than 7 kg in weight using a minimum oxygen flow rate of 200 ml/kg/minute for maintenance [12,17]. For induction with a nonrebreathing system, oxygen flow rates should be approximately double the maintenance rate to facilitate rapid induction.

A circle rebreathing system is used for birds over 7 kg with a minimum oxygen flow rate of 20–30 ml/kg/minute for maintenance. For induction of anesthesia with a circle system, flow rates should be at least twice the maintenance rate. One report suggested that

[d] Dinamap, Critikon, Tampa, FL.

TABLE 9.3. Suggested Concentrations for Inhalation Agents Used in Ratites

Agent	Induction	Maintenance
Isoflurane[3]	4.0–5.0%	2.0–3.0%
Halothane[3]	4.0%	2.0–3.0%
Methoxyflurane[13]	NA	0.5–1.0%

NA = Not applicable.

ostriches over 50 kg can be maintained more effectively on a large animal breathing circuit with a 15 l reservoir bag [14]. Either a small- or a large-animal circuit may be used to maintain inhalation anesthesia in adult ostriches. The size of the circuit probably has minimal influence on the quality of anesthesia. The maximum weight capacity for the small-animal circuit is 135 kg. If a large-animal circuit is used, the patient connection port of the Y-piece should be occluded and the circuit primed with anesthetic gas and oxygen prior to connecting to the patient.

Intubation

Ratites should be intubated for procedures requiring anesthesia of more than 10 minutes in duration. Intubation, which is easy to perform, protects the airway if regurgitation occurs, facilitates assisted ventilation, permits low oxygen flow rates, and minimizes environmental contamination with waste anesthetic gases.

When an adequate plane of anesthesia is achieved either with an injectable or inhalation method, the beak is held open, the glottis visualized, and the tube passed between the arytenoid cartilages into the trachea (Figures 9.3 and 9.4). This stimulation may excite some individuals especially when the plane of anesthesia is inadequate. In this situation, maintaining a high anesthetic concentration for an additional 1 to 5 minutes will usually result in smooth transition to an adequate anesthetic depth.

A wide range of endotracheal tube sizes are indicated for ratites due to their extreme variation in size (Table 9.4). Tubes should never be forced into the trachea, and if the "fit" seems tight, it is best to select the next smaller size to avoid damage to the tracheal mucosa. Some authors recommend the use of noncuffed tubes or noninflated cuffs. This is due to

TABLE 9.4. Endotracheal Tube Selection for Ratites

Patient	Tube Size (mm internal diameter)
Ratite chick (< 5 kg)	4–6
Ratite juvenile (5–20 kg)	7–11
Adult rhea/emu (25–50 kg)	12–14
Adult ostrich (>50 kg)	14–18

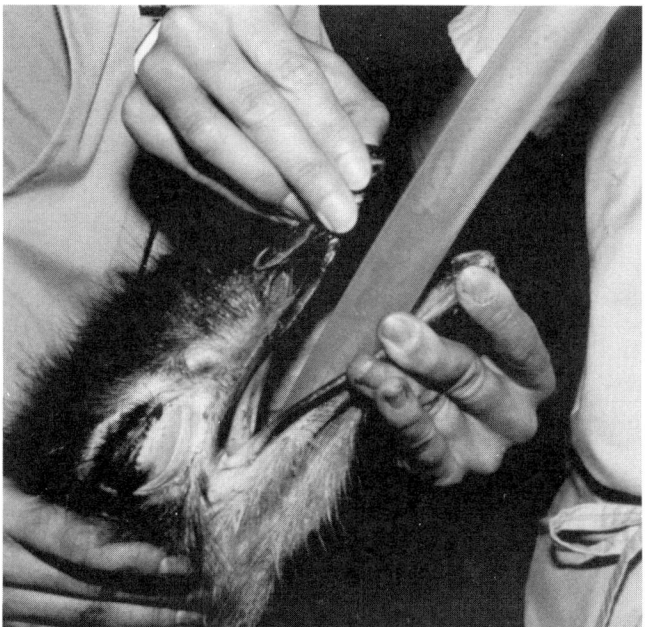

Figure 9.3 Adult ostrich positioned for intubation with a size 14 mm ID endotracheal tube.

the presence of complete tracheal rings in ratites with the increased risk of mucosal damage associated with overinflation of the cuff [10,17]. Inflation of the cuff is beneficial when expanded to the point of sealing the trachea. This is achieved by injecting small amounts of air into the cuff, then delivering a breath by closing the relief valve, and squeezing the reservoir bag to a peak inspiratory pressure of 12–15 cm H_2O. Adequate inflation is indicated by the absence of anesthetic gas odor and failure to detect air flow around the cuff. It is important to minimize waste gas contamination of the surgery suite and to protect the airway if regurgitation occurs. The inflated cuff facilitates effective mechanical ventilation which may be required for ratites, especially when the anesthetic period is prolonged.

PERIANESTHETIC SUPPORTIVE THERAPY

Padding and Positioning

Adult ratites, especially ostriches should be well padded during the anesthetic period to minimize neuropathy and myositis. The ostrich is susceptible to peroneal nerve paralysis when placed in lateral recumbency for as little as 1 hour [10]. Even juveniles should be placed on padding to avoid direct contact with the cold surgical table. Sternal recumbency does not interfere with normal respiration as it does in other avian species because ratites breath by lateral movement of the sternum [2,5,6]. Lateral recumbency may have a detrimental effect on expansion of the dependent air sacs especially in ratites anesthetized for a prolonged period [2]. Avian species may not ventilate effectively when placed in dorsal recumbency [27]. Either manual or mechanical ventilatory support is recommended for ratites placed in either dorsal or lateral recumbency, especially with prolonged duration of anesthesia.

Because the tracheal rings are flexible and easily compressed dorsoventrally, it is important that the neck be kept straight to avoid occluding the trachea or the endotracheal tube. The trachea should be free of the weight of the musculoskeletal portion of the neck to avoid compression and occlusion [3,10]. The head should be supported slightly above heart level to prevent edema in the structures of the head and to discourage passive regurgitation (Figure 9.4).

Thermal Support

Birds experience more rapid changes in body temperature during anesthesia than mammals [12]. Heat loss is significant during anesthesia, especially in juveniles, debilitated subjects and in smaller ratites. Even the adult ostrich is susceptible to hypothermia and several factors, including a cold surgery suite, administration of IV fluids at room temperature, and opening of a body cavity, exacerbate heat loss. Ideally, ratites should be placed on a circulating warm water heating pad and IV fluids should be warmed to body temperature.

Conversely, adult ratites may become hyperthermic, especially when exposed to a hot and humid ambient temperature, which might occur in a field situation. Both the injectable and inhalation agents impair thermoregulatory regulation. If a ratite has a rough induction and is maintained in a hot environment, hyperthermia may occur [9]. It is important to monitor temperature, especially under extreme environmental conditions. Adult ratites should not be anesthetized in hot weather unless shade and adequate methods for cooling can be effectively applied.

Fluid Administration

Intravenous fluids should be administered to anesthetized ratites especially during prolonged procedures or if significant blood loss is anticipated. Correcting dehydration prior to anesthetic induction will greatly improve the outcome of the anesthetic period. Lactated Ringer's solution is suggested for adult ratites, but Ringer's solution, 0.9% sodium chloride, and other commercial polyionic isotonic fluids are appropriate alternatives [13,14,26]. In juveniles and in

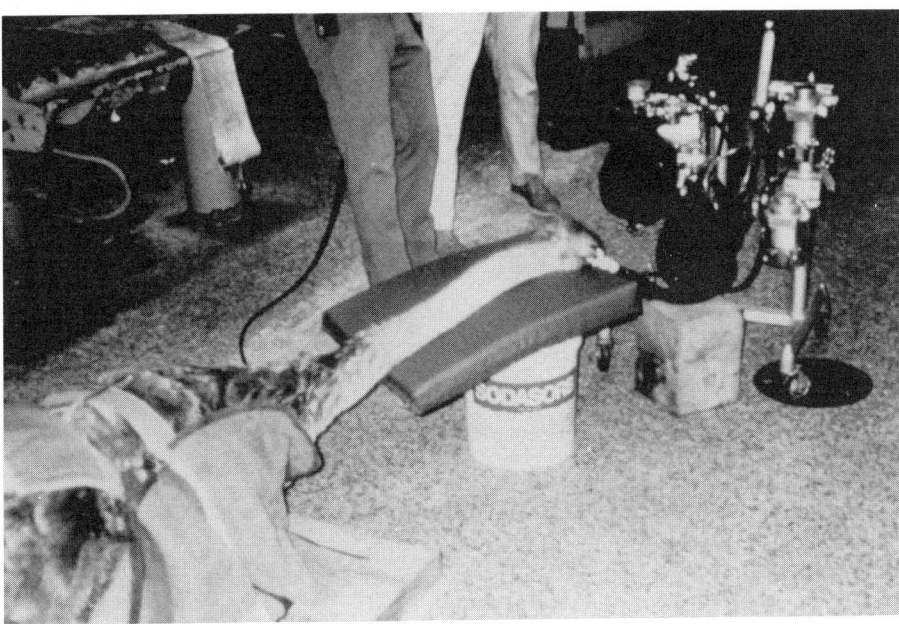

Figure 9.4 Adult ostrich placed in lateral recumbency. Note how the head is supported slightly above heart level and the neck is straight.

debilitated anorexic birds, fluids should be supplemented with 2.5% or 5.0% dextrose to prevent hypoglycemia. Commercial preparations which contain dextrose, such as 0.45% sodium chloride with 2.5% dextrose, are available for this purpose or dextrose may be added to isotonic fluids at a rate of 25–50 mg/ml. Five percent dextrose is not appropriate for prolonged administration because the free water load may promote edema. A flow rate of 5–10 ml/kg/hour is suitable for most patients [13,14]. This may be increased to compensate for blood loss or to aid in correcting hypotension.

The brachial vein is recommended for catheterization. In the ostrich, the brachial vein is large enough to place a 14-gauge, 13-cm catheter (Figure 9.5). In the emu, the right jugular vein is most commonly used for catheterization (Figure 9.6), and this site may be more appropriate in the ostrich and rhea when either positioning or the surgical site prevents access to the brachial vein. The medial metatarsal vein may also be catheterized for administration of fluid during the operative period. The skin over the limb is very thick and difficult to penetrate and either cut-down with a scalpel blade or puncture with a hypodermic needle one gauge larger than the catheter will facilitate entry through the skin. If a catheter is to be maintained after surgery for fluid or drug administration, the right jugular vein should be catheterized with a 13 cm catheter. A longer catheter is more likely to be maintained in the vein for a prolonged period of time [2]. The catheter should be secured in place with both adhesive glue and sutures. Wraps should be removed after recovery. Even a loose wrap may interfere with passage of a food bolus down the esophagus. Catheters should be flushed every 4–6 hours and changed every 72 hours.

When intravenous catheterization is impossible due to venous collapse or the size of the bird, an intraosseous catheter may used for rehydration. The ulna is the most commonly used site in companion animal birds [28].

Monitoring

Monitoring the depth of anesthesia in ratites must be performed with the same standard of care as in other avian and mammalian species. The plane of anesthesia is difficult to assess in ratites. Depth should be evaluated on the basis of a combination of observations including pulse rate and rhythm, respiratory rate and character, qualitative or quantitative assessment of arterial blood pressure, wing and neck tone, and the presence or absence of pedal, palpebral, and corneal reflexes [12]. Assessing anesthetic depth is difficult when injectable agents are used because the "traditional" planes of anesthesia may not be present [12]. When using an injectable protocol, the presence or absence of voluntary movement and frequent assessment of respiration, pulse rate, temperature, and mucous membrane color are the most reliable criteria [6].

The cardiovascular system is evaluated by continuously assessing pulse quality and rate, tissue perfusion

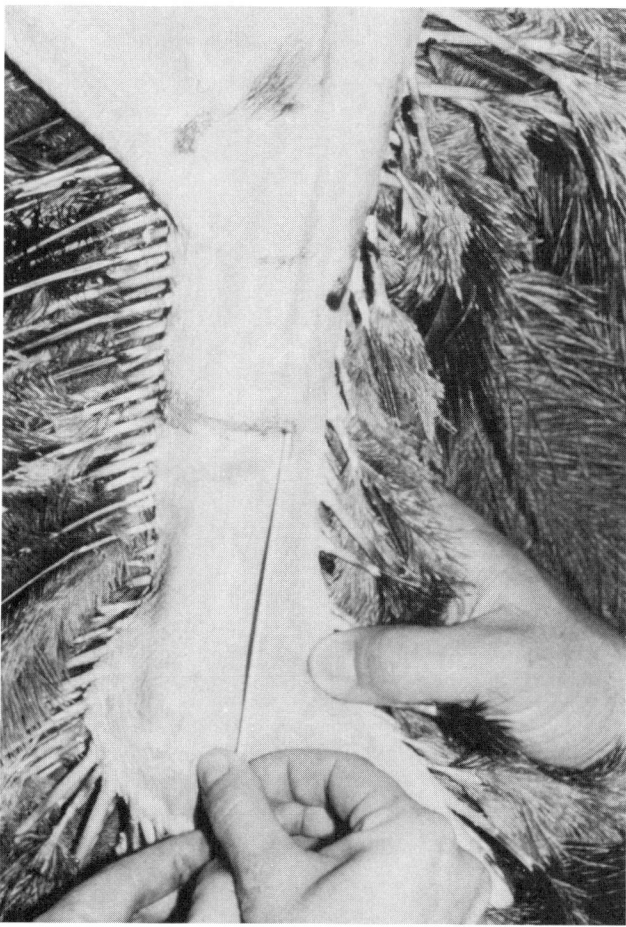

Figure 9.5 Catheter positioned in the right jugular vein of an ostrich.

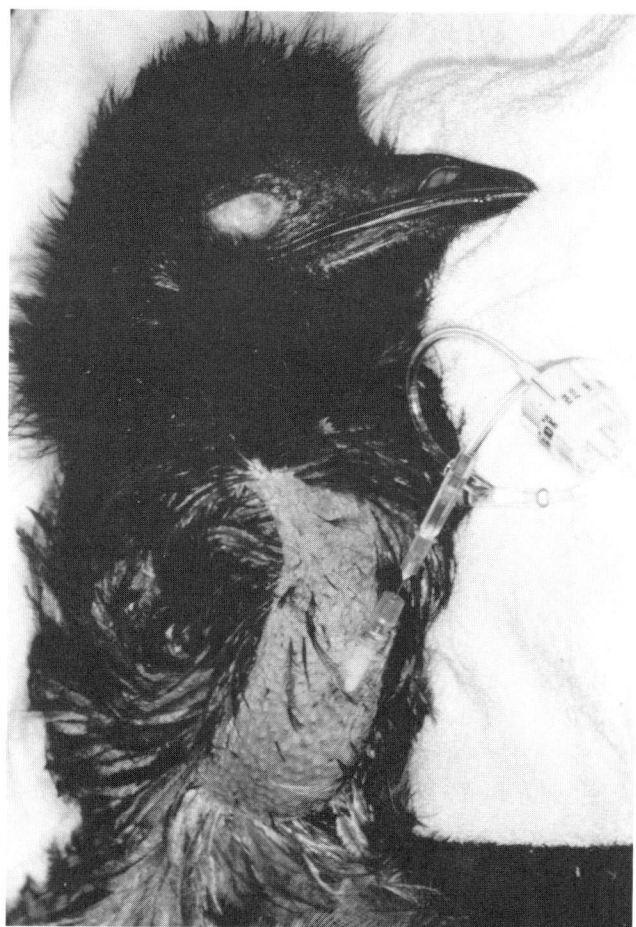

Figure 9.6 Fourteen-gauge, 13-cm catheter positioned in the right jugular vein of an emu.

(mucous membrane color), electrocardiogram display, and arterial blood pressure. Usually, a lead II electrocardiogram is recorded with the right arm and left arm leads placed on the right and left wing webs, respectively. The left leg lead is placed on the thigh of one leg (Figure 9.7).

The heart may be auscultated with a stethoscope over the lateral aspect of the ribs. The heart beat may also be readily detected by watching movement of the feathers over the sternum and wings. Significant decrease in the intensity of the heart beat is indicated by absence of feather movement. It has been suggested that heart rate may increase in ostriches during isoflurane anesthesia [3], but this is not a consistent finding and may be influenced by selection of preanesthetic and induction agents [14]. An esophageal stethoscope may be useful in ratite chicks to auscultate heart beat and respiration.

Pulse quality may be subjectively assessed by palpation of peripheral pulses including the brachial, medial metatarsal, and digital arteries. These arteries (except for the brachial artery in emus) may be catheterized, and the catheter can be connected to a pressure transducer for continuous direct assessment of arterial blood pressure. The brachial artery runs parallel to the brachial vein. The medial metatarsal artery runs along the craniomedial aspect of the tibiotarsal bone. The digital artery courses along the lateral aspect of the third digit parallel to the phalanges. Two other alternative sites have been identified if other arteries are inaccessible. An artery runs along the craniolateral aspect of the tibiotarsal bone, which is most accessible at the junction of the middle and distal thirds of the tibiotarsal bone. This can be used in the emu when the medial metatarsal artery cannot be successfully catheterized (Figure 9.8). A second artery, located in the oral cavity parallel to the mandibular beak, has been catheterized in an ostrich for intraoperative blood pressure monitoring. An indirect blood pressure monitor has also been used successfully on ratites [26]. The cuff should be placed around the distal tibiotarsal bone to monitor the artery adjacent to the craniolateral aspect of the bone (Figure 9.8). Arterial blood pressure values reported from ostriches during isoflurane anesthesia

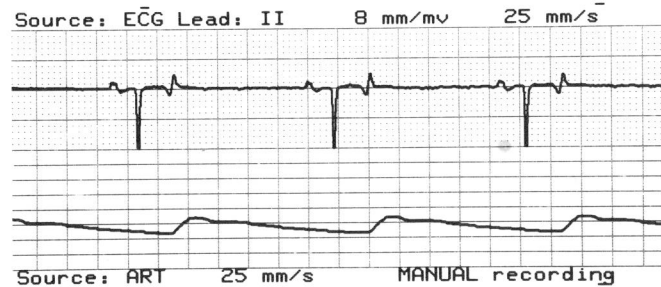

Figure 9.7 Normal electrocardiographic and arterial blood pressure tracings recorded from an anesthetized ostrich.

appear to be higher than values reported for mammalian species [14,26,29]. Arterial blood pressure values for ambient ratites have not been reported, but values reported during isoflurane anesthesia correlate well with values from conscious chickens [29]. Until more information is available regarding normal arterial blood pressure in conscious ratites, the same parameters used to evaluate pressure in mammals should be applied. A minimal mean arterial pressure of 60–70 mm Hg is desirable.

Respirations should be slow, deep and regular and remain at or above 8 breaths per minute during a surgical plane of anesthesia. Free-ranging ratites immobilized in the field should maintain a respiratory rate above 10 breaths per minute [6]. A rapid respiratory rate (25 to 40 breaths per minute) has been reported in isoflurane-anesthetized ratites [3,15], which could erroneously be interpreted as an inadequate depth of anesthesia. Any sudden change in the rate or character of respiration requires immediate assessment of anesthetic depth and airway patency.

Avian species are susceptible to respiratory arrest during general anesthesia. Apnea during general anesthesia is not uncommon in ratites [14]. When apnea occurs, anesthetic depth should be decreased and ventilation assisted manually or mechanically until spontaneous ventilation is reestablished. Movement of the reservoir bag, auscultation with an esophageal stethoscope in ratite chicks, and a respiratory monitor will all aid in assessing respiratory rate and rhythm. Respiratory rate may also be monitored by observing movement of the ribs and sternum. This method may be more important in ratite chicks that do not move the reservoir bag discernibly with respiration. Visualization may be facilitated using clear surgical drapes.

The most accurate assessment of adequacy of ventilation is measurement of arterial blood gases and pH. This is rarely practical in a clinical or field situation, and normal values have not been documented for conscious ratites. Capnography measures the end-tidal carbon dioxide concentration and provides a method to assess adequacy of ventilation. Reports on usefulness in ratites are unavailable at this time. Pulse oximetry, which is a noninvasive measure of pulse rate and the percent of oxygenated hemoglobin in the blood, has been reported in small birds. The probe is attached to a wing web, tongue, or the tibiotarsal area [12]. Reports on accuracy in ratites are unavailable; although, the author has had limited success by placing the probe on the mandibular beak.

Cloacal temperature should be monitored during the anesthetic period. Hypothermia occurs rapidly in anesthetized birds and is a problem in both young and thin patients. Hyperthermia may also occur in adult ratites

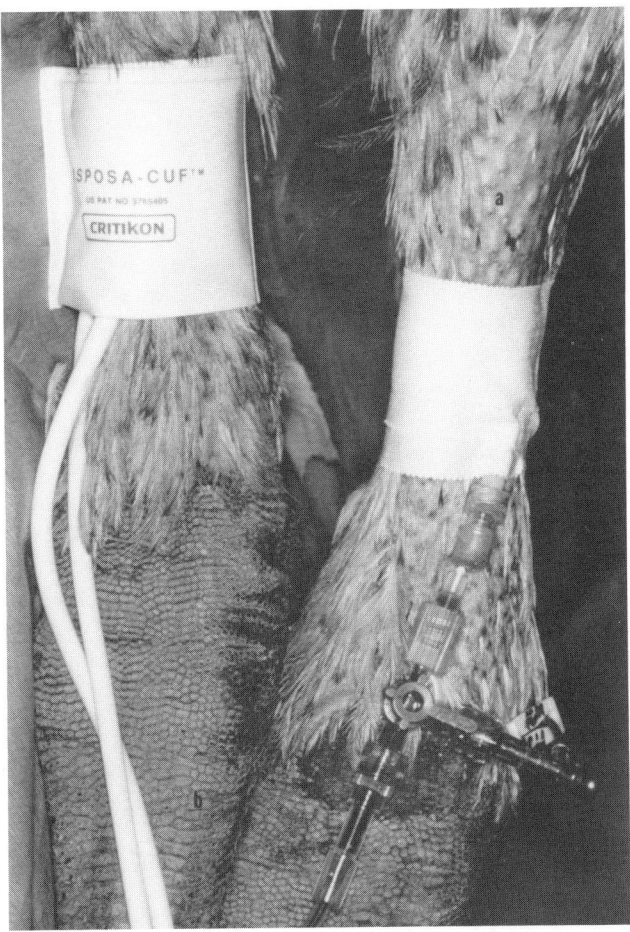

Figure 9.8 Twenty-two-gauge catheter postioned in an artery coursing along the craniolateral aspect of the tibiotarsal bone of an emu for direct arterial blood pressure measurement and collection of arterial blood. Blood pressure cuff positioned on the opposite limb at the level of the distal tibiotarsal bone for indirect blood pressure measurement.

following a rough induction and when the ambient temperature is high.

Muscle tone of the neck is a useful indicator of anesthetic depth in ratites. The neck should be relaxed during surgical anesthesia. The development of tension in the neck indicates that the bird is light and that voluntary movement is likely. This occurs during inhalation anesthesia but is less reliable when injectable anesthesia is used. Anesthetic depth in birds can be assessed using the palpebral, pedal and corneal reflexes [12]. The usefulness and dependability of these reflexes to evaluate anesthetic depth in ratites has not been reported. Loss of the corneal reflex indicates an excessive depth of anesthesia. The other reflexes are usually absent when an adequate surgical depth is achieved, however these reflexes are not reliable.

Ventilatory Support

Arterial blood gases recorded from spontaneously breathing ostriches during isoflurane anesthesia yielded $PaCO_2$ values ranging from 58 to 129 mm Hg, indicating that ratites hypoventilate during inhalation anesthesia [14]. Spontaneously ventilating emus breathing room air and anesthetized with guaifenesin, xylazine and ketamine maintained normal ventilation with $PaCO_2$ values ranging from 24 to 40 mm Hg. However, these emus developed mild to moderate hypoxemia. Based on these limited observations, it is advisable to provide ventilatory support for ratites during inhalation anesthesia, especially for prolonged procedures. Ventilation settings for ratites are similar to those used in small mammals and include a tidal volume of 10–20 ml/kg, respiratory rate of 6–12 per minute, and a peak inspiratory pressure of 10–15 cm H_2O.

The emu presents a unique problem when positive pressure ventilation is applied. The cartilaginous tracheal rings of the caudal cervical area are interrupted ventrally for a distance of 6–8 cm and an expandable pouch everts from this opening [10]. To effectively deliver positive pressure ventilation, this area of the neck should be wrapped to prevent inflation of the pouch.

Ideally, arterial blood gases should be reassessed after mechanical ventilation is initiated because it appears that hypocapnia can be rapidly induced [14]. For birds anesthetized with injectable anesthetic agents, oxygen supplementation may be beneficial, especially for prolonged procedures.

Complications

Apnea.

The respiratory rate should remain regular, deep and stable during anesthesia. Apnea is not uncommon during general anesthesia in ratites [14] and may be caused by hypocapnia associated with excessive ventilatory assistance, but is more likely due to extreme depth of anesthesia. When apnea occurs, depth of anesthesia should be assessed and decreased as indicated. Ventilation should be supported at 2–4 breaths per minute to a peak inspiratory pressure of 10–15 cm H_2O until spontaneous ventilation returns. The cardiovascular system should be critically evaluated because apnea may indicate an impending cardiac arrest. If spontaneous ventilation does not return within 15 minutes following discontinuation of delivery of anesthetic gas, administration of doxapram (5 mg/kg IV) [6] may be beneficial.

Cardiac Arrest.

Cardiac arrest is associated with a poor prognosis and efforts at resuscitation are often unsuccessful in avian species [12]. Cardiac arrest may be preceded by apnea, hypotension, arrhythmias or a sudden decrease in the intensity of the heart beat. If the bird is being appropriately monitored, there will be some indication of excessive anesthetic depth which will alert the anesthetist to correct the situation before an arrest occurs. Premature ventricular depolarizations have been reported in an ostrich during isoflurane anesthesia [26]. The same criteria used to evaluate and treat premature ventricular depolarizations in mammals should be applied to ratites. A dosage of 1.0–2.0 mg/kg IV of lidocaine has been used to treat ventricular arrhythmias in ostriches. Hypotension, as defined by a mean arterial pressure less than 60 mm Hg, is uncommon in ratites but has been observed in critically ill birds during isoflurane anesthesia. Hypotension should be treated by increasing the rate of IV fluid administration and decreasing anesthetic depth. Dobutamine (2–5 ug/kg/minute) or dopamine (5 ug/kg/minute) infusion may be beneficial for refractory hypotension. If cardiac arrest occurs, epinephrine should be administered (0.02–0.2 mg/kg IV or intratracheal) as soon as airway ventilation with 100% oxygen and chest compressions are instituted. Cardiac compressions should be applied laterally in ratites because the chest is more compressible in this plane. In reality, open cardiac massage would be more effective but there are no studies in ratites to describe the approach. Ideally, anesthetic complications should be detected and corrected before cardiac arrest occurs. Corticosteroids, such as prednisolone sodium succinate (20–30 mg/kg IV), may be beneficial as part of the treatment for a hypotensive crisis or for cardiopulmonary arrest.

Voluntary Movement.

A frustrating aspect of anesthetic management of ratites is the difficulty in maintaining an adequate depth of anesthesia, even when a surgical plane is achieved. This is more likely to occur when the patient is healthy. Intraoperative use of butorphanol (0.02–0.05 mg/kg IV) may contribute to analgesia and help prevent voluntary movement. Ketamine (0.2 mg/kg IV) may be administered intraoperatively to stop voluntary movement as needed.

Regurgitation.

Ratites may regurgitate during general anesthesia especially when fasting is inadequate or if abnormalities of the gastrointestinal tract exist. The cuff of the endotracheal tube should be adequately inflated to assure protection of the airway if regurgitation occurs. Slight elevation of the neck may help to prevent passive regurgitation. The oral cavity should be thoroughly cleaned prior to extubation and the endotracheal tube removed with the cuff partially inflated after onset of swallowing.

ANESTHETIC RECOVERY

Environment

A small, dark, and quiet environment, free of sharp objects and corners, should be used for recovery of adult birds. As noted above, the bird should be extubated when swallowing is observed. Hoods may help to smooth the recovery of ostriches [2,25]. Ideally, the bird should be positioned in sternal recumbency and the head supported until the bird is able to maintain normal head posture unassisted (Figure 9.9). This is not always possible because some birds will have a rough recovery and be too dangerous to assist. Diazepam (0.1–0.2 mg/kg IV) may be indicated when this occurs [2,14,17]. Once the bird is in sternal recumbency and can hold its head steady, the bird may be left unattended to stand on its own. One author suggests placing the bird in sternal recumbency inside a small closed crate (1.3 m × 1.3 m × 1.3 m for an adult ostrich) padded with straw until fully recovered [25]. An anesthetic recovery system[e] for ratites is now available (Figure 9.10). This device is wrapped around the bird with the limbs flexed and secured tightly with quick release straps (Figure 9.11). The bird is held in sternal recumbency until fully recovered to prevent self-trauma. Two sizes are available for ostrich and emus, respectively.

Young ratites may be wrapped in a towel with their limbs flexed to allow recovery in the arms of a restrainer. Emus, rheas, and young ostriches may be held in sternal recumbency by the handler straddling the bird and applying sufficient pressure to prevent

[e] Ratite Rap, Jorgensen Laboratories Inc, Loveland, CO.

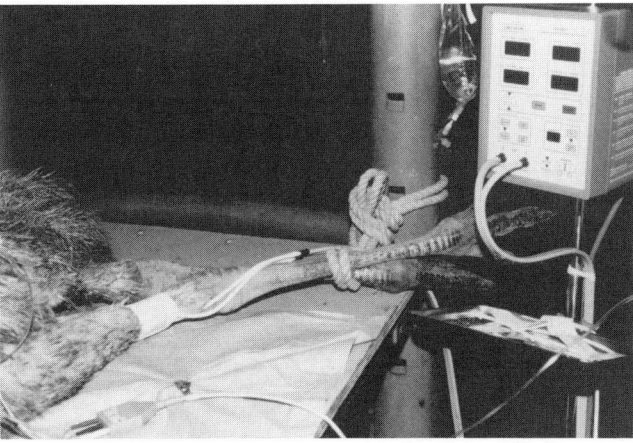

Figure 9.9 Blood pressure cuff positioned at the level of the distal tibiotarsal bone and connected to an oscillometric blood pressure monitoring device for indirect measurement of blood pressure.

Figure 9.10 The Ratite Rap.

the bird from rotating into a lateral position (Figure 9.12).

Potential Complications

The most common postoperative complication is self-trauma during recovery. This can be avoided by applying appropriate precautions. In very young and debilitated birds, hypoglycemia can occur especially if glucose was not administered intraoperatively. A postrecovery blood glucose measurement may be indicated in high-risk patients. Some birds may develop excessive oral and respiratory secretions intraoperatively. The oral cavity should be checked prior to extubation and a clear airway assured prior to leaving a bird unattended.

Exertional rhabdomyolysis may occur rarely in ratites and the risk increases if induction was rough and

Figure 9.11 The Ratite Rap positioned on an ostrich to facilitate safe anesthetic recovery.

Figure 9.12 If possible, ratites should be placed in sternal recumbency during recovery from anesthesia and monitored until the bird can maintain normal head position unassisted.

prolonged and if the ambient temperature is high [6,8]. Myopathy is indicated by failure of the bird to rise after a recovery period of 24–48 hours [8]. Treatment is largely unrewarding but supportive therapy, including administration of bicarbonate-containing IV fluids, dextrose, corticosteroids, vitamin E-selenium, calcium, antibiotics, and flunixin meglumine, may be beneficial in mild cases. Dantrolene,[f] a skeletal muscle relaxant used in the treatment and prevention of malignant hyperthermia in several species [30], may also be beneficial. One author systematically administered methylprednisolone succinate[g] (3 mg/kg IV), Vitamin E (8 mg/kg IM), and selenium (0.1 mg/kg IM) to ostriches immobilized in the field following the occurrence of fatal myopathy in one adult bird [6].

Postoperative Analgesics

Analgesics are used infrequently for the relief of pain in birds [31]. Although it has been incorrectly stated that birds do not suffer from pain to any significant degree, the majority of literature states that analgesics are recommended for avian species. Unfortunately, there is a paucity of information regarding dosages and efficacy of analgesic agents. Studies performed in chickens and pigeons indicate that opioids may be useful [32,33]. Two studies which investigated the use of butorphanol in cockatoos [31] and parakeets [34] indicated that this opioid agonist-antagonist may be a safe and effective analgesic agent. There have been no clinical studies involving the use of analgesic agents in ratites to date. Based on clinical use, butorphanol appears to be safe for ratites and may provide adequate analgesia in the postoperative period following surgical procedures.

SUMMARY

General anesthesia should be avoided in very young ratites. When necessary, isoflurane anesthesia is the method of choice for both induction and maintenance. Young ratites may be more susceptible to isoflurane-induced apnea and should be intubated unless the procedure is less than 10 minutes.

Selection of an anesthetic protocol for juvenile and adult ratites depends on the procedure to be performed, the ability to safely restrain the bird and the physical status of the bird.

- For invasive and painful procedures, such as orthopedic surgery and laparotomy (proventriculotomy, hysterotomy), birds should be intubated and maintained with an inhalation agent, preferably isoflurane. Induction of anesthesia will vary with the individual bird. For juveniles and sick adults, mask induction is feasible if the bird can be safely restrained. The preferred induction for adults is diazepam-ketamine administered IV. When IV induction is not possible, IM injection of zolazepam-tiletamine will produce immobilization for maintenance with an inhalation agent. Diazepam can be administered concurrently with the zolazepam-tiletamine and also prior to

---

[f] Dantrium, Eaton Laboratories, Norwich, NY.

[g] Solumedrol, Upjohn Co, Kalamazoo, MI.

recovery to improve the quality of induction and recovery, respectively. Induction and recovery may be rough when zolazepam-tiletamine is used and patients should be placed in a dark, padded, and hazard-free area.

- General anesthesia of ratites under field conditions presents a challenge to the practitioner. These animals are usually healthy and often excited. Xylazine and ketamine administered IM or zolazepam-tiletamine IM may provide adequate immobilization for short and minor procedures, such as repair of laceration, only. A combination of guaifenesin-xylazine-ketamine may provide a method to maintain anesthesia in emus but has not been used in ostriches.

Regardless of the protocol used, ratites must be closely monitored during anesthesia, including observation of respiratory rate and character, and frequent assessment of pulse rate and quality. Cloacal temperature should be measured periodically especially when ratites are being anesthetized in the field.

REFERENCES

1. Jensen J. Ratite restraint and handling. In: Fowler ME, ed. *Zoo and wild animal medicine: Current Therapy III*. Philadelphia: WB Saunders Co, 1993; 198–200.
2. Martin JA, Speer BL. Restraint and techniques of ratites for the veterinary technician, in *Proceedings*. Assoc Avian Vet 1993;369–380.
3. Kimminau KM. Introducing the ostrich. *Vet Tech* 1993;14:459–467.
4. Robinson PT, Fairfield J. Immobilization of an ostrich with ketamine HCl. *J Zoo Anim Med* 1974;5:11.
5. Jensen J, Johnson JH, Weiner S. Husbandry and medical management of ostriches, emus, and rheas. *Texas Wildlife and Exotic Animal Teleconsultants* 1992;1–114.
6. Ostrowski S, Ancrenaz M. Chemical immobilization of red-necked ostriches (*Struthio camelus*) under field conditions. *Vet Rec* 1995;136:145–147.
7. Samour JH, Irwin-Davies J, Faraj E. Chemical immobilisation in ostriches (*Struthio camelus*) using etorphine hydrochloride. *Vet Rec* 1990;127:575–576.
8. Raath JP, Quandt SK, Malan JH. Ostrich (*Struthio camelus*) immobilisation using carfentanil and xylazine and reversal with yohimbine and naltrexone. *J S Afr Vet Assoc* 1992;63:138–140.
9. Stoskopf MJ, Beall FB, Ensley PK, Neely E. Immobilization of large ratites: blue-necked ostrich (*Struthio camelus australis*) and double-wattled cassowary (*Casuarius casuarius*)—with hematologic and serum chemistry data. *J Zoo Anim Med* 1982;13:160–168.
10. Fowler ME. Comparative clinical anatomy of ratites. *J Zoo Wildl Med* 1991;22:204–227.
11. Oelafsen BW. The renal portal valves of the ostrich, *Struthio camelus*. *S Afr J Sci* 1977;73:57–58.
12. Sinn LC. Anesthesiology. In: Ritchie BW, Harrison GJ, Harrison LR, eds. *Avian medicine: Principles and application*. Lake Worth, FL: Wingers Publishing, Inc, 1994;1066–1080.
13. Fowler JD, Bauck L, Cribb PH, Presnell KR. Surgical correction of tibiotarsal rotation in an emu. *Companion Anim Pract* 1987;1:26–30.
14. Cornick JL, Jensen J. Anesthetic management of ostriches. *J Am Vet Med Assoc* 1992;200:1661–1666.
15. Jensen J. Ratite anesthesia and surgery, in *Proceedings*. Ostrich Medicine Seminar for Veterinarians, Texas A&M University, College Station, TX, 1990;6–7.
16. Van Heerden J, Keffen RH. A preliminary investigation into the immobilising potential of a tiletamine/zolazepam mixture, metomidate, a metomidate and azaperone combination and medetomidine in ostriches (*Struthio camelus*). *J S Afr Vet Assoc* 1991;62:114–117.
17. Ramsey E. Ratite restraint, immobilization and anesthesia. *Avian/Exotic Anim Med Symposium* 1991;176–178.
18. Jensen J. Husbandry, medical and surgical management of ratites: Part I, in *Proceedings*. Am Assoc Zoologic Med 1988;113–118.
19. Gandini GCM, Keffin RH, Burroughs REJ, Ebedes H. An anesthetic combination of ketamine, xylazine, and alphaxalone-alphadolone in ostriches (*Struthio camelus*). *Vet Rec* 1986;118:729–730.
20. Bruning DF, Dolensek EP. Ratites. In: Fowler ME, ed. *Zoo and wild animal medicine*. 2nd ed. Philadelphia: WB Saunders Co, 1986;277–291.
21. Altman RB. Avian Anesthesia. *Comp Cont Ed Pract Vet* 1980;11:38–43.
22. Blackshaw GD, Wakeman B. Immobilization of an adult ostrich for surgery. *J Zoo Anim Med* 1971;2:11–12.
23. Mikaelian J. Intravenously administered propofol for anesthesia of the common buzzard (*Buteo Buteo*), the tawny owl (*Strix aluco*), and the barn owl (*Tyto Alba*), in *Proceedings*. Europen Chap Assoc Avian Vets 1991;97–101.
24. Fitzgerald G, Cooper JE. Preliminary studies on the use of propofol in the domestic pigeon (*Columba livia*). *Res Vet Sci* 1990;49:334–338.
25. Stewart JS. Husbandry, medical and surgical management of ratites: Part II, in *Proceedings*. Am Assoc Zoologic Med 1988;119–122.
26. Matthews NS, Burba DJ, Cornick JL. Premature ventricular contractions and apparent hypertension during anesthesia in an ostrich. *J Am Vet Med Assoc* 1991;198:1959–1961.
27. Fedde MR. Avian respiratory physiology, in *Proceedings*. 8th Annual Vet Midwest Anes Conf 1992; 1–10.
28. Ritchie BW, Otto CM, Latimer KS, Crowe DT. A technique of intraosseous cannulation for intravenous therapy in birds. *Comp Cont Ed Pract Vet* 1990; 12:55–58.
29. Detweiler DK. Control mechanisms of the circulatory system. In: Swenson MJ, ed. *Dukes' physiology of domestic animals*. 10th ed. London: Cornell University Press Ltd, 1984;163–191.
30. Waldron-Mease E. Correlation of post-operative and exercise-induced equine myopathy with the defect malig-

nant hyperthermia, in *Proceedings*. 24th Ann Mtg Am Assoc Eq Pract 1978;95–99.
31. Curro TG, Brunson DB, Paul-Murphy J. Determination of the ED50 of isoflurane and evaluation of the isoflurane-sparing effect of butorphanol in cockatoos (*Cacatua spp.*) *Vet Surg* 1994;23:429–433.
32. France CP, Woods JH. Effects of morphine, naltrexone, and destrorphan in untreated and morphine-treated pigeons. *Psychopharmacology* 1985;85:377–382.
33. Bardo MT, Hughes RA. Shock elicited flight response in chickens as an index of morphine analgesia. *Pharmacol Biochem Behav* 1990;35:567–570.
34. Bauck L. Analgesics in avian medicine, in *Proceedings*. Assoc Avian Vet 1990;239–244.
35. Jones JH. Pulmonary blood flow distribution in panting ostriches. *J Appl Physiol* 1982;53:1411–1417.

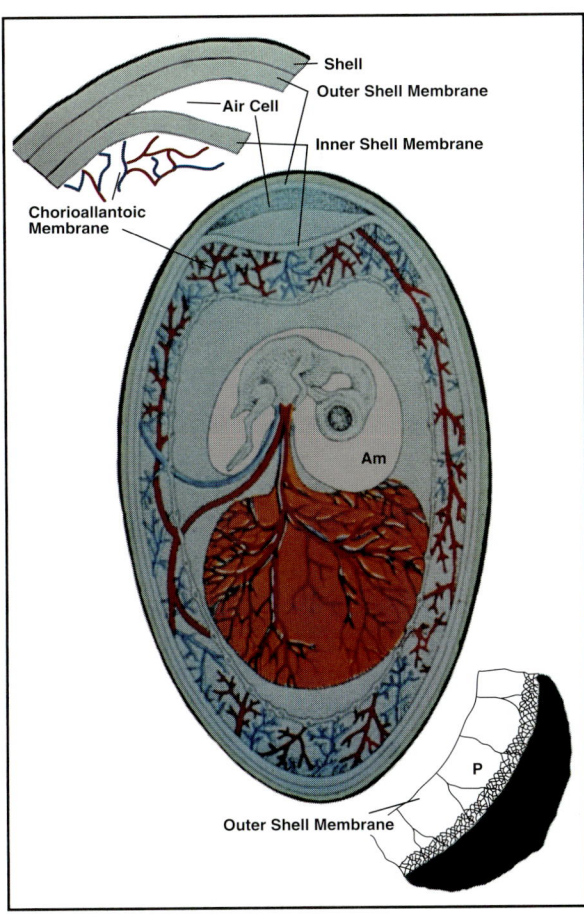

Figure 6.11b. Diagram of avian egg showing principal structures including the shell, membranes, albumen, and yolk. Courtesy: Inner-Vision Bio-Technologies Corp.

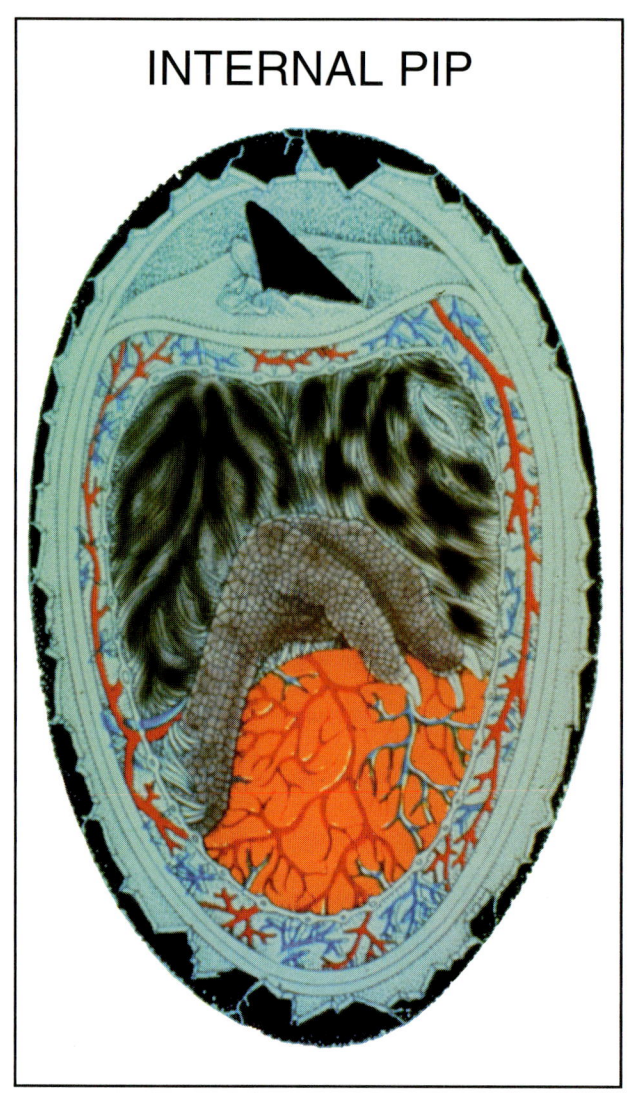

Figure 7.3. Normal position of the ratite embryo at the time of pipping. Courtesy: Inner-Vision Bio-Technologies Corp.

Figure 8.3. Oxygen requirements of the developing embryo. Courtesy: Inner-Vision Bio-Technologies Corp.

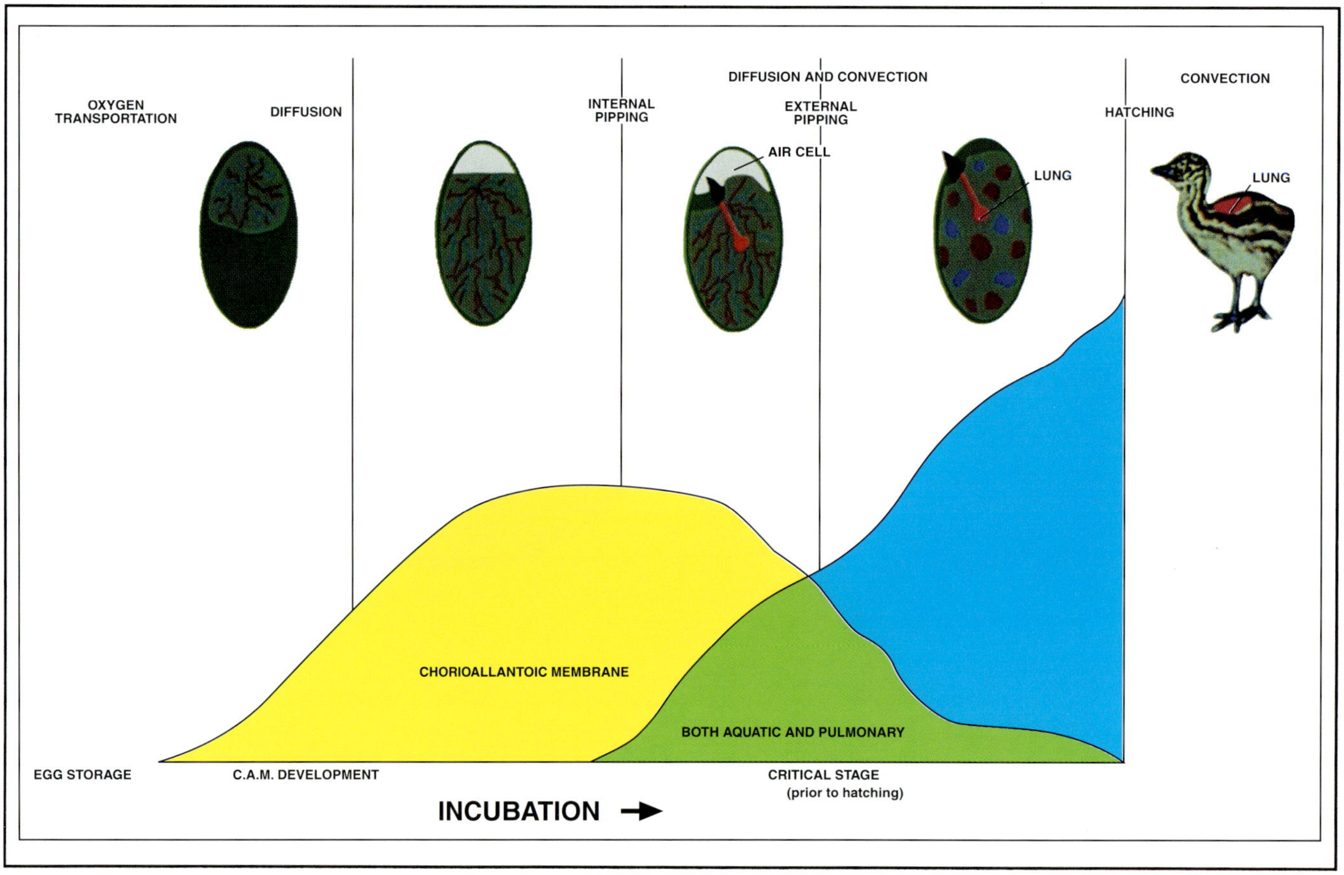

Figure 7.1. Bacterial infection of an ostrich egg.

Figure 7.4. Ostrich egg at the time of external pipping.

Figure 12.1. *Argus persicus*, soft-shelled tick.

Figure 12.2. *Struthiolipeurus struthionis* louse.

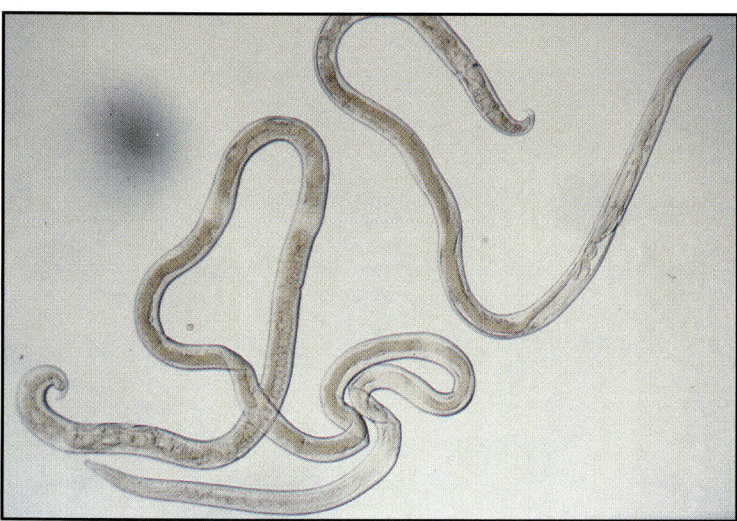

Figure 12.3. *Libyostrongylus douglassi*.

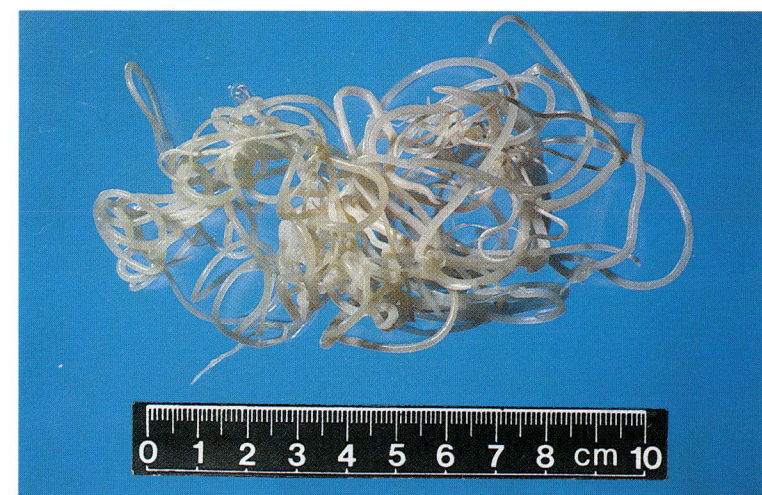

Figure 12.4. *Dicheilonema* sp.

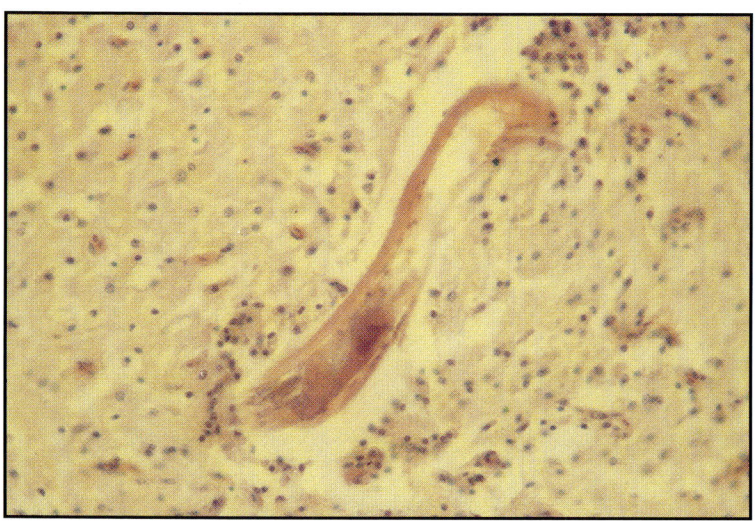

Figure 12.5. *Baylisascaris* sp. in cerebral tissue of emu.

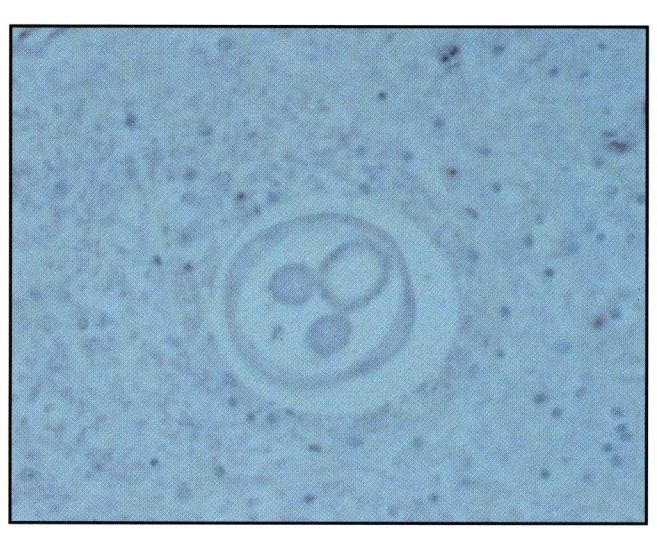

Figure 12.6. *Chandlerella quiscali*.

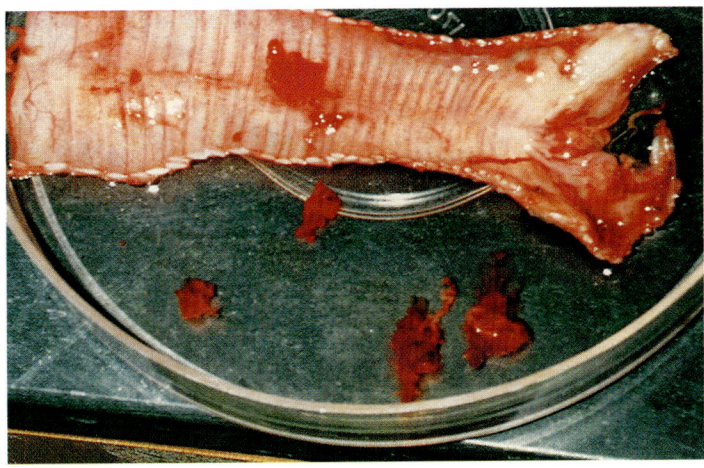

Figure 12.7. *Syngamus trachea* (gapeworm).

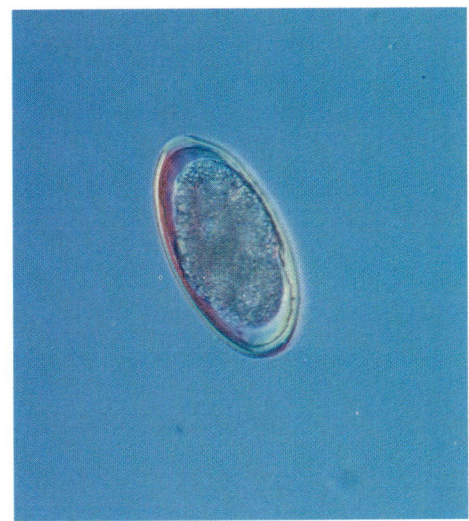

Figure 12.8. Ovum of *Syngamus trachea* on fecal flotation.

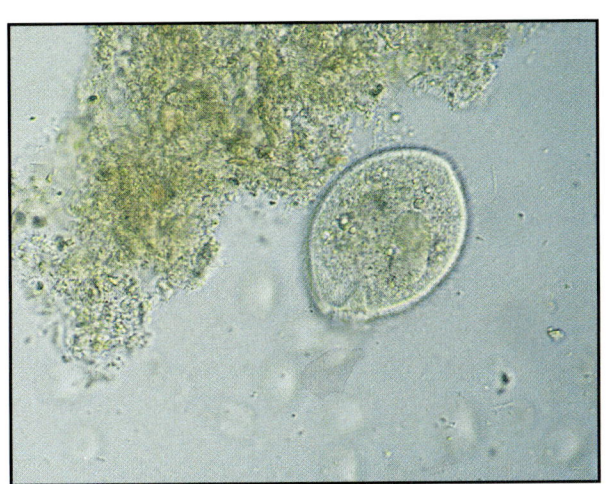

Figure 12.9. *Balantidium* sp. in fecal suspension.

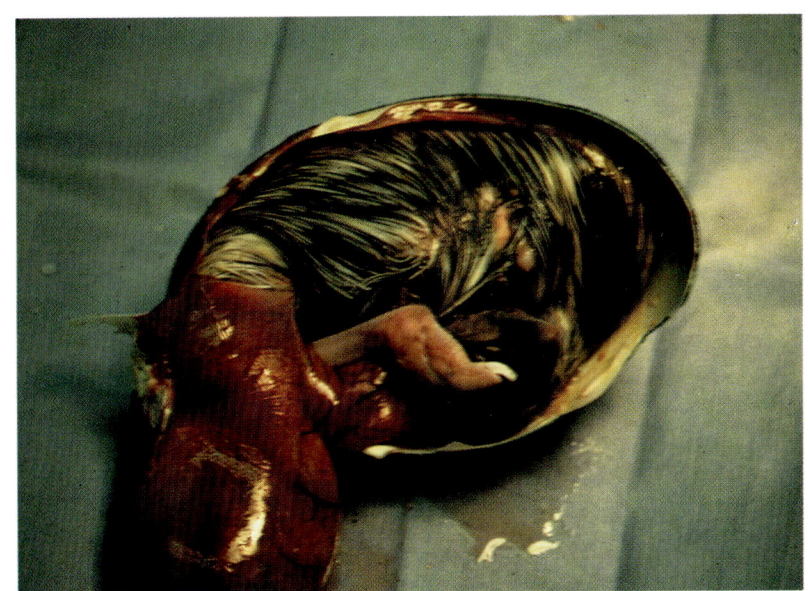

Figure 13.1. Dead embryo in shell showing omphalitis due to congenital infection.

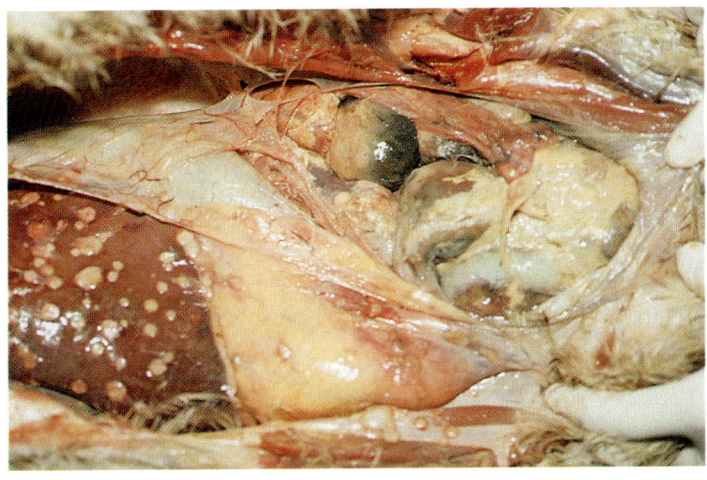

Figure 13.2. Lesions of tuberculosis in a juvenile emu showing numerous granulomas of the peritoneum and the capsule of the liver.

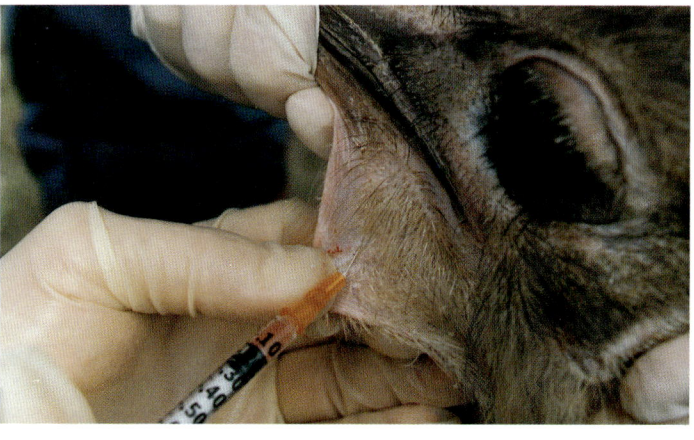 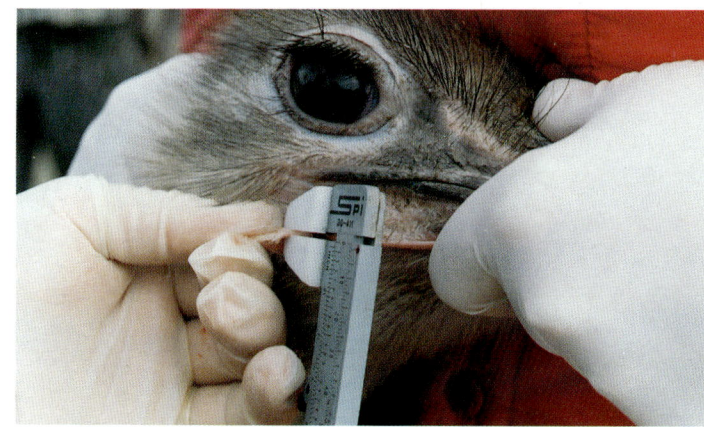

Figure 13.3. Site for intradermal mycobacterium sensitivity test in ostriches.

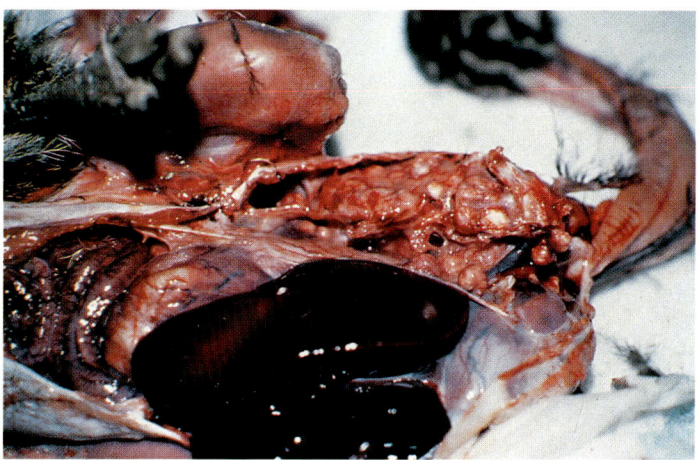

Figure 13.4. Severe aspergillosis in emu chick showing numerous pulmonary granulomas.

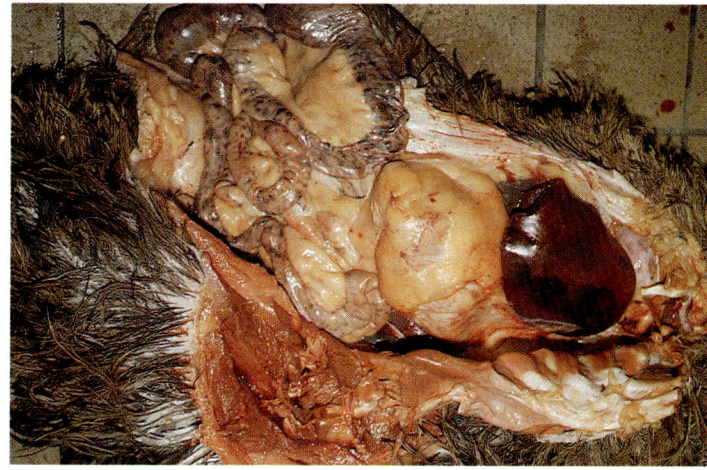

Figure 13.5. Characteristic hemorrhages associated with eastern equine encephalitis in emus.

Figure 13.6. Superficial pox lesion cranial to the eye in a young ostrich.

Figure 13.7. Severe chronic pox lesions in ostrich chick.

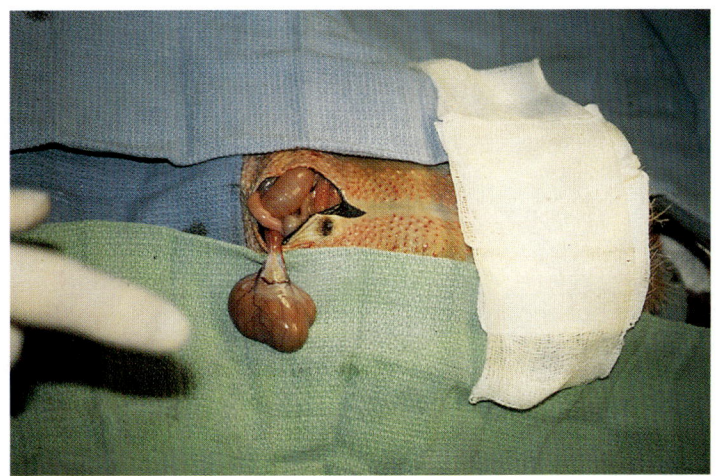

Figure 14.2. Surgical removal of the retained yolk sac in ostrich chick.

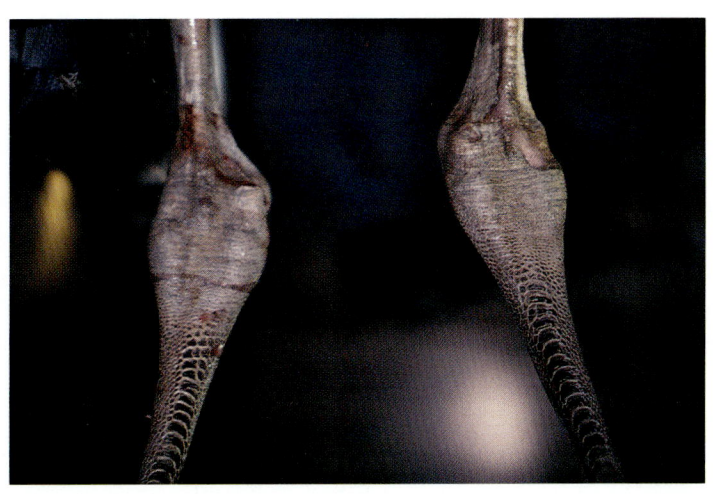

Figure 14.1. Enlarged yolk sac and unhealed navel associated with water retention in an ostrich chick.

Figure 14.3. "Tumbler" emu chick showing retraction of the head and incoordination.

Figure 14.4. Angular leg deformity in young emu chick.

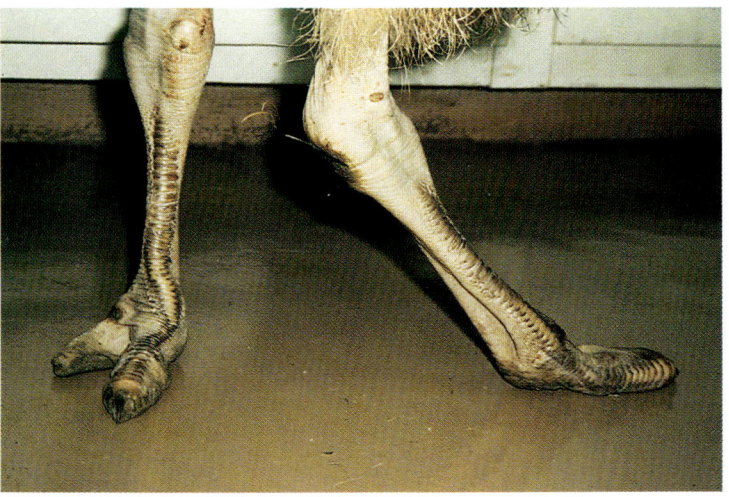

Figure 14.5. Angular limb deformity in a juvenile.

Figure 14.7. Untreated rolled toe in a juvenile ostrich.

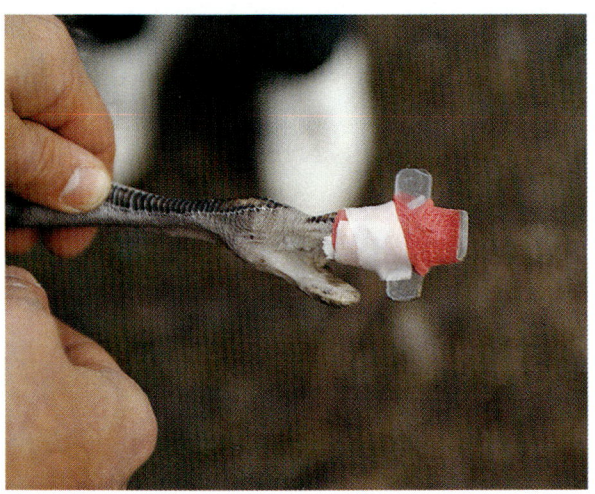

Figure 14.8. Splinting of rolled toe in young ostrich chick.

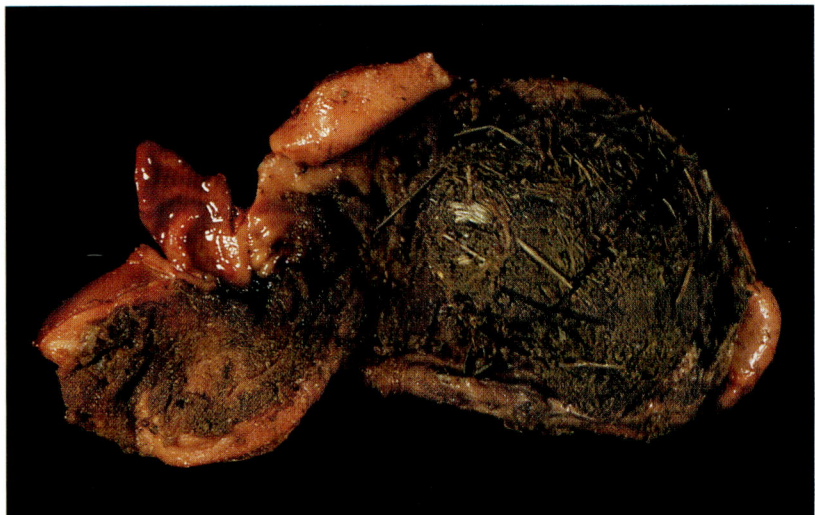

Figure 14.9. Impaction of the proventriculus and ventriculus with fibrous material.

Figure 14.10. Prolapse of the cloaca in an emu chick.

Chapter 10

Surgical Conditions of Ratites

Mark R. Crabill and Clifford M. Honnas

IMPACTION OF THE PROVENTRICULUS

Indiscriminate ingestion of foreign material by ratites leads to impaction of the proventriculus [1–3]. The normal foraging behavior of ratites combined with the stress of captivity and access to foreign material may result in this syndrome. Rocks, sand, leaves, grass, and other fibrous material commonly result in impaction of the proventriculus [2,3]. Nails and other metallic objects are also commonly ingested (Figure 10.1).

Clinical signs associated with impaction of the proventriculus include anorexia, weight loss, changes in volume or consistency of feces, loss of reproductive performance, depression, and lethargy [2,3]. Loss of reproductive performance results from chronic anorexia and debilitation. While impaction is often a chronic condition, some birds may exhibit acute clinical signs resulting in death within 24 to 48 hours [2]. Ingestion of sharp foreign objects leads to perforation of the gastrointestinal tract with consequential peritonitis.

Palpation of the abdomen caudal to the keel bone (sternum) and to the left of ventral midline facilitates diagnosis of impaction [1–3]. With sand and rocks, a gritty, firm, distended viscus is often palpable [4]. Impaction with leaves or hay is less discernable by palpation but usually an enlarged viscus can be appreciated. Endoscopy can be used as a diagnostic aid to confirm impaction with fibrous material that cannot be visualized radiographically. Dorsoventral and lateral radiographs of the abdomen will provide evidence of impactions caused by radiodense material (metal, sand, or rocks). In adult birds, lateral oblique radiographs taken cranial or caudal to the thigh are useful to diagnose an impaction and can be obtained without anesthesia. A complete blood cell count and serum biochemistry panel are useful to detect concurrent problems of dehydration, electrolyte imbalance, and

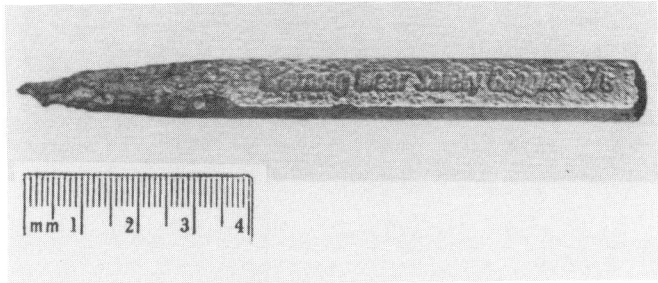

Figure 10.1 Foreign body (metallic punch) removed from the proventriculus of an ostrich.

infection, thereby aiding the veterinarian in perioperative management of the bird.

Medical treatment with laxatives is often ineffective in resolving severe impactions. In these cases, proventriculotomy is required for successful resolution. Proventriculotomy is also indicated in birds that have ingested sharp foreign objects capable of penetrating the gastrointestinal tract (Figure 10.2).

Preoperative treatment comprises insertion of a catheter in the left brachial vein and initiation of antimicrobial therapy (amoxicillin (7 mg/kg, IM, q 6 h) or enrofloxacin (2 mg/kg, IM or PO, q 12h)) [3]. Following induction of anesthesia, the bird is positioned in right lateral recumbency with the left pelvic limb abducted and supported caudally in a stand (Figure 10.3). The ventral abdomen and left paramedian areas are plucked and prepared aseptically for surgery. A left paramedian approach is used for access to the proventriculus, placing the cutaneous incision in the middle of the quilled band on the left of ventral midline (Figure 10.4). In adult ostriches, the incision should start approximately 15 cm caudal to the keel. The cranial aspect of the abducted left leg serves as the

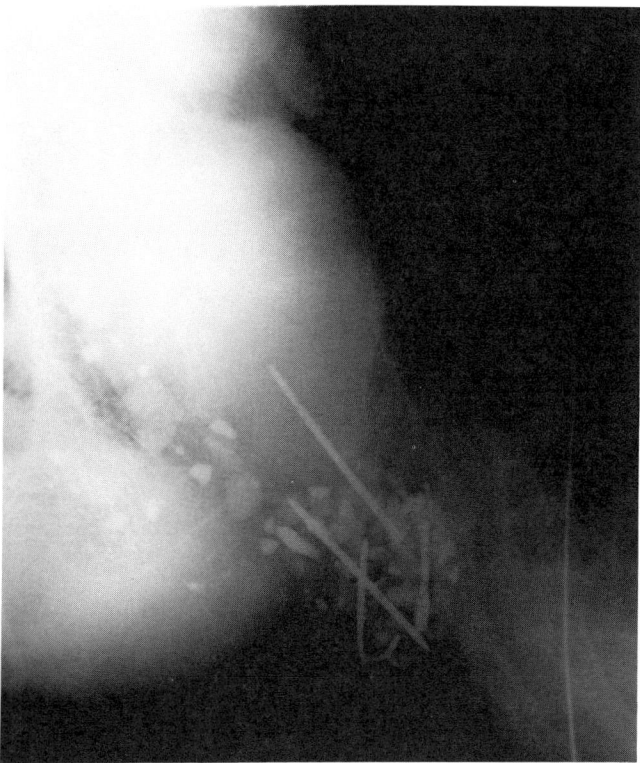

Figure 10.2 Lateral radiograph of the abdomen of an ostrich showing the presence of nails and rock within the proventriculus.

Figure 10.3 Proper positioning of the ratite for a left paramedian approach to the abdomen. (From Honnas CM, Jensen J, Cornick JL, et al. Proventriculotomy to relieve forign body impaction in ostriches. *J Am Vet Med Assoc* 1991;199:461–465, reproduced with permission).

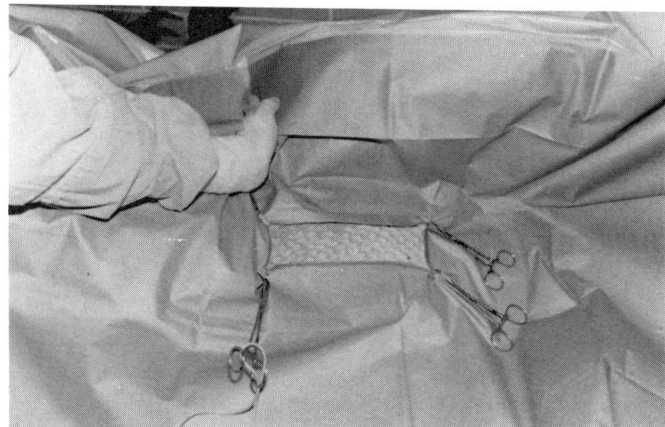

Figure 10.4 The cranial landmark for a left paramedian approach to access the proventriculus is the cranial aspect of the abducted left leg, indicated by the surgeon's hand.

landmark for the cranial extent of the incision. The incision is extended caudally for 12–18 cm. Abdominal air sacs will be encountered after the skin, subcutaneous tissue and rectus abdominus muscle are incised. The air sac is bluntly separated to gain access to the abaxial surface of the proventriculus (Figure 10.5). If the skin and muscle incisions are placed too far caudally, the caudal air sac will be incised, exposing the intestines. If this occurs, the caudal air sac should be closed, and the incisions through the skin and rectus abdominus muscle extended cranially to gain access to the abaxial surface of the proventriculus.

To decrease contamination of the retroperitoneal tissue surrounding the proventriculus, size 0 monofilament suture can be used to suture the proventriculus to the skin. An alternative is to isolate the proposed incision in the proventriculus from surrounding exposed tissues using saline-moistened laparotomy sponges. This option is preferred because it is quicker, and minimal contamination of these tissues does not appear to increase postoperative morbidity. When the wall of the proventriculus is identified, size 0 to 1 monofilament stay sutures or Allis tissue forceps are placed at the ends of the proposed proventriculotomy incision. A stab incision is made into the lumen of the proventriculus and a suction tip introduced to evacuate fluid from within the organ (Figure 10.6). The incision is then lengthened to 8–10 cm (Figure 10.7). Depending on the character of the material causing the impaction, a variety of instruments can be used to evacuate the viscus. A large spoon is useful for the removal of sand and rocks; sponge forceps for roughage (leaves, hay, twigs, and grass) [1]. In larger birds, the surgeon can also introduce a hand through the proventriculotomy incision to remove foreign material. The lumen of the ventriculus can also be explored via the proventriculotomy incision. Instru-

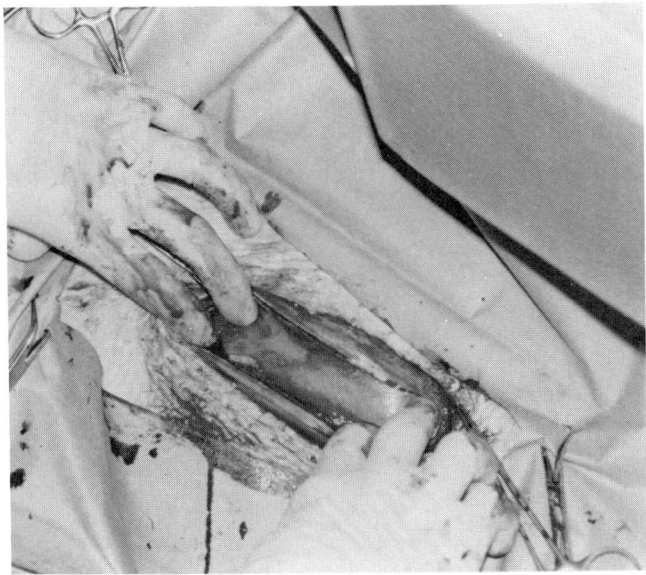

Figure 10.5 Blunt separation of the air sac with the fingers to access the abaxial surface of the proventriculus.

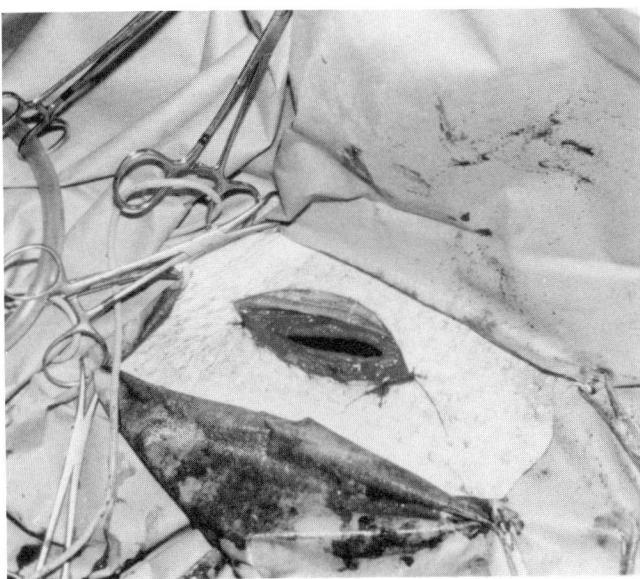

Figure 10.7 Lengthening of the proventriculotomy incision allows room for removal of foreign material.

Figure 10.6 Fluid is aspirated from the proventriculus through a stab incision prior to lengthening the incision for removal of foreign material.

ments or a hand are passed in a cranial direction along the ventrolateral aspect of the proventricular lumen to access the ventriculus. When the surgeon uses a hand to explore and remove foreign material from the proventriculus and ventriculus, the incision should be sufficiently long to accommodate entry of the hand to avoid tearing of the tissue at the limits of the incision, which may rupture blood vessels within the wall of the proventriculus, complicating closure of the incision.

Once the contents of the proventriculus and ventriculus have been removed, the proventriculotomy incision and surrounding tissues are cleaned with gauze sponges moistened with saline. The proventriculotomy is closed in two layers using #0 polydioxanone or polyglactin 910. The first layer is closed in a simple continuous pattern, the suture line is lavaged with saline, and the first layer is then oversewn with a continuous Cushing or Lembert pattern. The surgical site is again lavaged and cleaned with saline-moistened gauze sponges, the laparotomy sponges are removed, and the surface of the proventriculus and surrounding air sacs are lavaged and suctioned free of debris. The rectus abdominus muscle is closed with #1 polydioxanone or polyglactin 910 in a simple continuous pattern. Because the subcutaneous tissue is sparse in the ostrich, this layer is not sutured. The skin is closed with #0 PDS in a simple continuous pattern.

After completion of the proventriculotomy, an esophagotomy is recommended so that young and debilitated birds can be force-fed after surgery [4]. This contributes to nutritional requirements during the postoperative and recovery period because many birds are anorexic for several days after surgery. To perform an esophagotomy, an equine nasogastric tube is passed orally and advanced into the distal third of the cervical portion of the esophagus. The esophageal lumen is entered by incising over the end of the tube. Care should be taken to avoid incision of major vessels in the area. After entering the esophageal lumen, the tube is withdrawn from the mouth and

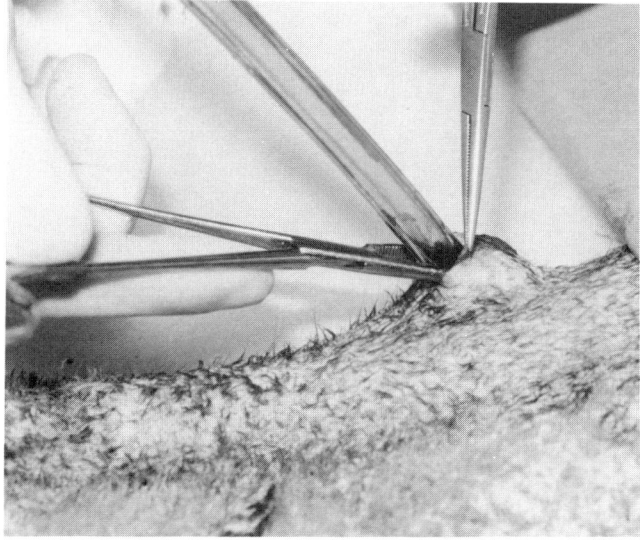

Figure 10.8 Placement of a tube into the proventriculus through an esophagotomy for postoperative feeding of debilitated patients.

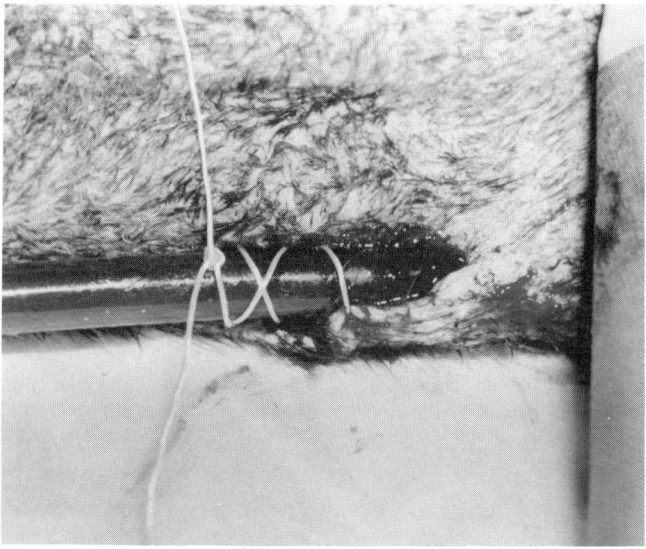

Figure 10.9 Esophagotomy tube is secured to adjacent skin with suture material.

inserted through the esophagotomy incision and passed into the proventriculus (Figure 10.8). The tube is sutured to the neck using either a finger trap (Figure 10.9), multiple half hitches, or 2 in. white tape placed in butterfly fashion over the tube that is sutured to the skin.

Birds are allowed access to water and feed within several hours of surgery. Psyllium can be fed to birds impacted on sand or rock. Debilitated birds with esophagotomy tubes in place are fed a commercially available diet[h] until they are observed to eat on their own. Esophagotomy tubes are removed when no longer needed and the esophagotomy is allowed to close by second intent. After removal of the esophageal tube, the esophageal incision generally seals within 24 hours. Complications such as cellulitis or fistulas do not result from esophagotomy.

If clinical signs consistent with impaction of the proventriculus are observed, affected birds should be palpated and radiographic examination of the abdomen should be performed. Affected birds should be surgically treated as soon as possible to avoid cachexia and debilitation that develops subsequent to chronic anorexia. In a retrospective study of proventriculotomies in ostriches [3], less than 50% of the birds undergoing surgery for impaction survived long term. The high mortality rate was attributed to presentation of ostriches in an advanced state of debilitation following chronic proventricular impaction. Since the reported retrospective study [3], prompt surgical correction of impactions coupled with aggressive perioperative management (e.g., IV fluid therapy and force-feeding) has significantly improved the success rate at Texas A&M University.

Education should be an integral aspect of preventing impactions of the proventriculus. Clients should be advised to change the substrate contributing to impactions on farms with recurrent cases, to reduce stresses on the flock by minimizing movement of birds within groups and within enclosures on the farm, and to feed psyllium to birds suspected of having an impaction or that have been observed to eat foreign material. Birds that become anorexic or show a change in fecal consistency, or become weak, should be evaluated to determine if a chronic impaction exists and whether removal of foreign material via proventriculotomy is indicated [3].

HYSTEROTOMY TO RELIEVE EGG RETENTION

Egg retention in ratites occurs infrequently and has been seen exclusively in emus submitted to Texas A&M University. Clinical signs associated with egg retention include cessation of laying activity, straining, passing fragments of egg shell, lethargy, and anorexia [5]. A diagnosis of egg retention is usually made from the combined findings of abdominal palpation and radiography (Figure 10.10). Because nonmineralized eggs cannot be detected radiographically, ultrasonography is a useful diagnostic aid.

Prior to surgery, antimicrobials are administered because salpingitis and peritonitis often occur as a result of egg retention. The bird is placed on a padded

[h] Exact, Kaytee Products Inc., Chilton, WI 53014.

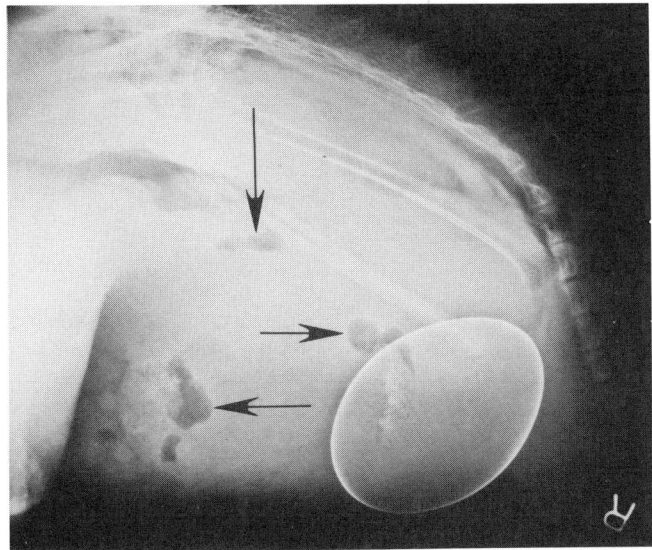

Figure 10.10 Lateral abdominal radiograph of an emu with egg retention. Arrows indicate free gas within the abdomen. At surgery, the uterus had ruptured and the egg was free in the abdomen.

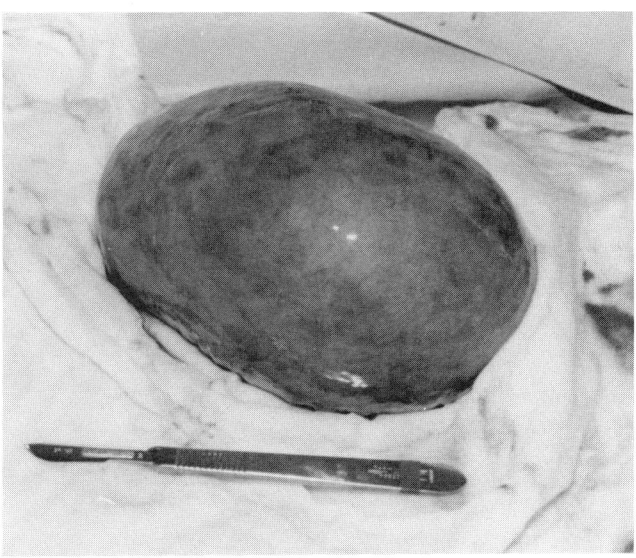

Figure 10.11 Exteriorization of the uterine segment containing the egg in preparation for hysterotomy. The uterus is isolated from surrounding tissue with laparotomy sponges.

table in right lateral recumbency; the left pelvic limb is positioned in abduction as described for proventriculotomy in ostriches [2]. The ventral and left paramedian abdomen is plucked and aseptically prepared for surgery. A left paramedian incision is made over the palpable egg. Once the abdomen is entered, the uterine segment containing the egg is exteriorized and isolated from the other abdominal viscera with saline-moistened sponges (Figure 10.11). A small stab incision is made just proximal to the egg, and accumulated fluid that has been trapped by the egg is aspirated. The incision in the uterus is then lengthened so that the egg can be removed (Figure 10.12). Any additional fluid accumulation distal to the removed egg is then aspirated by suction. The hysterotomy is closed in 2 layers using size 0 or 1 polydioxanone or polyglactin 910 in a simple continuous pattern, followed by a continuous Cushing oversew with the same size suture material (Figure 10.13). The uterine surface is lavaged and the uterus is replaced into the abdomen. The abdomen be explored and any nonmineralized eggs within the abdomen removed. Use of warm isotonic fluid to lavage the abdominal cavity is indicated, especially in birds with previous intraperitoneal ovulation. The rectus abdominus musculature is closed with size 1 absorbable suture material using a simple continuous pattern and the skin is closed with 0 polydioxanone in a simple continuous pattern.

Information relating to restoration of successful breeding status and a description of the pathogenesis of egg retention and egg peritonitis in ratites has not been

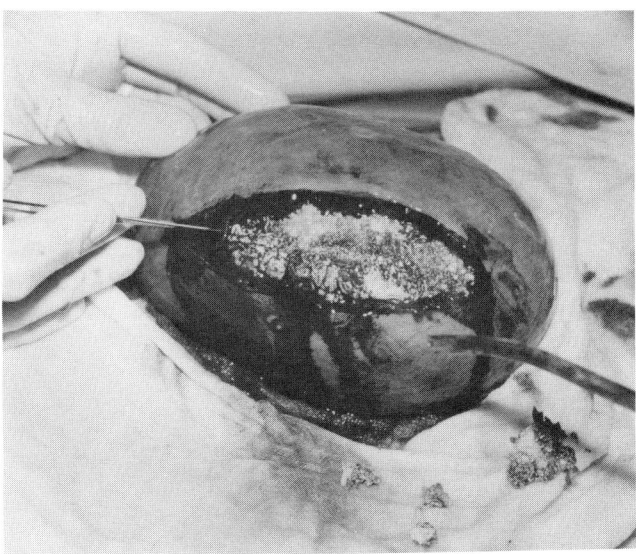

Figure 10.12 An incision is made directly over the retained egg and the egg removed.

published. The isolation of various bacterial species from the uterine lumen suggests that salpingitis may initiate egg retention and subsequent egg peritonitis in birds that ovulate intraabdominally [5]. Birds that have a history of laying abnormal eggs or that have had retained eggs can be examined endoscopically to identify bacterial infections [6].

In caged birds egg retention is a frequent problem. Spasm or atony of the oviduct, interrupting normal

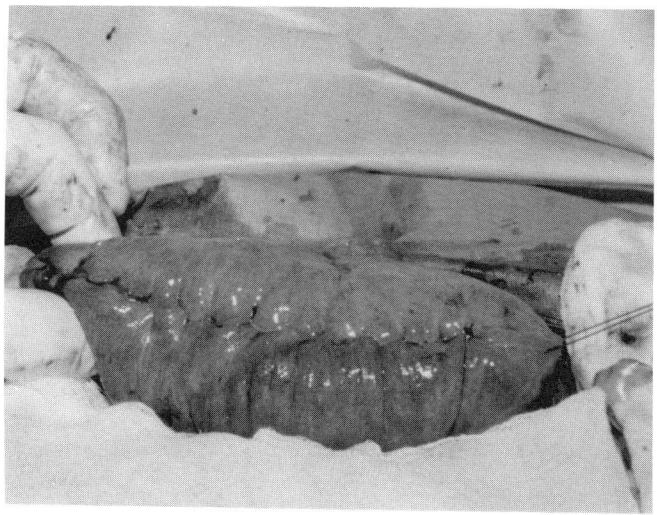

Figure 10.13 Closure of the hysterotomy is accomplished with a simple continuous pattern oversewn with a continuous Cushing pattern (shown).

progress of the ovum in the oviduct, may result from salpingitis, lack of muscle tone in the oviduct wall, exhaustion due to overbreeding and out of season breeding or environmental chilling. These factors may also contribute to egg retention in ratites.

Continued ovulation after oviductal obstruction leads to egg peritonitis, which is common in ratites with egg retention [5]. Failure of ova to enter the infundibulum of the oviduct or rupture of the organ result in egg peritonitis [5]. Treatment of egg peritonitis consists of removal of egg material from the abdomen, peritoneal lavage, and systemic antimicrobial therapy.

YOLK SACCULECTOMY

The egg yolk, which provides nourishment to the developing chick during incubation, is internalized into the abdomen just prior to hatching [7]. The yolk sac, which is a diverticulum of the intestine, is attached via a yolk stalk and accompanying vasculature to the intestine. Yolk that is absorbed provides nourishment during the first few days of life. After it is consumed, the remnant of the yolk sac is converted to scar tissue identified as Meckel's diverticulum [8].

Clinical signs of yolk sac retention or infection in ratite chicks are observed 10 to 14 days after hatching and include failure to grow, or weight loss, anorexia, depression, abdominal enlargement, and isolation from other chicks. Affected chicks may also hold their heads down or against their body, display ataxia, or tuck their heads between their legs [9]. Yolk sac retention may occur as a result of high temperature and humidity during incubation or brooding, food being offered too early, gastrointestinal ileus, or infection of the yolk sac [8–10].

Abdominal palpation of chicks with clinical signs suggestive of a retained yolk sac will reveal the presence of a resilient abdominal mass that may range in size from 0.5 cm to 6 cm. Abdominal radiography may also be employed as a diagnostic aid. This is not necessary in most cases, as clinical signs and palpation provide sufficient evidence of a retained yolk sac [9]. Abdominal ultrasonography would likely provide more valuable information than radiography in birds with obscure clinical signs or inconclusive findings on abdominal palpation.

Prior to surgery, antimicrobial therapy [enrofloxacin (2 mg/kg, IM or PO, q 12h)] should be initiated and continued for 5 days postoperatively. Anesthesia is induced with xylazine (2 mg/kg) and ketamine (20 mg/kg), both administered intramuscularly, and the plane of anesthesia is maintained with isoflurane [9]. The chick is positioned in dorsal recumbency, the feathers plucked from the ventral region of the abdomen which is prepared for aseptic surgery. A 2- to 3-cm fusiform skin incision is made around to the umbilicus. Because the rectus abdominus muscle is extremely thin (0.05 mm) in young chicks, it is appropriate to tent the musculature with thumb forceps prior to incision. This avoids inadvertent penetration of the yolk sac which will lead to egg peritonitis and death. After exposing the abdomen, the yolk sac is exteriorized by applying gentle bilateral pressure. Depending on the size of the yolk sac, it may be necessary to enlarge the incision to avoid rupture of the yolk sac. Once the yolk sac has been exteriorized, a hemostat is clamped around the yolk stalk next to the yolk sac to prevent discharge of yolk material when the stalk is severed. Using 3–0 polydioxanone or polygalactin 910, an encircling ligature is placed around the yolk stalk adjacent to the intestine. The yolk stalk is severed between the encircling ligature and the hemostat and the yolk sac is sampled for bacterial culture and is discarded [9]. The abdominal wall and skin are closed with 2–0 absorbable suture applying the surgeons preference of suture patterns.

The prognosis for survival is estimated to be approximately 80% for emus and 50% for ostriches [9]. Early recognition of candidates for surgery will result in higher survival rates. Management practices that may reduce the occurrence of yolk sac retention include witholding feed from newly hatched chicks for 4 to 5 days and allowing moderate amounts of exercise to deplete the yolk that remains after hatching [9].

Surgical Conditions of Ratites

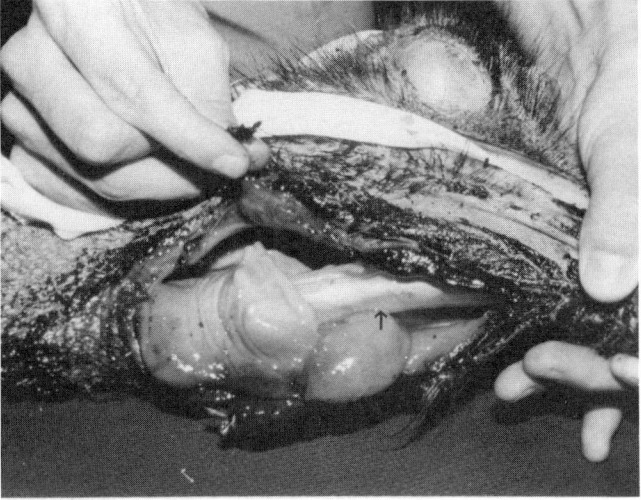

Figure 10.14 Laceration into the oral cavity and pharnyx of an ostrich. An endotracheal tube is in place (arrow).

MISCELLANEOUS ABDOMINAL PROCEDURES

Surgical exploration of the abdomen may be indicated in birds with intestinal disease from diverse causes. Exploratory celiotomies have been performed requiring intestinal resection and anastomosis due to foreign body penetration of the intestine or an infarcted bowel. A left paramedian approach is used to gain access to the abdomen, making the incision in the quilled band caudal to that described for proventriculotomy. This approach provides sufficient exposure to allow exploration and surgical treatment of most gastrointestinal tract conditions.

WOUND MANAGEMENT

Principles of wound management employed in mammalian species are applicable to ratites. Initial physical examination to assess the status of vital functions and immediate therapy are directed at supporting the respiratory and cardiovascular systems. Profuse hemorrhage is addressed with pressure bandaging, application of hemostatic forceps or ligatures, when possible.

The head, neck and pelvic limbs are the more common sites of trauma in ratites (Figures 10.14 and 10.15). Wounds are evaluated to ascertain involvement of vital structures such as vessels, nerves, and regional organs. Wound debridement and tissue closure are implemented as with other animals. Efforts are made to close areas of dead space or provide drainage of regions that cannot be sutured effectively.

Figure 10.15 Longitudinal neck laceration in an ostrich.

Wounds having a large area of the integument missing or that will not heal by contraction and epithelization, may benefit from autografting [11]. This technique may be used for fresh wound beds or on healthy granulation tissue. Full-thickness autografts may be harvested from the axilla or inguinal area. Subcutaneous tissue is sharply excised from the harvested skin, which may then be expanded with a mesh graft dermatome or using multiple incisions with a scalpel blade. The subcutaneous tissue of the donor site is closed with 0 polydioxanone or polyglactin 910 and 2–0 polydioxanone or polypropylene may be used to close the skin. Harvested skin is attached to the recipient site with cyanoacrylate glue around the periphery; scattered simple interrupted sutures of 2–0 catgut are used to attach the autograft directly to the wound. Care of the graft site includes protection from excessive motion through the use of bandaging, and the application of topical antimicrobials (neomycin-polymyxin-bacitracin ointment, gentamicin, ceftiofur, or ticarcillin) to prevent loss of the graft from infection [11].

Esophageal lacerations heal rapidly in ratites. Following debridement of wound edges, 2–0 absorbable suture material is used to close the esophagus in 2 layers (mucosal and seromuscular).

Use of tracheal ring prostheses for repair of a collapsed trachea secondary to trauma has been reported [12]. Rings are fashioned from polypropylene syringe cases and 4 to 5 holes are drilled into the rings. A ventral midline cervical approach is used to access the affected area. Once the trachea is exposed, gentle blunt dissection is used to allow placement of prosthetic rings around the affected tracheal segment. There are no major nerves (i.e., recurrent laryngeal nerve) close to the trachea as in mammals. The prosthetic rings are sutured to the trachea with 2–0 polydioxanone to expand against the prostheses. Care is taken to avoid penetration of the tracheal mucosa, which may result in infection. In cases where deformity of the tracheal rings prevents expansion of the trachea, chondrotomies of the tracheal rings may be performed with a #15 blade, taking care to avoid mucosal penetration. Closure of surrounding soft tissue and skin is routine. Birds are placed on broad spectrum antimicrobials perioperatively.

Orthopedic Surgery

Developmental orthopedic diseases are common in young growing birds and can cause significant morbidity in some flocks. Acquired orthopedic diseases (including fractures) are encountered as a result of panic or fights between birds, collisions with fence posts or objects in their pens and improper handling and restraint.

Developmental Orthopedic Diseases.

Many of the developmental orthopedic diseases of ratites are associated with rapid growth or insufficient exercise in young chicks. Deformities observed include curled and rolled toes, splayed legs, angular limb deformities, osteochondrosis, and rotational deformities of the tibiotarsus [13].

Angular limb deformity (ALD) is multifactorial in etiology. The condition is seen in fast-growing chicks and in birds bearing weight on one limb because of contralateral pain. Most of these cases present as a tarsal valgus deformity. Surgical correction involves transection and elevation of the periosteum near the growth plate on the concave side of the deformed leg. The origin of the deformity is ascertained radiographi-

callyd prior to surgery as described for the foal. Most ALD's occur at the proximal physis of the tarsometatarsus [13]. A T-shaped incision is made through the periosteum, just distal to the proximal physis, on the concave side of the limb. Care is taken to not transect the lateral digital extensor tendon on the lateral aspect of the limb. The corners of the periosteum formed by the incisions are then elevated with a periosteal elevator. The periosteal flaps are then repositioned in their original location and the skin is closed with 2–0 polydioxanone or polypropylene. The surgical site is wrapped for 48 hours and chicks are confined to small pens [13].

If the ALD has been caused by pain in the contralateral limb, the source of pain in that limb should be identified, if still present, and treatment with nonsteroidal anti-inflammatory agents is indicated. If the bone responds to the periosteal elevation, correction of the ALD should occur within 3 to 4 weeks of surgery.

Valgus deformities originating in the diaphysis of the tarsometatarsus, as determined by radiography, require wedge osteotomy for correction. Results of this surgery are often unsatisfactory because of the difficulty encountered in attaining rigid fixation sufficient to withstand weightbearing while allowing healing to occur. Following wedge osteotomy, fixation is best attempted with a transfixation cast in young chicks and with bone plates and screws in older birds.

Rotational deformities of the tibiotarsus requires osteotomy and derotation of the bone if rotations progress past 45°-60°. The type of fixation employed depends on the age of the bird and the implant holding quality of the bone. Repair of these deformities is generally unrewarding, as many legs will re-rotate following successful repair.

Acquired Orthopedic Diseases

Because ratites are often excitable and unpredictable in their response to environmental stimuli, fractures of pelvic limbs occur frequently. Fractures occur secondary to trauma from running into fences, barns, trees, during fights, or when a hen is kicked by an aggressive male [10].

Palpation of the affected bone is usually sufficient to diagnose long bone fractures. Radiographic assessment of the fracture is imperative to assess the option of surgical repair. Open and comminuted fractures complicate treatment. The goal of initial triage of the fracture patient is to prevent further trauma to the soft tissues surrounding the fracture site, preserve blood flow to the region and to prevent closed fractures from becoming open. To accomplish this, the fractured limb should be stabilized in some manner. This can be very difficult to accomplish, especially in adult birds with fractures of long bones of the pelvic limb. Birds will attempt to use the limb during transport to a surgical facility. Application of a Robert-Jones type bandage is probably the best method to provide some stability for fractures of the tarsometatarsus and tibiotarsus. Femoral fractures cannot be bandaged, but muscle mass will provide limited inherent stability.

Most fractures of long bones of ratites carry a guarded to poor prognosis. A major problem with long bone fractures in ratites is that birds need to bear weight on the fractured limb soon after surgery to avoid complications, such as rhabdomyolosis, in the contralateral limb. Fracture management with internal fixation is complicated by sparse soft tissue coverage of implants in the tarsometatarsus, and the thin, soft cortical bone in young birds may preclude fixation with pins and screws. In some instances, bone fails around the implants. Long bone fractures have been reported to be more easily managed in chicks younger than 3 months of age because of their relatively small mass and reduced strength [13].

Fractures of the femur are difficult to repair in young and adult ostriches because the femur is pneumatized and, in the chick, the bone has a thin cortex and epiphysis [13]. Currently, there is no recommended method of surgical management of femoral fractures.

Fractures of the tibiotarsus are best repaired in birds under 6 months of age using plates and screws. The fracture is approached from the medial aspect of the bone because the limited soft tissues over the bone in this region allows easy access to the distal metaphysis and diaphysis. The implant should span the bone from metaphysis to metaphysis to avoid failure of the bone where the end of a short plate corresponds to the thin diaphyseal region. In birds less than 6 months of age, application of a transfixation cast may be successful to repair fractures. It has been reported that intramedullary pins along with Figure-8 hemicerclage wire for rotational stability is successful in chicks younger than 2 months [13]. In general, the prognosis for fractures of the tibiotarsus is guarded. Breakdown of the supporting limb is always a concern, and in chicks, overuse of the supporting limb may result in the development of an angular limb deformity.

Fractures of the tarsometatarsus can be managed with open reduction and internal fixation with plates and screws, or with transfixation casts. The tarsometatarsus appears to have more favorable bone density and thicker cortices to hold implants. The bone is easily accessible because of sparse soft tissue coverage. This may be counterproductive when internal fixation is employed as it is often difficult to adequately cover the implant with soft tissue following repair. Open reduction and internal fixation of fractures of the tarsometatarsus in adult ostriches is further complicated by standard plate lengths that are too short to span from metaphysis to metaphysis, thereby avoiding stress concentration at the end of plates that end in the diaphyseal region of the bone. Staggering two plates on the bone (one dorsomedial and one dorsolateral) avoids this complication, but soft tissue closure over the implants can be extremely difficult. Careful management of the skin overlying a tarsometatarsal fracture is critical for success [13].

Repair of long bone fractures in ratites is associated with numerous potential complications. The requirement for immediate weightbearing following surgical repair can result in failure of fixation. Recovery from anesthesia presents unique challenges in adult birds and hyperexcitation may result in failure of the fracture repair. Rhabdomyolysis is likely to develop in ratites that remain recumbent after surgery. Peroneal nerve paralysis or ischemic edema may occur due to pressure imposed during lateral recumbency. Slings may be used postoperatively for docile birds, however, some birds will not tolerate a sling and may struggle excessively, resulting in injury.

REFERENCES

1. Gamble KC, Honnas CM. Surgical correction of impaction of the proventriculus in ostriches. *Compend Contin Educ Pract Vet* 1993;15:235–244.
2. Honnas CM, Jensen J, Cornick JL, Hicks K, Kuesis BS. Proventriculotomy to relieve foreign body impaction in ostriches. *J Am Vet Med Assoc* 1991;199:461–465.
3. Honnas CM, Blue-McLendon A, Zamos DT, Parson E, Jensen J. Proventriculotomy in ostriches: 18 cases (1990–1992). *J Am Vet Med Assoc* 1993;202:1989–1992.
4. Honnas CM. *Ratite surgical techniques*. 21st Annual Surgical Forum, San Francisco, 1993;21:395–397.
5. Honnas CM, Jensen JM, Blue-McLendon A, Zamos DT, Light GS. Surgical treatment of egg retention in emus. *J Am Vet Med Assoc* 1993;203:1445–1447.
6. Jensen JM, Schumacher J. Endoscopic examination of the distal uterus of ostriches and emus, in *Proceedings*. AAZV 1994;138–139.
7. Harrison GJ, Harrison LR. *Clinical Avian Medicine and Surgery*. Philadelphia: W.B. Saunders Co., 1986;634–635.
8. Kenny D, Cambre RC. Indications and technique for the removal of the avian yolk sac. *J Zoo Wildl Med* 1992;23:55–61.
9. Lamar Crossland, DVM, Sunset Canyon Veterinary Clinic, Dripping Springs, TX: Personal Communication, 1995.
10. Gilsleider E. Common abdominal surgeries in ratites, in *Proceedings*. Amer Assoc Zoo Vet 1993;272–274.

11. Johnson JH, Schumacher J, McClure SR, Jensen JM. Skin grafting in an ostrich, in *Proceedings*. Amer Assoc Avian Vet 1993;138–141.
12. McClure SR, Taylor TS, Johnson JH, Heisterkamp KB, Sanders EA. Surgical repair of a traumatic induced collapsing trachea in an ostrich (*Struthio camelus*).
13. Gilsleider E. Ratite orthopedics. *Seminars in Avian and Exotic Pet Med* 1994;3:81–91.

Chapter 11

Clinical Hematology and Chemistry of Ratites

Alan M. Fudge

The application of avian hematology and biochemical analysis in avian diagnostics has its origin in poultry and mammalian physiology. This diagnostic approach has become most advanced clinically among companion and zoological avian species [1–3]. Examination of ratite blood is frequently employed by veterinary clinicians as an aid for health certification and breeding evaluation and for diagnosis in obviously diseased birds. Examination of blood will sometimes assist the veterinary clinician to quantify the effects of management deficiencies, which account for a large proportion of flock problems [4–8].

Little information can be found in the literature correlating changes in ratite blood with specific diseases. Published reference ranges vary greatly among studies, with difference in subjects, sampling, methodologies, and analyses tabulated [9–14] (Tables 11.1 and 11.2).

BLOOD COLLECTION AND PROCESSING

Diagnostic laboratories experience problems with sample quality which can affect the validity of many determinations. Problems can be minimized if standard procedures are followed [15].

Blood collection techniques should minimize stress and patient contact time. Blood volumes of up to 1% of patient's body weight can be collected, which vastly exceeds sampling requirements in ratites. Avian patients do not possess contractible spleens, although stress hemograms reflected in leukocytosis, and heterophilia, appear to occur.

Proper restraint of patients is important for optimal results and skill in capture will minimize problems [14].

The jugular or wing vein should be used to obtain blood using a 3- to 10-ml syringe and a 20- to 22-gauge needle.

Ostrich blood is adversely affected by EDTA. Many clinicians employ heparinized green-top tubes to collect whole-blood samples for ratite hematology. The author's laboratory requests whole blood from ratites in citrated blue-tops, which yields the best results when analyzed by laser flow-cytometry (Abbott Cell-Dyn).

A blood film should be made at the time of collection when the needle is withdrawn from the vein. An air-dried, thin smear provides superior cellular morphology to samples stored in EDTA or heparin. In the author's laboratory, the coverslip smear technique contributes to a low occurrence of "smudge cells." A drop of blood is placed on the center of a clean 22 × 22 mm glass coverslip. A second coverslip is immediately moved along the surface of the lower coverslip before the blood spreads out. Routine slide to slide smears or long coverslip to slide smears produce unacceptable rupture and damage, resulting in "smudge" cells. These cells are recorded as unrecognizable granulocytes, resulting in inaccurate hematology data [15].

Blood collected for biochemical assay should be immediately transferred to an appropriate container. Plasma harvested from centrifuged, heparinized green-top tubes, provides the best results. Silicone plugs enhance separation, but centrifugation should be completed within 2 hours of collection. Samples for chemical analyses should never be collected or submitted in containers with EDTA. This will invalidate assays of calcium and possibly affect other determinations.

Unseparated, clotted, or hemolyzed samples frequently result in spurious values. Serum potassium level drops in samples which are not separated for 2 or more hours. Samples shipped unseparated for up to 2

TABLE 11.1. Literature Values for Ostriches

Ostrich Blood Values

Reference	13 Fudge 1995 Ostrich Adults Mean	13 Fudge 1995 Ostrich Adults Range	10 Okotie-Eboh 1992 Ostrich Pooled Range	12 Palomeque 1991 Ostrich Adults	12 Palomeque 1991 Ostrich Juv	11, 26 Levy 1989 Ostrich Pooled Mean	9 Van Heerden 1985 Ostrich Adult Range
WBC × 1000	18.65	10–24	ND	21.0 ± 8.0	19.5 ± 13.7	5.5 ± 1.9	ND
Heterophils %	72.9	58–89	ND	ND	ND	62 ± 7.6	ND
Lymphocyte %	24.2	12–41	ND	ND	ND	34.1 ± 7.0	ND
Monocyte %	2.64	0–4	ND	ND	ND	2.8 ± 1.3	ND
Basophil %	0.2	0–2	ND	ND	ND	0.2 ± 0.5	ND
Eosinophil %	0.035	0–2	ND	ND	ND	0.3 ± 0.5	ND
RBC M/μl	1.8		ND	2.42 ± 0.37	1.91 ± 0.28	1.5 ±	ND
Hematocrit %	45	41–57	ND	48 ± 2.4	37 ± 2.1	32 ± .03	ND
Hemoglobin g/dl	16.92		ND	15.6 ± 0.89	13.3 ± 0.39	12.0 ± 2.0	ND
MCV fl	212		ND	201.1 ± 25.1	196.9 ± 31.2	174 ± 42	ND
MCH pg	82.19		ND	65.3 ± 7.0	70.5 ± 9.0	61 ± 16	ND
MCHC g/l	37.65		ND	32.5 ± 3.0	35.9 ± 3.1	33 ± 5	ND
RDW %	11.11		ND	ND	ND	ND	ND
AST(SGOT)IU/L	447.9	226–547	110–449	190.5 ± 39.4	152.7 ± 35.2	131	100–892
LDH IU/L	970	408–1236	224–1245	514.9 ± 286.1	463.3 ± 168	1565	240–2187
CPK (CK) IU/L	3702	800–6600	0–6588	933.0 ± 269.0	570.8 ± 172	688	394–2500
Uric Acid mg/dl	8.62	1–14.5	1.1–17.1	11.17 ± 2.11	9.81 ± 1.26	8.2	ND
Calcium mg/dl	10.7	8.0–13.6	6–13	18.07 ± 1.62	10.61 ± 0.63	9.2	7.54–11.2
Glucose mg/dl	217	164–330	153–337	207.4 ± 33.6	263.5 ± 109.9	250.2	64.8–262.8
Creatinine mg/dl	0.26	0.1–0.7	ND	0.144 ± 0.04	0.621 ± 0.138	ND	ND
Phosphorus mg/dl	5.33	2.9–7.7	3.1–7.4	13.71 ± 2.92	11.95 ± 2.40	4.96	3.0–11.3
Total Protein mg/dl	3.93	2.4–5.3	2.4–5.2	3.87 ± 0.58	6.22 ± 0.92	3.7	2.6–5.0
Albumin g/dl	1.72	1.1–2.3	1.2–2.4	ND	ND		1.4–2.4
Globulin g/dl	2.21	1.4–3.1		ND	ND		
Cholesterol mg/dl	103	39–172	24–177	116.2 ± 27.2	148.2 ± 65.1	ND	ND
Bile Acids μmol/l	21	2–30	ND	ND	ND	ND	ND
ALP IU/L	ND	ND	0–420	171.5 ± 45.9	339.1 ± 135.8	ND	150–1384
Sodium mmol/L	ND	ND	ND			147	ND
Potassium mmol/L	ND	ND	ND			3	ND
Chloride mmol/L	ND	ND	ND			100	ND

Note: Urea, GGT, Bilirubin, ALT are omitted

ND = Not Done

days can result in elevated potassium, lactic dehydrogenase and protein levels, decreased glucose and variable changes in calcium and phosphorus [15].

AVIAN HEMATOLOGY

Leukocytes

Until recently, it was not possible to quantify avian leukocytes by automated methods, due to the presence of nucleated erythrocytes and thrombocytes. Variability in leukocyte count can occur due to collection and preparation of the sample, shipping time and temperature, and assay procedure.

The direct manual hematocytometer method involves dilution of the blood sample with Natt and Herrick's solution. This stains cells types differentially, leaving the technician to determine the total leukocyte number in nine squares. The total leukocyte count is derived by adding 10% and multiplying by 200. The indirect manual hematocytometer method requires dilution of the blood sample with phloxine (Unopette™#5877-Becton-Dickinson) which stains only heterophils and eosinophils. Cells are counted in 18 squares. Total leukocyte count is calculated from the differential count applying a formula. The indirect estimate of total leukocyte count from a blood film involves counting all leukocytes in ten fields at 40x magnification and multiplying the average value by 2000. This method presumes a constant erythrocyte to leukocyte ratio. The estimated leukocyte count must be adjusted in cases with out of range PCV measurements [1,2,17].

Hematocytometer methodology, when performed under controlled conditions applying good techniques and with whole blood collected within 12 hours, provides reproducible results. An estimate of the leukocyte count from a blood film provides the best results when:

- the smear is uniform,
- is made with a 22 × 22 mm coverslip resulting in a monolayer of cells, and

TABLE 11.2. Literature Values for Emus

Emu Blood Values

Reference	13 Fudge 1995 Emu Adults Mean	13 Fudge 1995 Emu Adults Range	10 Okotie-Eboh 1992 Emu Pooled	14 Costa 1993 Emu Adult	14 Costa 1993 Emu 8 weeks	14 Costa 1993 Emu 4 weeks	14 Costa 1993 Emu 1 week
WBC × 1000	14.87	8–21	ND	ND	ND	ND	ND
Heterophils %	78.8	54–88	ND	ND	ND	ND	ND
Lymphocyte %	19.8	10–44	ND	ND	ND	ND	ND
Monocyte %	0.1	0–1	ND	ND	ND	ND	ND
Basophil %	0.2	0–1	ND	ND	ND	ND	ND
Eosinophil %	2.58	0–6	ND	ND	ND	ND	ND
RBC M/μl	1.85		ND	ND	ND	ND	ND
Hematocrit %	47.4	39–57	ND	ND	ND	ND	ND
Hemoglobin g/dl	16.04	ND	ND	ND	ND	ND	ND
MCV fl	219	ND	ND	ND	ND	ND	ND
MCH pg	86.51	ND	ND	ND	ND	ND	ND
RDW %	10.9	ND	ND	ND	ND	ND	ND
AST(SGOT)IU/L	227.2	80–380	57–151	151	258.7	234.5	216.8
LDH IU/L	778.1	318–1243	60–422	ND	ND	ND	ND
CPK (CK) IU/L	428.8	70–818	0–603	222.7	330.7	339.4	295.3
Uric Acid mg/dl	6.3	1–13.7	0.7–8.7	10.6	2.47	10.72	16.76
Calcium mg/dl	11.1	8.8–12.5	7.9–13.2	11.48	10.36	10.4	11
Glucose mg/dl	134.1	101–243	114–203	285.4	216.2	376.7	531.5
Creatinine mg/dl	0.22	0.1–0.4	0.1–0.4	ND	ND	ND	ND
Phosphorus mg/dl	5.7	3.8–7.2	3.3–7.4	6.91	9.32	6.97	5.42
Total Protein mg/dl	3.93	3.4–4.4	3.1–5.3	4.526	3.895	2.986	2.635
Albumin g/dl	1.7	1.2–2.4	1.9–3.2	2.19	2.04	1.49	1.29
Globulin g/dl	2.23	1.4–3.2					
Cholesterol mg/dl	122	68–170	42–166	106	115.7	97.1	318.1
Bile Acids μmol/l	18	2–34	ND	ND	ND	ND	ND
ALP IU/L	ND	ND	0–172	278.9	549.8	665.7	577.9

Note: Urea, GGT, Bilirubin, ALT are omitted
ND = Not Done

- exhibits minimal presence of ruptured or "smudge" cells.

The author's laboratory is now using automated laser flow-cytometric methods to tabulate avian blood cells. Results have been validated using blood from ratite species. Citrated chilled blood, shipped by overnight air express, will provide consistent hematologic analyses for ratites.

Leukocytosis occurs in ratites as a result of either infection or stress. This is common in ratites since the avian spleen is not associated with blood pooling. Clinical observations indicate that leukocyte counts can increase markedly in excited birds compared to samples taken when the patient is quiescent [17]. Stress hemograms can also be observed in patients recently treated with corticosteroids [15].

Juvenile birds can demonstrate great variability in total leukocyte count. Leukocytosis in ratites is common and results must be interpreted with caution, as the bird may in fact be normal [17]. Clinicians should review differential counts and cell morphology in relation to other information.

Leukocytosis in birds occurs as a result of inflammation caused by infectious agents [18]. The clinician cannot easily differentiate among diseases based solely on leukocytosis, but certain inferences are valid. Leukocytosis is associated with infection of wounds, aspergillosis, septicemia, and tuberculosis.

Aspergillosis affects the leukogram in a variable manner, but leukocytosis generally occurs in ratites [19]. Since most cases of aspergillosis in ratites are diagnosed at postmortem examination, hematologic assessment is precluded [8,20].

Leukograms associated with tuberculosis are not extensively documented [21,22]. Hematologic changes with chlamydiosis in ratites have not been quantified [23].

Leukopenia must be interpreted in relation to standard reference ranges for ratites. Sample artifacts should be first eliminated when evaluating leukopenia. Whole blood that is incompletely mixed with an anticoagulant will clot or yield a low count. Lysis of leukocytes before analysis, due to prolonged shipping and storage can result in pseudoleukopenia. Poor quality blood films often display a high proportion of ruptured leukocytes or "smudge cells" which will falsely indicate a leukopenia when using the slide technique.

Leukopenia usually arises from an overwhelming bacterial infection, severe viral disease [1,2], or occasionally toxins. Bacterial septicemia will cause a degenerative left shift, progressing to heteropenia with only a few lymphocytes present. Depending on the duration of the condition, leukopenia may present as a non-regenerative anemia. The presence of intracellular bacteria suggest a grave prognosis.

Normal leukocyte counts can occur in the medically-compromised bird. Chronic, low-grade infections or degenerative disorders may present with a normal leukocyte count. The clinician should assess changes in leukocyte morphology and differential counts in relation to concurrent medical data and observations [17].

DIFFERENTIAL LEUKOCYTE COUNT AND MORPHOLOGY

Heterophils

The heterophil is the most frequently observed leukocyte in the ratite hemogram. This cell is the analogue to the mammalian neutrophil but displays some differences in cytochemistry [1,2,17]. Heterophilia occurs frequently in the abnormal avian hemogram. Absolute heterophilias often are the primary contributors to leukocytosis. Heterophilia occurs in acute inflammatory and infectious processes [18].

Immature heterophils occur occasionally in peripheral blood. Bands and mesomyelocytes indicate a guarded prognosis.

Eosinophils are present in higher numbers in the normal emu compared to the ostrich. Differentiating eosinophils from heterophils requires experience. Eosinophils are distinguished by their round shape, distinct contrast of nuclear and cytoplasm, and uniform cytoplasmic color. The shape and color of granules varies greatly. Clinicians should question avian hemograms showing high eosinophil counts, as an inexperienced technologist can easily confuse the pink-staining heterophil with an avian eosinophil. Eosinophilia is seen inconsistently in association with parasitism.

Basophils are uncommon in ratite blood. In most cases, avian basophils resemble their mammalian counterparts. Basophils can be confused with toxic heterophils, which often have large amounts of variable basophilic cytoplasmic inclusions. Basophilia is observed in respiratory infections and during resolution of tissue damage. Avian respiratory infections can be associated with basophilia, although this change may not be prominent in ratites.

MONONUCLEAR CELLS

Lymphocytes occur at a higher frequency than all other leukocytes, except for heterophils. There are two to three unique populations of lymphocytes, based on size. These groups are not generally differentiated by clinical laboratories [1,2,18].

Lymphocytosis is not common. Lymphopenia is largely a function of heterophilia due to stress, excitement, inflammation, or steroid administration [17].

Monocytes are uncommon in ratite blood. Relative or absolute monocytosis is characteristic of chronic infection. In the avian patient, monocytosis can indicate bacterial, mycobacterial, fungal, and granulomatous bacterial infections [17,18]. Leukocytosis has been reported in an ostrich with aspergillosis [1,2,17,24]. It is likely that monocytosis is common, based on experiences with other avian species [1,2, 17,25].

THROMBOCYTES

Thrombocyte counts are not routinely performed by avian clinical laboratories. One to two thrombocytes per 100X field is considered to be normal in companion species [2]. Laboratories will often report that this cell as present, or noted in low or high numbers. The thrombocyte derives from a stem cell precursor in avians and not a megakaryocytic cell, as in mammals.

Thrombocytopenia occurs in pancytopenic forms of some psittacine viral diseases, such as psittacine circovirus but is poorly documented in ratites.

ERYTHROCYTE COUNTS

Automated erythrocyte counting is possible using standard equipment. Manual counts can be complicated by the presence of clots and other nucleated cells. Avian blood is very labile, necessitating whole cell counts within several hours of collection. Ostrich blood, submitted in EDTA for hematologic analysis, will rapidly hemolyze.

Appropriate indices can be calculated from erythrocyte count, hemoglobin, and hematocrit. Ratite erythrocytes are relatively large, with a mean corpuscular volume (MCV) greater than 200 femtoliters, compared to 125–179 for most other avian species. A large erythrocyte correlates with a lower total erythrocyte count, usually less then 2 million cells per microliter. Adult ostrich and emu hematocrits are similar to other avian species. The lower mean hematocrit and erythrocyte values reported by Levy [22] are probably influenced by the number of young ostriches in the sampled populations. Studies on four ostriches, evaluated as juveniles and again at one year of age, showed age-related increases in erythrocyte count, hematocrit, hemoglobin and mean corpuscular volume. When analyzed, the values for adult ostriches were higher than in juveniles. This observation has been confirmed by studies which showed changes in cell size and count, increasing with age until maturity [12]. Similar erythrocytic changes have been observed in clinically normal ostrich, emu, and rhea.

Erythrocytic Polychromasia

Polychromasia refers to variation in erythrocyte coloration, associated with maturation of the cell. Younger forms demonstrate bluer cytoplasm using most stains. There is no standardized index of avian polychromasia for veterinary diagnostic laboratories. Polychromasia suggests an increased bone marrow response [2,17].

Erythrocytic Anisocytosis

Mature avian erythrocytes are oval. Less mature cells are more round in shape and basophilic in color. Reticulocytes are normally present in peripheral blood at approximately 1–2% of the total erythrocyte count [2]. The red cell distribution width percentage (RDW%) provides a quantitative method to report the degree of anisocytosis and is based on the distribution of erythrocytes and mean corpuscular volume. The RDW% is generated by some automated hematology analyzers.

Anemia

Anemia is evidenced by a decrease in the erythrocyte count, packed cell volume or hematocrit [2,17]. Deficiency anemias have been reported experimentally in poultry [23], but have not been documented in exotic birds or ratites. Anemias can be classified as regenerative or nonregenerative, blood-loss related or hemolytic [17].

Nonregenerative Anemia

Nonregenerative anemias (NRA) are the most common type observed in ratites and are caused by factors which reduce erythropoiesis. These include infectious disease [27], cachexia, neoplasia, and certain toxins [28,29].

Hemolytic Anemia

Lead toxicosis denatures hemoglobin and results in premature lysis and removal of affected erythrocytes [28,29]. These are observed in peripheral blood smears showing hypochromatic cytoplasm and erythrocytic ballooning. Vigorous regenerative response occurring within hours of the onset of toxicity is a hallmark of lead toxicosis. Basophilic stippling is not a typical finding with lead toxicosis in avian species.

Oil ingestion is well documented as causing hemolytic anemia and is observed in the blood of shore birds located in areas where oil discharges occur [28].

Blood Loss Anemia

Trauma, rupture of organs, aneurysm, and iatrogenic causes can result in severe blood loss. Rapid recovery from blood loss occurs in healthy birds. No significant blood storage pool exists in birds, but erythropoiesis will increase dramatically within hours [2,28]. Recovery from blood loss anemia is characterized by an elevated PCV and RBC count and an increase in immature erythrocytes, anisocytosis, and polychromasia [28].

Polycythemia

Polycythemia is characterized by elevated hematocrit and erythrocyte counts. Hypoxia is the principal cause of avian polycythemia. Relative polycythemia is caused by hemoconcentration due to dehydration [29]. The clinician should assess hydration of the patient, in addition to reviewing laboratory parameters.

Absolute polycythemia is signaled by an increase in erythrocyte number in the absence of clinical dehydration or laboratory evidence of hemoconcentration [29].

Clinical Biochemistry

Chemical analysis provides information on organ function and occasionally can suggest an organ-specific diagnosis.

Iatrogenic artifacts frequently result in out-of-range biochemical values in ratite blood specimens submitted to laboratories. Causes include intramuscular injections, bacterial contamination during sampling and shipping, unseparated blood, clots, hemolysis, and improper anticoagulants [13].

Uric Acid

Avian species excrete nitrogenous waste through hepatic synthesis of uric acid. This compound is eliminated only by tubular secretion and is independent of glomerular filtration.

There is no sensitive and effective analyte for avian renal function. When moderate-to-severe elevation in plasma uric acid occurs, significant tubular damage has already occurred. Veterinarians presented with iatrogenic aminoglycoside toxicosis may not see significant uric acid elevation until the patient is in extremis.

Diagnoses of renal disease are usually established at postmortem examination.

Slight elevations in uric acid are commonly observed in dehydrated birds due to a prerenal increase in solute concentration. Values in the range of 10–20 mg/dl should always be confirmed after clinical stabilization of the patient. Studies in emus have shown that reduced renal ability to concentrate urine is compensated by efficient cloacal-rectal reabsorption [30].

Creatinine

Creatinine is a normal constituent of avian urine but is formed in very small quantities compared to creatine. Although creatinine can be removed by tubular secretion in birds, clearance is variable. At normal plasma levels, this analyte is reabsorbed. Certain chemicals can inhibit reabsorption, but diuresis will increase elimination. Spurious elevations in creatinine can be caused by nonspecific chromogens in avian plasma. Creatinine is not a reliable indicator of avian renal function.

Urea

Urea is synthesized in very small quantities in avian patients. This analyte is completely filtered by the kidney and can be reabsorbed. Urea is not useful in assessing avian renal function, but may be a valid indicator of prerenal dehydration.

Calcium

Nutritional and metabolic dysfunctions involving calcium imbalance are known to occur widely in avian species. Calcium assay has limitations in assessing clinical disorders. Nutritional history is important when considering the differential diagnoses of possible disorders of mineral metabolism.

Females uniquely develop intramedullary bone at the onset of egg production stimulated by estrogen. Large quantities of calcium are transferred from the plasma to the shell by the uterine gland during egg formation. Elevation in total blood calcium level occurs during egg production by the hen. Excess calcium is principally bound to a specific carrier protein [25].

Hypocalcemia in Ratites

Artifactual hypocalcemia is common. Blood for avian chemical assays should never be collected into tubes containing EDTA, as calcium will be chelated by EDTA *in vitro* [15]. Suboptimal serum quality, including hemolysis and bacterial contamination, can reduce calcium values. Because most blood calcium is protein-bound, the clinician should be careful to assess total protein level in relation to the results of calcium assay.

Phosphorus

Phosphorus metabolism in birds is similar to mammals, although elevation is not commonly noted in birds. Hyperphosphatemia, associated with apparent renal disease, is rarely encountered. Hypophosphatemia is also uncommon [25] but occurs in rheas with rickets. Artifactual alteration in plasma phosphorus level can be caused by collection of blood into tubes containing EDTA. Hemolysis can falsely depress values, and hemolysis and lipemia can elevate serum phosphorous [15].

Glucose

Ratites maintain higher plasma glucose levels than mammals. Moderate hyperglycemia can occur due to stress but tends to be transient in nature [25]. Diabetes mellitus has not been reported in ratites.

True hypoglycemia in young ratites is typically due to primary or secondary starvation, and values less than 100 mg/dl can be life-threatening.

Artifactual decreases in plasma glucose are common. Shipping unseparated avian blood can result in a significant decrease in glucose level. Bacterial contamination of samples during collection and processing can also markedly lower glucose values [15].

ELECTROLYTES

The cellular functions of sodium and potassium in birds are similar to mammals. Regulation of avian electrolytes by the kidney is modulated by a variety of hormones. Corticosterone appears to be the primary avian glucocorticoid.

Electrolyte values are not routinely requested by avian clinicians. Imbalances due to organ dysfunction, alimentary tract disorders, and iatrogenic causes parallel changes in mammals.

PROTEIN

Serum protein is important in avian homeostasis by maintaining osmotic pressure, preventing extravasation, and maintaining blood pH. Protein fractions include albumin and the globulins. Albumin serves as a reserve protein and as a carrier for metabolites. Globulins include inflammatory and clotting proteins and immunoglobulins [25].

No reliable biochemical method exists to quantify avian albumin, resulting in abnormally low values. Globulin and albumin:globulin (A:G) ratios are usually calculated using total protein and albumin measurements. This results in reproducible but inaccurate results. Protein electrophoresis (EPH) remains the only reliable method to quantify the major protein fractions in avian blood [31].

Cholesterol

Reference ranges for ratites have been reported but little clinical importance is attached to this analyte [9–14].

Bile Acids

Bile acids are synthesized in the liver from cholesterol. In seed-eating birds, chenodeoxycholic acid is the primary bile acid produced [32]. Bile acids are stored in the gallbladder of the emu and rhea and are released after a meal. Ostrich do not have a gallbladder. Bile acids facilitate digestion and absorption of fats. A majority of bile acids are absorbed by the small intestine, transported to the liver by portal veins, and reclaimed by the liver.

Elevation in bile acids is associated with reduced hepatic function. Clinical experience indicates that bile acid values are a more sensitive indicator of hepatic function. Enzymes measure hepatocellular damage and often return to normal after acute insult. Persistent loss of hepatic function is reflected in elevation in bile acids. This is uncommon in ratites, reflecting a low prevalence of chronic liver disease compared to companion avian species.

ENZYMES

Lactic Dehydrogenase

Lactic dehydrogenase (LDH) activity occurs in cardiac and skeletal muscle, liver, kidney, and erythrocytes. Cellular damage results in increased serum LDH. This enzyme rises most rapidly after an acute insult and is not liver-specific. Increases in nonhemolyzed samples originate primarily from skeletal muscle and the liver, with occasional contributions from cardiac muscle. Renal LDH will be excreted in the urine and will not be detected in the blood [3].

Sampling and processing artifacts probably represent the most common causes of elevated LDH in ratites. Hemolysis of erythrocytes releases intracellular LDH isoenzyme. Appropriate blood collection techniques avoid hemolysis and clotting. Blood samples must be centrifuged promptly, and serum or plasma should be immediately decanted from the cell fraction [15].

Aspartate Transaminase

Aspartate transaminase (AST, SGOT) is regarded as a nonspecific indicator of liver disease in avian species. As with LDH, elevated AST levels are due to damage to liver tissue and skeletal and cardiac muscle. Since AST has a longer half-life than LDH, levels will remain elevated for a longer period following cellular insult [3]. Nonspecific muscle damage, including trauma, surgery, debilitation, and intramuscular injections are the most common cause of elevation.

Alanine Transaminase

Alanine transaminase (ALT, SGPT) is a liver-specific enzyme in mammals. Elevation of ALT following experimentally induced liver disease in psittacine birds is inconsequential [3], suggesting minimal significance in ratites.

Creatine Kinase

Creatine kinase (CK, CPK) occurs in smooth, skeletal and cardiac muscle, and in nervous tissue. Clinical or physiologic elevations in CK are primarily attributed to degeneration in skeletal or cardiac muscle [3]. Myocarditis can result in cellular damage and leakage of CK, resulting in elevated serum values. Muscle wasting, vigorous physical activity, rough handling, irritating injections, surgery, and trauma can cause marked elevations in CK. Mechanical impaction of the proventriculus in ratites inconsistently results in elevation in this enzyme [33]. No smooth muscle specific avian CK isoenzyme is commercially available. The half-life of psittacine and pigeon CK falls between LDH and AST, and clinical observations suggest a similar relationship among the half-life of enzymes in ratites. The importance of CK is to assist the clinician in differentiating between liver and muscle damage in the presence of elevated AST [3].

Alkaline Phosphatase

Alkaline phosphatase (AP) is not useful to diagnose liver disease in psittacines. Elevated AP occurs following fracture, neoplasia and osteomyelitis [3]. Studies in farm-raised Australian emus showed that juvenile birds have higher values than adults, which may be attributed to rate of bone growth [14]. Elevation in alkaline phosphatase occurs in ratites due to bone accretion, female reproductive activity, and iatrogenic hemolysis.

Gamma Glutamyl Transferase

Gamma glutamyl transferase (GGT) is not relevant in liver diseases in avian species.

Amylase and Lipase

Amylase and lipase activities have been documented in chickens. A variety of pathological changes have been reported in the avian pancreas, but clinical studies have not been completed to assess the value of pancreatic enzymes in ratites.

RATITE BLOOD PANELS IN HEALTH AND DISEASE

Retained yolk sacs often become infected, resulting in an acute, septic response.

Bursal depletion is a frequent pathologic finding associated with mortality in ostrich chicks [8]. A number of management factors such as physical stress may be associated with bursal depletion. Ostrich chicks may be exposed to infectious bursal disease virus with subsequent, secondary bacterial infection. Blood panels vary from normal to a severe disturbance in cellular and enzyme parameters.

Starvation in young chicks may result in anemia with stress leukocytosis. Elevation in enzymes (CK, LDH, AST) occurs with cachexia. Elevation of LDH and AST suggests hepatocellular damage but is not specific.

Dehydration due to improper incubation or environmental management may result in prerenal elevation in uric acid. Marked neonatal dehydration may result in visceral gout preceding death.

Congenital and acquired leg abnormalities (slipped tendons, rotations of long-bones) occur frequently in commercial ratites. Generally, blood panels are not helpful in the diagnosis, management and prevention of leg problems. In a small number of cases, deficiencies or imbalances in dietary nutrient content may lead to developmental abnormalities which may be reflected in altered plasma calcium and phosphorus values.

PURCHASE EXAMINATIONS

Veterinarians are frequently requested to perform prepurchase, postpurchase, and insurance examinations. Subjects may include juveniles, yearlings near breeding age birds, and adults. Veterinary clinicians vary in their approach to assessing health. Parameters will vary according to the age, sex of bird and season. Many insurance companies require some type of blood analysis in addition to clinical evaluation of the birds.

Ideally, new birds should be examined and blood samples obtained before shipment, rather than after delivery to the receiving farm. During the immediate postshipment period of up to several days, stress hemograms will result in addition to elevated enzymes and prerenal dehydration, resulting in abnormal biochemistry values.

Insurance companies may require various blood determinations for health certification. Rejection of healthy ratites occurs, based on artifactual elevation in laboratory values. This situation is evidenced by panels showing high levels in analytes which have not been documented as being elevated in avian species occurring as a result of sample hemolysis.

The analytes to be included in a ratite panel should include:

- CBC
- Protein
- Glucose
- Uric acid
- CPK
- AST (SGOT)
- LDH

Analytes which have no relevance in assessing the health of ratites include: ALT(SGPT), SGT, urea, and bilirubin. Analytes which are rarely useful to evaluate health but which are frequently elevated due to improper handling of samples or rise under normal physiological conditions include creatinine, cholesterol, and alkaline phosphatase.

PREBREEDING PHYSICAL EXAMINATION

A thorough description of recommended procedures is discussed in Chapter 6. Male birds may require a CBC. Blood may be submitted from the hen for a complete blood panel, including CBC, and biochemical parameters. Ovulating hens will show elevated serum calcium.

BACTERIAL INFECTIONS

Clinicians have probably overemphasized bacterial screening of ratites. These birds are geophagic and when young, coprophagic isolation of coliforms from feces is of no significance. Bacterial infections are a secondary cause of death, particularly in chicks and juveniles. Occasionally, bacteria, such as *Mycobacterium* spp., are the primary cause of death or illness [21,22,34]. *Bordetella* infection in ostriches is associated with a mild leukocytosis. Except for peracute infections, systemic and organ-reactive bacterial infections will show predictable changes in the leukogram, including leukocytosis and heterophilia. Toxic heterophils, left shift, and monocytosis occur in more advanced or chronic infections.

FUNGAL INFECTIONS

Immunosuppressed chicks can become infected with yeasts and other fungal pathogens, resulting in nonspecific leukocytosis. This inconsistent finding in aspergillosis depends on the degree of immunosuppression.

VIRAL INFECTIONS

A number of viral infections and diseases have been diagnosed in ostriches and emus. In most reported outbreaks of viral infection in ratites, hematologic and biochemical parameters were not determined. Togavirus infection in emus result in little or no specific blood changes. Eastern equine encephalitis in emus is too acute to show changes related to intestinal bleeding. Western equine encephalitis in emus will show nonspecific blood changes associated with cachexia.

PARASITISM

Helminths have been reported in the brain of emus. Filariid infection results in an elevation in leukocyte count and plasma AST. *Balylisascaris* infections, resulting in verminous encephalitis in emus, can produce moderate leukocytosis and elevation in plasma creatinine phosphokinase.

Wireworms (*Libyostrongylus douglassii*) occur in the ostrich with variable clinical impact. There are no published reports on blood changes in this infection. *Syngamus* sp. infections in asymptomatic emus show no blood changes. Typhlitis in rheas associated with combined spirochete and flagellated protozoan infection has not be documented hematologically, as the condition involves sudden death.

TRAUMA

Acute trauma is typically associated with profound elevation in LDH, CK, and AST. Enzyme elevation in enzymes occurs following laparotomy.

REFERENCES

1. Campbell T. Hematology. In: Ritchie B, Harrison GJ, and Harrison LR, eds. *Avian medicine, principles, and application*. Lake Worth, FL: Wingers Publishing, 1994; 176–198.
2. Campbell TW. *Avian hematology and cytology*. Ames, IA: Iowa State University Press, 1994.
3. Lumeij J. Avian clinical enzymology, in *Seminars*. Avian and Exotic Pet Medicine 1994;3:14–24.
4. Dhillon AS. High mortality in young ostriches related to management. In: Kornelsen MJ, ed. *Research education networking opportunities—Main Conference Proceedings*. Assoc Avian Vet 1994;423–424.
5. Parsons B. Emu farming in Florida. In: Kornelsen MJ, ed. *Research education networking opportunities—Main Conference Proceedings*. Assoc Avian Vet 1994;438–439.
6. Raines AM. How to evaluate a ratite facility to aid in diagnosing chick mortality. In: Kornelsen MJ, ed. *Research education networking opportunities—Main Conference Proceedings*. Assoc Avian Vet 1994;97–102.
7. Smith CA. Ostrich chick survival presents challenge. *J Amer Vet Med Assn* 1993;203:637–643.
8. Terzich M, Vanhooser S. Postmortem findings of ostriches submitted to the Oklahoma Animal-Disease Diagnostic Laboratory. *Avian Dis* 1993;37:1136–1141.
9. van Heerden J, et al. Blood chemical and electrolyte concentrations in the ostrich (*Struthio camelus*). *J So Afr Vet Assoc* 1985;56:75–79.
10. Okotie-Eboh G, et al. Reference serum biochemical values for emus and ostriches. *Am J Vet Res* 1992;53:1765–1768.
11. Levy A, et al. Reference blood chemical values in ostriches. *Am J Vet Res* 1989;50:1548.
12. Palomeque J, Pinto D, Viscor G. Hematologic and blood chemistry values of the Masai ostrich (*Struthio camelus*). *J Wildl Dis* 1991;27:34–40.
13. Fudge A. *Ratite reference ranges*. Citrus Heights, CA: California Avian Laboratory, 1995.
14. Costa ND, McDonald DE, Swan RA. Age-related changes in plasma biochemical values of farmed emus (*Dromaius-Novaehollandiae*). *Aust Vet J* 1993;70:341–344.
15. Fudge AM. Blood testing artifacts: Interpretation and prevention, in *Seminars*. Avian Exotic Pet Med 1994;3:2–4.
16. Blue-McLendon A. Basic techniques of ratite restraint and handling. In: Kornelsen MJ, ed. *Research Education Networking Opportunities—Main Conference Proceedings*. Orlando, FL: Assoc Avian Veterinarians, 1994;309–311.
17. Van der Heyden N. Evaluation and interpretation of the avian hemogram, in *Seminars*. Avian Exotic Pet Med 1994;3:5–13.
18. Montali R. Comparative pathology of inflammation in the higher vertebrates (reptiles, birds, and mammals). *J Comparative Path* 1988;99:1–27.
19. Marks SL, Stauber EH, Ernstrom SB. Aspergillosis in an ostrich. *J Amer Vet Med Assn* 1994;204:784–785.
20. Rousseaux CG, Dalziel JB. Aspergillus pneumonia in an ostrich (*Struthio camelus*). *Aust Vet J* 1981;57:151–2.

21. Sanford SE, Rehmtulla AJ, Josephson GKA. Tuberculosis in farmed rheas (*Rhea-Americana*). *Avian Dis* 1994;38:193–196.
22. Shane SM, et al. Tuberculosis in commercial emus (*Dromaius-Novaehollandiae*). *Avian Dis* 1993;37:1172–1176.
23. Grimes JE, Arizmendi F. Case-reports of ratite chlamydiosis and update on the chlamydias. In: Kornelsen MJ, ed. *Research education networking opportunities—Main Conference Proceedings.* Assoc Avian Vet 1994;133–140.
24. Marks SL, Stauber EH, Ernstrom SB. Aspergillosis in an ostrich. *J Amer Vet Med Assoc* 1994;204:784–785.
25. Jenkins J. Avian metabolic chemistries, in *Seminars.* Avian and Exotic Pet Medicine 1994;3:25–32.
26. Levy A, et al. Haematological parameters of the ostrich (*Struthio camelus*). *Avian Pathol* 1989;18:321–327.
27. Randolph JF, et al. Bacterial endocarditis and thromboembolism of a pelvic limb in an emu. *J Am Vet Med Assoc* 1984;185:1409–10.
28. Dein F. *Avian clinical hematology: Erythrocytes and anemia.* San Diego, CA: Assoc Avian Vet, 1983.
29. Jain, NC. Clinical and laboratory evaluation of anemias and polycythemias. In: J NC, ed. *Schalm's Veterinary Hematology.* Philadelphia, PA: Lea & Febiger, 1986;563–576.
30. Dawson T, Herd R, Skadhauge E. Osmotic and ionic regulation during dehydration in a large bird, the emu (*Dromaius novaehollandiae*): An important role for the cloaca-rectum. *Qtrly J Exper Physiol* 1985;70:423–436.
31. Spano J, et al. Comparative albumin determinations in ducks, chickens, and turkeys by electrophoretic and dye-binding methods. *Amer J Vet Res* 1988;49:325–326.
32. Lumeij JT, Meidam M, Wolfswinkel J. Changes in plasma chemistry after drug induced liver disease or muscle necrosis in racing pigeons (*Columbia livia domestica*). *Avian Pathol*, 1988. 17:pp 865–874.
33. Gamble KC, Honnas CM. Surgical-correction of impaction of the proventriculus in ostriches. *Compend Cont Educ Pract Vet* 1993;15:235.
34. Bowes V. Avian tuberculosis in ostriches. *Can Vet J* 1993;34:758.

Chapter 12

Parasites of Ratites

Thomas M. Craig and P. Lea Diamond

Current literature contains many citations relating to parasites of ratites but with few details as to diagnosis, economic importance, pathogeneses, or practical control programs. This chapter will review current information on parasites of ostriches, emus, and rheas with special reference to North America.

In 1993, the United States Department of Agriculture, Animal and Plant Health Inspection Service, temporarily banned the importation of adult ratites to restrict the entry of exotic ectoparasites into the United States [1]. This moratorium on imports was introduced to prevent entry of exotic ostrich ectoparasites of ruminants. The internal parasites that accompanied these birds have a far greater economic potential to impact ratites than the *Amblyomma* ticks which led to federal action.

ECTOPARASITES

A number of ectoparasites have been identified in the ratite species (Table 12.1). Adult, female ixodid (hard) ticks feed upon their host until repletion then drop off to lay their eggs in the environment. Males generally remain on the host, searching for females. Most of the ticks found on imported birds were males, and exotic populations have not been established in North America.

Hard ticks do not transmit known disease-causing organisms to ratites, but they can cause slow growth, unthriftiness and reduced egg production. Tick paralysis caused by *Hyalomma truncatum* has been observed in young ostrich chicks in Africa (2) and this condition may occur in North America. Large mouth parts may damage hides, lowering market value.

The fact that most ixodid ticks are not host-specific has implications for infestation. Larvae and nymphs reside on a variety of small, ground-dwelling mammals and birds that can easily enter the most secure

TABLE 12.1. Ectoparasites of Ratites

Ixodid Ticks	
Amblyomma gemma	ostrich
A. lepidum	ostrich
A. variegatum	ostrich
A. hebraeum	ostrich
Haemaphysalis punctata	ostrich
Hyalomma albiparmatum	ostrich
H. luscitanicum	ostrich
H. rufipes	ostrich
H. truncatum	ostrich
H. impeltatum	ostrich
H. dromedarii	ostrich
Rhipicephalus turanicus	ostrich
Argasid Ticks	
Argus persicus	ostrich
Otobius megnini	ostrich
Mites	
Gabucinia (Pterolinchus) bicaudatus	ostrich, rhea
G. sculpturata	ostrich
G. abbreviata	ostrich
G. nouveli	ostrich
Paralges parchycnemis	ostrich
Lice	
Struthioliperirus struthionis	ostrich
S. rhea	rhea
S. nandu	rhea
S. renschi	rhea
Meinertzhageniella latus	rhea
Dahlemhornia asymetrica	emu
Flies	
Struthiobosca struthionis	ostrich

fenced enclosure. Adult ticks are frequently found on large mammals or birds such as ratites. Recommended treatment consists of direct application of 5% carbaryl dust or 2% malathion spray. It will be necessary to repeat the treatment, as it may be impossible to render the pens and surroundings free of ticks.

The soft tick, *Argus persicus* (Figure 12.1), is easily overlooked due to nocturnal, intermittent feeding. The *Argus* tick lives away from the host during the day, in shelters and nesting areas, feeding on hosts at night. Transmission of disease carrying organisms by *Argus* has not been recorded in the United States. In Africa,

A. persicus transmits aegyptianellosis from chickens to ostriches [3]. *Argus* can exsanguinate young birds, especially with intensive indoor rearing of chicks. Treatment for *Argus* spp. is the same as for hard ticks and lice. Application of the acaracide should be directed at the environment as well as the birds.

Both the ostrich and rhea may be infested with *Gabucinia bicaudatus*, the quill mite. Other species of feather mites (*Paralges parchycnemis, Gabucinia sculpturata, G. abbreviata* and *G. nouveli*) have also been described from the ostrich [4]. These mites may be found upon the skin, feathers or, within the calmus of the quill. Mites parasitizing the feather shaft require thorough microscopic inspection for diagnosis.

Pathogenicity associated with mites in the family Pterolichidae of ostriches include pruritus, excessive preening, and feather loss. Birds may show bacterial dermatitis and respiratory distress [5]. Three treatments of Ivermectin 200 ug/kg per os at 4-week intervals will control mites.

Several species of chewing lice occur on ratites, but only *Struthiolipeurus* spp. have been recorded in the United States. *Struthiolipeurus struthionis* occurs on ostriches, and *S. rhea, S. nandu,* and *S. renschi* have been identified on rheas in North America [6]. Lice are generally seen on domesticated or zoo-housed ratites and can be expected where there is transfer of birds among flocks (Figure 12.2).

Struthiolipeurus struthionis feed on body scales and feather debris. Diagnosis of louse infestation may be possible by observing the lice or their egg clusters (nits) attached to the feathers. Clinical signs generally associated with louse infestation include excessive grooming, ruffled and frayed feathers, and feather loss.

Placing the birds in a warm dark environment for 1–2 hours may facilitate diagnosis because the lice will migrate to the outer feathers of the body. Lice which desert anaesthetized or dead birds may be seen crawling on prosectors at postmortem examination but they do not parasitize humans.

A closed-flock policy and quarantine of new introductions will effectively reduce infestation. Transmission of lice occurs directly from one bird to another. Malathion spray [8] and carbaryl dust are safe and effective treatments for ectoparasites of ostriches.

In areas suitable for the breeding of blackflies (*Simulium* spp.), ratites may suffer from stress, anemia, or even death from anaphylactic shock caused by a reaction to bites. In Texas, during the spring of 1991, ostrich and emu ranchers lost over $1.5 million in birds due to blackfly swarms [9]. Supportive treatment included the use of corticosteroids and fluids. The control of blackflies is difficult because they breed in swift moving water.

Culicoides crepuscularis, biting midgets or "no-see-ums," can cause considerable irritation, especially to young birds. The midge serves as a vector of *Chandlerella quiscali* to emus. Culicoides breed in moist areas and in seasons with high rainfall and elevated temperature, midges swarm in tremendous numbers.

Mosquito, midge, and blackfly infestations are difficult to control when birds are kept in outdoor pens or open range. Chicks may be protected in screened enclosures. Because blackflies and midges are poor fliers, strategically located fans can prevent these pests from reaching and feeding on young birds. Applying products, such as Skin-so-Soft™, may discourage these insects from feeding on ratite chicks.

Leukocytozoon and *Plasmodium* occur in ostriches in Africa, but these genera have not been reported in the United States. Hemoparasites are transmitted by various species of gnats and mosquitoes, but the ability of North American insects to transmit these parasites is unknown.

The hippoboscid fly, *Struthiobosca struthionis,* has been reported on ostriches imported from Tanzania. This parasite does not appear to have naturalized and thus is not a problem in the United States at this time [10].

ENDOPARASITES

Although various helminth species have accompanied ostriches and rheas to North America, none appear to have been imported with the emu (Table 12.2). Generally, parasites with direct life cycles have established in North America, in contrast to species requiring an intermediate host. Each ratite family has undergone convergent evolution in geographic isolation, accounting for the limited number of common parasites. Where two or more ratite or other avian species are raised in proximity, parasites may infect birds that have no history of previous exposure. The accidental host may be more severely affected than the normal definitive host, which coevolved with the parasite.

Parasites with a direct life cycle require bird-to-bird transmission by contact with eggs or larvae passed in excreta. The most effective way to avoid infection and subsequent loss is to prevent introduction of the parasite into the flock. Parasites which require one or more intermediate hosts may be more difficult to control.

The "wireworm," *Libyostrongylus douglassii,* is the most deleterious helminth of young ostriches with mortality rates occasionally greater than 50%. This

TABLE 12.2. Helminths of Ratites

Ostrich
  Nematodes
Libyostrongylus douglassii	proventriculus
L. magnus	proventriculus
L. dentatus	proventriculus
Cyrnea colini	proventriculus
Codiostomum struthionis	caeca
Paronchocerca struthionus	lungs
Dicheilonema spicularia	air sac
Struthiofilaria megalocephala	abdominal cavity
Baylisascaris procyonis	brain

  Cestodes
Houttuynia struthionis	small intestine

  Trematodes
Philothalamus gralli	eye

Rheas
  Nematodes
Ascaridia orthocerca	small intestine
Deletrocephalus dimidiatus	small intestine
D. cesarpintoi	small intestine
Paradeletrocephalus minor	small intestine
Sicarius uncinipenis	small intestine
S. waltoni	small intestine
Vaznema zschokkei	small intestine
Dicheilonema rhea	abdominal cavity
Syngamus trachea	trachea

  Acanthocephala
Prosthorhynchus rheae	small intestine

  Cestodes
Houttuynia struthionis	small intestine
Chapmania tauricollis	small intestine

Emus
  Nematodes
Dromaestrongylus bicuspis	small intestine
Trichostrongylus tenuis	small intestine
Baylisascaris procyonis	brain
B. columnaris	brain
Chandlerella quiscali	brain
Syngamus trachea	trachea
Cyathostoma variegatum	bronchi

parasite was introduced into North America with imported ostriches and is now well established. The worms are small, blood-feeding nematodes which inhabit and physically occlude the ducts of the proventricular glands. This causes a compensatory production of thick mucous which impairs digestion and may lead to impaction [11]. They cause a diphtheritic proventriculitis, which is commonly called "vrotmaag" (rotten stomach) by producers in South Africa. Clinical signs of infection include general wasting, anorexia, anemia, and death [3, 12]. Healthy adult ostriches may be infected and are a source of infection to other birds.

Examination of the fresh proventriculus by stripping the koilin (mucosal) lining will yield delicate 4–6 mm long, red worm with a very small buccal cavity (Figure 12.3). The males are bursate with the dorsal ray cleft in their distal half forming three branches on each side, and spicules of 0.14–0.16 μm, each ending in a small and large spine. [12]

A positive fecal examination contains strongylid-type eggs approximately 72 × 41 μm, which cannot be differentiated from the eggs of *Codiostomum struthionis*. Larval culture to the L3 stage is necessary for definitive identification. The Texas A&M College of Veterinary Medicine laboratory has cultured suspect fecal material at 23°C in one week. Eggs will develop into infective larvae in 60 hours if incubated at 36°C. [11]

Infective larvae of *L. douglassii* are approximately 830 μm long, with 16 triangular cells and a 4 μm diameter knob approximately 23 μm from the tip of the sheath [11]. Species differentiation is based on the length of the tail of the sheath; *L. douglassii* has a tail about 85 μm long, whereas *C. struthionis* is much longer and whiplike, typical of strongylids. The knob at the end of the tail of the L3 is diagnostic for *L. douglassii*.

A second species of *Libyostrongylus*, *L. dentatus*, has been found in the proventriculus and gizzard of ostriches in North America. [13] This parasite is also found under the koilin lining. The adult worms are 7–13 μm long and are easily overlooked or confused with *L. douglassii*. The eggs of this species are shorter in length (52–62 μm long) than those of *L. douglassii*. A prominent dorsal esophageal tooth is the best characteristic for the identification of the adult stage of *L. dentatus*. Its importance is unknown, and thus far it has only been found in mixed infections where there were signs of libyostrongylosis. Larvae believed to be *L. dentatus* have been found in birds apparently infected with both species.

The best way to avoid introducing *L. douglassii* (or any other direct life cycle helminth) is to hold all new birds in quarantine for at least as long as the prepatent period of the parasite, then repeatedly test for the presence of helminth eggs or larvae [11]. The prepatent period for *L. douglassii* is approximately 30–35 days. During this time, feces should be collected within 48 hours and destroyed so that any larvae present do not have the opportunity to develop into the infective L3 stage. Infective larvae are very robust. Pastures will remain infective for several years. Considering the arid origins of the parasites, this is not surprising. Under ideal circumstances, the eggs and infective larvae can survive desiccation for 30 months. Direct sunlight is lethal to the unprotected L1 larvae. Pasture infection will be of longer duration if fecal pats remain intact than if rain or other mechanical forces expose the larvae before they develop to the L3 stage. Under natural conditions, infective larvae survive for up to 14 months [11].

In South Africa, L. donglassii infection is associated with ostrich chicks grazing on alfalfa pastures which previously held infected birds. It is suggested that pastures should be seeded to hay or be grazed by another species of livestock for at least one, but preferably two years, prior to reuse by ostrich chicks.

Care must be taken to ensure that the donors of fecal cultures are parasite free when establishing intestinal flora in chicks.

At present, there are no anthelmintics approved in the United States for ratites. However, fenbendazole, levamisole and ivermectin have been tested and found efficacious and safe. Levamisole, a drug registered in South Africa (30 mg/kg) for use against *L. douglassii,* is no longer as effective against some populations of this parasite due to the development of resistance [14]. Fenbendazole (15 mg/kg) was shown to be >99% effective against adults and >82% effective against the L4 larvae [15], and resistance has not been reported. Ivermectin at 0.2 mg/kg may also be effective.

The life cycle of *Codiostomum struthionis,* the cecal and colon parasite of ostrich, is unknown, but is probably simple and direct. The helminth has a large buccal cavity with a markedly thickened, chitinous capsular wall devoid of teeth. There is an external leaf-crown arising from the mouth collar and an internal leaf-crown arising from the anterior of the buccal capsule. The elements of the internal leaf crown are longer than the elements of the external leaf crown [12]. The vulva is located near the anus and has a prominent cuticular elevation just anterior. Once again, the eggs of *C. struthionis* are identical to those of *L. douglassii,* but the infective larvae have a long sheath. *Codiostomum struthionis* is a potentially pathogenic helminth [16] possibly causing anemia or impaired growth but is unlikely to be important in healthy, well-managed flocks. Benzimidazoles or levamisole should be efficacious against this worm.

Sicarius uncinipenis and *S. waltoni,* spiruids of the gizzard, were described from rheas in Brazil [17] and have been reported in zoological gardens. They resemble *Habronema* spp. of horses. *Sicarius uncinipenis* is 18–30 mm in length and 0.6–0.8 mm wide. The unequal spicules measure 3–3.7 mm and 0.7–0.8 mm in length. *Sicarius waltoni* is slightly smaller; females are approximately 25 mm and males 20 mm in length with spicules of 2.5 mm and 0.35 mm in length. Other characteristics are similar to those of *S. uncinipenis.*

Another spirurid, *Vaznema zschokkei,* is found in the submucosa of the proventriculus of rheas. The males measured 16–17 mm and females 17–25 mm in length. The spicules are long and filiform, 10 mm and 11 mm in length. The habitat and filiform spicules are adequate for identification of this species.

Members of the family Deletrocephalidae are an ancient, primitive group that evolved with ratites. *Deletrocephalus dimidiatus* is found in rheas. The worm has been diagnosed in young, domesticated rheas and appears to be well established in North America. Infections with *D. dimidiatus* have been associated with weak, diarrheic chicks with high flock mortality. The worm is apparently a blood feeder and may be found in large numbers in both the small and large intestines. Identification of the helminth is based on the presence of robust worms with a well-developed, cup-shaped buccal capsule with teeth at the base. The sinuous internal corona radiata is a unique feature of this parasite. Males are 11 to 18 mm long with slender spicules 0.9 to 1.025 mm long. Females are 17 to 24 mm long with a covered vulva close to the anus. The eggs are oval (120–125 μm by 70–75 μm), thin-walled, smooth shelled and segmented with a swelling at either pole. They are said to embryonate rapidly in the environment. The life cycle is unknown but is presumably direct. Birds ingesting infected larvae while foraging.

A related genus, *Paradeletrocephalus,* has not yet been identified in the United States. It can be differentiated from *Deletrocephalus* spp. based on the absence of an external corona radiata and sinuous internal corona radiata. The lining of the buccal capsule has vertical ridges and the esophagus is flask shaped.

An *Ascaridia* sp. specimen was found in the small intestine of a rhea submitted to the TAMU Diagnostic Laboratory in 1993. Species identification was speculative because the helminth was an immature specimen, but it was believed to be *A. orthocerca*. This is the only *Ascaridia* sp. that has been seen in ratites. Adults are 30–40 mm long × 1–2 mm wide with transverse cuticle striations. The oral opening is surrounded by three similar sized lips, with the dorsal lip having a divided pulp and two papillae. In the male, the caudal alae are very poorly developed; however, there is an oval preanal sucker with a chitinous ring and 12 pair of caudal papillae. The spicules are very long, alate, and slightly curved at the posterior ends. The eggs are oval with a thick, smooth shell and are probably hardy in the environment. Ascarididae are easily identified by fecal flotation. The eggs are relatively thick walled and single celled. The pathogenic effects and life cycle of this helminth are probably similar to *A. galli*. If ascarid eggs are found, treatment with piperazine (50–100 mg/kg) at 2-week intervals is recommended until fecal exams are negative. Fenbendazole at 15 mg/kg/os should also be effective.

An intestinal worm, *Trichostrongylus tenuis* (caecal threadworm), has been identified in the emu. This is a common parasite of several species, including the grouse, partridges, quail, and various other galliformes and anseriformes. The worm is delicate, transparent, and approximately 8 mm (males) to 10 mm (females) in length. It has a direct life cycle with a prepatent period of 7–8 days. Eggs are elliptical, 65 to 75 μm long by 35 to 42 μm wide, and segmented when passed. It takes approximately 6–7 days for the larva to develop to the infective stage, which migrates onto vegetation. When swallowed, the larva locate in the ceca and undergo two more molts to the adult stage [18]. In order to differentiate T. tenuis eggs from similar trichostrongle eggs, the larvae were fed to chickens, which were later euthanized, and the adult worms were retrieved for identification.

Pathology and treatment of T. tenáis in emus must be extrapolated from game bird research. Initially, the infection may cause a bloody, mucoid diarrhea which later resolves. A light infestation in the grouse does not result in clinical signs, but large numbers can cause anorexia, progressive anemia and toxemia. There appears to be little or no acquired immunity [19,20]. Older birds may accumulate sufficient numbers to cause disease. There were no clinical signs that could be directly attributed to T. tenuis in any of the infected emus observed in Texas.

Dromaestrongylus bicuspis was identified by Lubimov in 1933 in an emu from the Moscow zoo. The worms were found in the small intestine and no pathology was noted. The adults measured 6–8 mm in length and 150 μm in width. Adult worms have a rudimentary buccal capsule with an inflated cephalic vesicle. Females are ovoviviparous and males are bursate with a salient dorsal lobe and spicules 820 μm in length. The tail of these worms is conically elongate and terminated in a digitiform dorsal appendage and thin cuticular spine. It is impossible, at this time, to differentiate eggs of D. bicuspis from those of T. tenuis. This is an enigmatic parasite, as there are no reports of D. bicuspis in the literature from Australia.

CESTODES

Two tapeworms, Houttuynia struthionis and Chapmania tauricollis have been recorded in ratites in North America. Houttuynia struthionis has been reported both in the ostrich and the rhea, while C. tauricollis has been identified only in rheas [21]. After L. douglassii, Houttuynia struthionis is the most detrimental parasite seen in young ostriches in southern Africa. The disease in ostrich chicks is characterized by poor growth. Infection is recognized by the presence of proglottids or eggs in the feces of the chicks. The intermediate host(s) of these tapeworms are unknown, but insects are suspected.

Houttuynia is differentiated from Chapmania in that Houttuynia have unarmed suckers, unilateral genital pores, and a double circle of large hooks on the scolex separated from the suckers by a robust stock which can be seen grossly on occasion. The genital pores of Chapmania are irregularly alternating and suckers are unarmed [21]. Specific diagnosis requires the collection of the entire worm. Allowing tapeworms to relax in water in a refrigerator overnight prior to fixation, will facilitate recognition of morphologic features for identification. In South Africa, fenbendazole (15 and 25 mg/kg) exhibited mixed results against H. struthionis infections and was found to be more efficacious when used in combination with resorantel [15,22]. A second treatment may be necessary to eliminate heavy infections. Praziquantel at 7.5 mg/kg was highly effective against all stages of Houttuynia struthionis [3]. Chapmania probably can be successfully treated by the same drugs as Houttuynia.

ACANTHOCEPHALA

The rhea is the definitive host for the acanthocephalan, Prosthorhynchus rhea. This is a smooth bodied, elongate-cylindrical worm with a 0.8–0.9 mm long proboscis possessing 16 hooks in each of 18 longitudinal rows. The eggs of this worm are 0.07 mm by 0.01 mm with concentric membranes. It has been identified in rheas from South America. There is no evidence that this parasite occurs in North America, the life cycle is unknown. By the very nature of its attachment, Prosthorhynchus is likely to cause local inflammation and possibly peritonitis. There is no known treatment at this time.

EXTRAINTESTINAL HELMINTHS

The filariid Dicheilonema rhea has been found in rheas in Texas. Dicheilonema rhea is about 2–4 mm wide and up to 65 cm in length, however some reports indicate it may be much longer. A comparable worm, Contortospiculum rhea (80 cm long), has also been described in the rhea [23]. Dicheilonema rhea are found in the vicinity of the abdominal air sacs, although migration occurs frequently, resulting in location in the abdomen and between fascial planes including the muscles of the legs. Up to 30–40 worms have been found in a single bird during necropsy (Figure 12.4). Often, healthy birds appearing in good flesh harbor large numbers of this parasite. The length of the worm, dissimilar shape and size of the spicules, and the presence of eggs in the uterus are useful in distinguishing this parasite.

The life cycle of D. rhea has not been described, but given its location in the bird, the eggs or larvae would need to be coughed up and swallowed to be passed into the environment in feces. It is probable that an insect serves as an intermediate host. Anthelmintic treatment could be detrimental due to the immune response to dead worms in the air sacs and body cavity.

Paronchocerca struthionus was described from the lungs of a West African ostrich [24]. This filariid is believed to be extremely host specific, which is characteristic of the genus. This helminth is slender (males approximately 31 mm, females 35–45 mm) with a uniform body width, except for bluntly rounded extremities. The mature worm has a very long (>4 mm in female), broad glandular esophagus. In the male, the spicules are markedly dissimilar in length and morphology (left approximately 325 μm long, right approximately 163 μm long). Microfilaria from the anterior uteri were described as having a loose sheath, bluntly rounded anterior, and a markedly tapered to narrowly rounded tail. Information indicating pathogenicity or life cycle has not been published.

A worm identified as *Dicheilonema spicularia* was found in the serosa or air sac of an ostrich which originated from the Republic of South Africa.

Another species of filarial nematode, *Struthiofilaria megalocephala*, may also be found in the abdominal cavity [25]. The nematode is described as moderate size (females approximately 7 cm long and males approximately 4 cm long), with a laterally expanded head, small oral opening, and reduced buccal cavity. Many minute cuticular bosses or papillae stretch along the lateral fields of the body and at the cloacal aperture of both sexes. The tail is less than twice the body width and strongly coiled. Males have unequal and dissimilar spicules (about 199 μm and 118 μm long). Females have a didelphic uterus with an anterior vulva. Microfilariae are unsheathed, conforming tightly to the body which is about 247 μm long and 7 μm wide.

Early in 1990, an ostrich imported from Tanzania, was submitted for necropsy, and yielded 2 *Struthiofilaria megalocephala*. The parasite has not been identified in any birds raised in North America. There was no evidence that the parasites caused any damage to the ostrich. If microfilaria are detected in blood, benign neglect might be the best approach, as treatment may elicit an intense body reaction.

Three nematodes, *Baylisascaris procyonis*, *B. columnaris*, and *Chandlerella quiscali*, have been implicated as causal agents of cerebral nematodiasis in emus. Clinical signs noted with *Baylisascaris* spp. infection in ratites include ataxia and abnormal gait, muscle weakness, progressive recumbency, and death [26,27,28]. Diagnosis includes consideration of the age of the bird, geographic, and seasonal occurrence in relation to probable vectors. The condition must be differentiated from viral infections, such as western and St. Louis encephalitis.

The raccoon and the skunk are definitive hosts of *B. procyonis* and *B. columnaris*, respectively. Eggs of *Baylisascaris* spp. passed by raccoons or skunks may survive in the soil of enclosures for many years and birds of any age can become infected by ingesting the infective egg. Larvae migrate within the tissues of both emus and ostrich. A single larva in the brain may result in severe dysfunction and there is no effective treatment against baylisascariasis. Keeping facilities free of overhanging limbs, maintaining fences, and storing feed to avoid attracting raccoons are the only practical measures to control the parasite.

The common grackle, the definitive host of *Chandlerella quiscali*, is apparently not affected by the presence of filariids in the lateral ventricles of the brain. In some areas, up to 98% of grackles surveyed were found to be infected. Many other wild birds such as blue jays, brown-headed cowbirds and starlings also carry heavy parasite loads without any adverse effects. [29] Migration of this worm through the brain and spinal cord of young emus results in profound torticollis and incoordination leading to emaciation, scoliosis, and death.

During the spring and early summer of 1991, emu chicks in Louisiana, Mississippi, and Texas were presented to veterinarians with *C. quiscali* infection. These birds came from areas where soils were poorly drained and there had been excessive rainfall during preceding months. Farms were in rural or semirural areas where large numbers of wild birds and mammals were found along with an unusually high population of biting insects.

Chandlerella quiscali is transmitted by the bite of *Culicoides crepuscularis* [30], but other potential vectors may exist. If emus are infected during their first year, clinical signs occur within 30 days. Adult birds are apparently unaffected. Whether this is due to lack of exposure, which is unlikely, or due to maturation of the immune system is unknown. These neurotropic nematodes cannot mature and reproduce in emus as denoted by the absence of circulating microfilaria [28].

Morphology and histological criteria are used to differentiate *C. quiscali* from *Baylisascaris* spp. *Baylisascaris* larvae are approximately 60–65 μm in diameter with a pair of prominent lateral alae, and a centrally located intestinal tube lined with low columnar epithelial cells [27] (Figure 12.5). *Chandlerella quiscali* become adults during migration They are 70 to 100 μm in diameter, without lateral alae, and both intestinal and reproductive organs may be seen. (Figure 12.6).

Once these neurotropic parasites have entered the central nervous system, anthelmintic treatment is ineffective due to the blood-brain barrier. Practical field observations indicate that Ivermectin administered at a dose rate of 200 μg/kg by the subcutaneous route at 3- to 4-week intervals protects young emus from larval migration to the brain, effectively suppressing clinical signs [28].

Syngamus trachea (gapeworm) is a common parasite of galliformes, passeriformes, gruiforms and other birds, including emus. Wild birds such as blackbirds, crows, sparrows, and starlings serve as a continuous reservoir of infection. [16] Young birds are most seriously affected. Gasping, head shaking, and, in advanced cases, expulsion of froth and blood are the most common signs in emus. Worms are observed in the trachea on postmortem (Figure 12.7). Occasionally the double-capped egg (85–90 × 50) may be observed on fecal flotation (Figure 12.8). Earthworms are the most common intermediate host, but other invertebrates, such as slugs, beetles, snails, and cockroaches, may disseminate infection [31]. Eggs and larvae which develop to the third stage within the egg can survive nearly a year unprotected in soil. Once inside the earthworm the L3 stage can survive for the life (several years) of the vector. When a bird ingests either the eggs, free infective larvae, or encysted L3 within an intermediate host, the blood carries the L3 to the lungs where it undergoes two molts before moving to the trachea. The migration of these worms can result in secondary respiratory infection and possibly pneumonia. Clinical signs of infection occur about 7 days after ingestion of the L3. Adult worms appear blood-red upon fresh examination and will always be found paired in a state of permanent copulation. Treatment for *Syngamus* has been successful in a number of avian species using Fenbendazole at 60 ppm for 6 days or 120 ppm for 3 days. Mebendazole (15 mg/kg) and Chloramphenicol (0.3 g/100 kg) are also reported to be effective.

A related species, *Cyathostoma variegatum*, has been found in the bronchi of young emus where it caused severe respiratory distress [32]. The parasite is larger than *Syngamus trachea*, and the adults are not in constant copulation but are otherwise similar. The life cycle of *Cyathostoma* is thought to be the same as *Syngamus*. The normal hosts of *C. variegatum* are anseriformes (ducks), cinconiiformes (storks) and gruiformes (cranes). It appears that a number of intermediate hosts may serve to transport the larvae to the environment of the emu. In localized areas, this parasite appears to be an important pathogen. Several anthelmintics should be effective in killing the parasite; however, considerable mechanical damage and an inflammatory response occurs during larval migration. Infection often results in death of the bird despite treatment. Ivermectin at 200 μg/kg by the subcutaneous route may prove to be of prophylactic value.

FLUKES

The Oriental eyefluke, *Philophthalmus gralli*, is an unexpected parasite reported in large numbers from captive raised ostriches in Florida [33]. The fluke and is usually found in ansseriforms. requires a fresh water snail as an intermediate host.

Philophthalmus gralli is acquired by ingestion of the metacercaria attached to the shells of snails, crayfish, or other solid objects which may include pebbles ingested by ostriches. This parasite is unlikely to be a problem unless birds have access to fresh water such as a spring or free-running creek. The metacercaria rapidly excyst after entering the mouth of the birds and migrate up the esophagus to the pharynx or directly into the nasal passages and lacrimal ducts. Once in the eye, young adults develop under the nictating membrane and later migrate and attach to the outside of the nictating membrane in the lower portion of the eye [34]. Infestation with this small (3.2–3.9 × 0.1–1.4 mm) fluke results in a severe conjunctivitis which produces a blood-tinged fluid containing fluke eggs and active miracidia. The eggs of this fluke are yellow, oval, thin-shelled, and nonoperculate, averaging 0.07–0.11 × 0.03–0.05 mm. The prepatent period is approximately 30 days. [35] Repeated topical treatment with 5% carbaryl dust reportedly eliminated the parasite from ostriches [33].

The ostrich is most probably an accidental host for this parasite. The eggs of this fluke are unable to pass through the digestive system and must be dispersed into water from the exudate from the eyes, mouth and nasal cavities. The ostrich does not typically immerse its head while drinking and uses sand for bathing, which disfavors transmission.

PROTOZOA

Protozoan parasites of ratites have been only studied superficially (Table 12.3). In some instances, protozoa observed in a sick or dead bird have been implicated as the cause of disease with no further evidence other than that their presence. To prove that these agents are the cause of disease is not always simple, as many of the agents cannot be propagated outside of the body. In some cases, coinfection with a second pathogen or environmental stress is required to cause disease. Protozoa that are found in clinically normal birds may cause disease under specific circumstances and be of no significance in others. The following protozoa are reviewed in relation to the tissue or organ in which they occur.

Leukocytozoon struthionis is a common parasite of ostrich chicks in South Africa. It is apparently transmit-

TABLE 12.3. Protozoans of Ratites

Hemoparasites	
Leukocytozoon struthionis	ostrich
Plasmodium struthionis	ostrich, rhea
Aegyptianella pullorum	ostrich
Intestinal Sarcomastigophora	
Giardia sp.	ostrich
Histomonas meleagridis	ostrich, rhea
Trichomonas sp.	ostrich, rhea
Hexamita sp.	rhea
Amoeba	ostrich
Ciliates	
Balantidium struthionis	ostrich
Balantidium sp.	rhea
Coccidia	
Cryptosporidium sp.	ostrich
Isospora sp.	ostrich
Others	
Blasocystis hominis	ostrich, rhea
Tissue Protozoans	
Toxoplasma gondii	rhea

ted by blackflies, *Simulium* spp., and under some conditions may cause disease. Marked anemia was reported during early parasitemia in ostrich chicks as the gamonts invade erythrocytes and are destroyed by the immune response of the host. A severe inflammatory reaction associated with megaloschizonts in the myocardium occurred in an outbreak that resulted in high mortality in 3-week-old chicks. The surviving chicks had large numbers of gamonts in the erythrocytes at 5–6 weeks of age [36]. Identification of either gamonts in blood cells or megaloschizonts in tissues is diagnostic.

The gamonts of *L. struthionis* are round and are most often diagnosed in 4- to 7-month-old chicks. Gamonts are similar in morphology to *L. schoutedeni*, a common parasite of backyard chickens in southern Africa. The fact that parasites have not been found in older birds suggests that ostriches are abnormal hosts for this parasite, which has not been reported in North America.

Plasmodium struthionis has been identified by the presence of gamonts and schizonts in erythrocytes. It is still to be determined if this species is specific to ostriches or whether this or a similar species occurs in rheas [37].

A rickettsia, *Aegyptianella pullorum* was reported in young ostrich chicks housed with chickens in Chad. The chicks were febrile with paresis and somnolence before death. *Argus* ticks were implicated in the transmission of the agent from chickens to ostriches [3]. *Aegyptianella* has not been reported in ostriches in North America.

There have been several reports of coccidia in ratites. However, without information regarding size, shape, sporulation, or other morphologic criteria for speciation, or examination of the intestines of the birds for evidence of meronts or gamonts, it is impossible to determine whether the coccidia were parasites of ratites or passing through the digestive tract. Fecal samples from 14 of 165 adult ostriches imported into Canada from Botswana yielded *Cryptosporidium* spp. oocysts [38]. The sporulated oocyst measured 4.0–6.1 × 3.3–5.0 µm and contained four sporozoites. Nomarski interference contrast optics was used to facilitate examination and photography of these coccidia. Oocysts reacted positively to Merifluor-Cryptosporidium (Meridian Diagnostics) immunofluorescent stain which confirmed the infection. No clinical signs were noted in any of the parasitized ostriches. Attempts to infect other avian and mammalian species with the oocysts obtained from ostriches were unsuccessful [39]. Reports from South Africa indicate *Cryptosporidium* as the cause of disease in chicks suffering from phallic or cloacal prolapse [40,41]. Cloacal *Cryptosporidium* have been diagnosed in diarrheic chicks in Texas.

The status of small flagellates in ratites is confusing. *Giardia,* Trichomonads, *Hexamita,* and possibly other genera have been found in diarrheic feces, but there is no experimental evidence to confirm pathogenicity. A wet mount preparation will reveal organisms, if present, in fresh diarrheic feces. Clinically normal in-contact birds may be infected, although with lesser numbers of trophozoites. Trichomonads of cecal origin are considered normal inhabitants of the bowel. Necrotizing typhlitis in rhea chicks with high flock mortality was associated with a *Serpulina*-like organism and *Trichomonas* sp. in the ceca and colon [42].

Flagellates of the small intestine are more likely to be pathogens than protozoa of the cecum. Some of these organisms have been associated with the fading chick syndrome in ostriches. Attempts to cultivate these organisms have met with poor results, as they are extremely fragile and die quickly after being removed from the host. The morphologic characteristic of these organisms have not been well described. Treatment of ostrich and rhea chicks with metronidazole or Fenbendazole have met with varying success, but no controlled experiments have been published as to the pathogenicity of isolates or the value of chemotherapy.

A common protozoal parasite of many gallinaceous species, *Histomonas meleagridis,* has been reported in rheas and ostriches [43,44]. This parasite is transmitted either directly in the feces of contaminated birds or via the eggs of a paratenic host, the cecal nematode, *Heterakis gallinarum*. Earthworms also harbor this parasite. This parasite can survive within cecal worm eggs, healthy earthworms, or in soil, for at least 2 years. After ingestion protozoan parasite begins replication within the cecal tissues. These parasites migrate in the cecal tissues, and produce severe ulceration with inflammation and extensive necrosis. Infected rheas show depression, anorexia, and yellowish diarrhea with death occurring 3–5 days after the onset of signs [45]. This disease is controlled by rigorous sanitation. Carbarsone and dimetridazole have been used to treat turkeys and peafowl and may be beneficial in ratites. Unfortunately, both drugs have been withdrawn from the market. Metranidazole, although expensive, may be substituted.

Amoebae have been identified in the feces of both clinically normal and diarrheic ostriches. Amoeba vary considerably in size but are approximately 15–30 um in diameter with a single nucleus, similar to those of the family Entamoeidae. There is no evidence that these organisms cause disease as neither invasion of tissues nor ulceration of the intestinal tract have been observed.

Balantidium spp. are ciliates which inhabit the digestive tract of many reptiles, fish, and mammals, including man. Pathogenicity in ratites is uncertain but has been reported in dogs, man, and other primates. *Balantidium struthionis* was first described and named in 1934 when it was found in the intestines of an ostrich held in captivity in the zoological garden in Druid Hill Park, Baltimore [46]. The ciliates are relatively small (45–70 μm × 35–48 μm), with a comparably large macronucleus with a deep depression in one side, resembling the letter "B." These organisms have been seen in young birds, especially associated with a diarrhea. In southern Africa, *Balantidium* spp. has been associated with typhlitis, but this has not been reported in North America. *Balantidium* spp. may be opportunistic organisms since large numbers are observed in association with diarrhea. Cysts are commonly seen and have been mistaken for coccidia or other organisms (Figure 12.9). *Balantidium* spp. have been found in the ceca and stool of young rheas about 4–5 weeks of age but their importance as a pathogen is unknown. These are much larger (80 × 55 μm) than other *Balantidium*.

The protozoan, *Blastocystis,* has been found in the ceca of ostriches [47,48] and rheas, but the significance of this protozoan is unknown. The organisms were round to oval and 6–10 μm in diameter with a thick (0.5 μm) fibrillar coat. The organisms contained a single, small round nucleus with an elliptical concentration of condensed chromatin. A single cyst form was seen by light microscopy. More cysts may be seen in older samples as encystation appears to take place in the environment.

Toxoplasmosis was reported in an 8-month-old rhea with bloody diarrhea and anorexia. The bird was serologically positive for *Toxoplasma gondii,* but treatment with trimethoprim sulfadiazine (50 mg/kg) bid for 9 days led to an uneventful recover [49].

CONCLUSIONS

Newly acquired birds must pass through a quarantine facility where they can be screened for ectoparasites. An examination of feces (not urates) should be carried out to determine the presence of helminths. In addition to eggs, the cysts of some protozoa can also be observed on fecal flotation. Ratites should be examined on arrival and at 3 and 6 weeks thereafter. Only birds free of parasite eggs or ectoparasites should be mixed with the flock. If the birds are infested or infected, they should be treated and remain in quarantine for an additional 6 weeks before reexamination to confirm freedom from infection.

Ratites, especially emus, should be raised in an environment separated from contact with either wild, exotic or domesticated birds or animals, and vermin. Areas where gnats and black flies congregate should not be used to rear ratites, as effective screening of the flock from swarms of these insects is not practical.

It is necessary to conduct a thorough necropsy on birds, irrespective of the reason for death. Some significant parasites have been found on careful examination of cases submitted for routine or insurance postmortem. Parasites have been identified and described even though they did not contribute to the death of the bird.

The submission of several sick or dead birds, especially when immature, for necropsy will increase the probability of obtaining an accurate diagnosis. This is less expensive than random treatment with a wide range of anthelmintics and antibiotics. Often when birds are submitted in extremis or for necropsy, parasites or primary pathogens cannot be identified. It is important to actually identify parasites in order to implement successful control measures.

As a general rule, it is best to house birds of a similar age together but not to follow a set pattern of moving birds to new pens as they get older. Younger birds will acquire parasites carried by the preceding batch of older birds. Leaving pens vacant for extended periods, at least 6 months in winter or 3 months in summer, will assist in elimination of many parasites. Grass in vacated areas should be mowed or grazed by sheep which may act as biological parasite control agents. Rearing a new batch of young birds on ground not used the preceding year is recommended. If this is not possible, soil should be turned and exposed to direct sunlight.

Environmental sanitation is a key to parasite control. The prophylactic use of drugs may be helpful to control specific parasites but only if administered to susceptible birds at the appropriate time. Many of the programs used by ratite owners are ineffective and money is wasted by administering drugs for parasites that are not present or to the wrong birds at an incorrect time.

SPECIMEN COLLECTION AND DIAGNOSTIC METHODS

The species and age of the bird and a relevant history should be provided with all submissions to a diagnostic laboratory. When gross specimens are collected, the organ or tissue where the specimen was collected can assist in identification.

Ratite helminths can be diagnosed using fecal flotation in sugar or salt solutions. A specific gravity of 1.2 is sufficient to identify the category of worm present. Centrifugal flotation should be used with sugar due to the viscosity of the solution. Simple precipitation or centrifugation is satisfactory when using salt solutions. Sodium chloride, zinc sulfate or sodium nitrate are also satisfactory flotation media.

A major problem in diagnosing ratite parasites is the presence of fungal spores and protozoan cysts (especially *Balantidium*) in ostriches and rheas. In addition, a variety of nonpathogenic and exogenous parasite eggs are consumed by these omnivorous birds. Coccidia oocysts, ascarid and capillarid eggs are likely to pass through the intestinal tract as soil contaminants. It is important to examine only the fecal portion of excreta (as opposed to urates) for the presences of cysts or eggs. Multiple sampling may help determine if the structures are actual parasites or contaminants.

Certain eggs, especially stronglid nematodes, are almost certain to be parasites and not artifacts. It is important to differentiate *Libyostrongylus* from *Codostomum* in ostriches. This can be close only by identifying the eggs and identifying larvae. The tail of the infective larvae is unique to *Libyostrongylus douglassii*.

Larvae can be cultured by placing a few grams of fecal material in a 10 cm square of gauze. The ends are folded together to create a bag containing the feces. The bag is secured with string and is suspended in a humidity chamber comprising a 1 liter jar with ± 1 cm water in the bottom. The bag is suspended a few centimeters above the water level, and the chamber is incubated at or slightly above room temperature for a week. The fecal bag is then placed in a Baerman apparatus (funnel with a collecting tube attached) filled with enough warm water to immerse the bag. Larvae will exit the feces and gradually descend into the collection tube during the succeeding 10 hours. The supernatant is decanted and the larvae in the sediment on bottom of the collecting tube are pipetted onto a microscope slide. The larvae should be inactivated and stained using a drop of Lugol's iodine solution which diffuses under the coverslip. Larvae can be evaluated for size and morphology.

If tapeworms are observed on postmortem examination, tissue where the rostellum is located should be removed together with the worm and placed in normal saline solution in a refrigerator overnight to allow relaxation. The next morning, an equal amount of 10% formalin solution can be added to the medium for preservation. This method can also be used for flukes and acanthocephalans. Nematodes should be placed in warmed 70% ethanol to straighten worms and preserve them for subsequent examination.

Ectoparasites should be placed in 70% ethanol. In all cases, more than 3 specimens should be provided, if available, because some mites and lice are identified by characteristics of one sex. There may be considerable sexual and stage-specific polymorphism, requiring examination of a number of individual parasites. Drawings of motile protozoans are helpful in diagnosing an infection. A description of characteristic movements, such as "jerky," "progressive, corkscrew motion through the medium," or "remain in place but rapid flagellar or ciliar motion," can be helpful to a parasitologist.

Feces can be submitted to a laboratory for egg or cyst examination by mixing fecal material with 10 times the volume of 5% formalin. Motile protozoa will not survive shipment. Air-dried smears on microscopic slides, can be submitted which can be stained on arrival at the laboratory.

REFERENCES

1. Anon. Importation of ostriches and other ratites. *USDA APHIS Federal Register* 1991;56:31858–31868.
2. Burger WP. Role of management in the epidemiology of disease: Therapeutic approaches in the sick bird, in *Proceedings*. South African Veterinary Association Biennial National Congress 1992;141–148.
3. Huchzermeyer FW. *Ostrich diseases*. Republic of South Africa, Onderstepoort Veterinary Institute, 1994;8–45.
4. Andre M. *Sarcoptides plumicoles* parasites des autruches. *Acarolgia* 1960;2:556–567.
5. Hoover JP, Lochner FK, Mullins SB. Quill mites in an ostrich with rhinitis, sinusitis and airsacculitis. *Companion Animal Practice* 1988;2:23–26.
6. Weisbroth SH, Seelig AW. Struthiolipeurus rhea (Mallophaga:Philopteridae), an Ectoparasite of the Common Rhea (Rhea americana). *J Parasitol* 1974;60:892–894.
7. Dolensek E, Bruning D. Ratites. In: Fowler ME, ed.

Zoo and wild animal medicine. Philadelphia: WB Sanders Co, 1978;167–180.
8. Suedmeyer WK. Treating feather lice. *J Assoc Avian Vet* 1992;6:202.
9. Anon. Black flies cause costly losses in East Texas ostriches and emus. *Vet Qtrly Rev,* Texas Ag Extension Service, 1993;9:3.
10. Mertins JW, Schlater JL. Exotic ectoparasites of ostriches recently imported into the United States. *J Wildlife Dis* 1991;27:180–182.
11. Barton NJ, Seward DA. Detection of *Libyostrongylus douglassi* in ostriches in Australia. *Aust Vet J* 1993;70:31–32.
12. Reinecke RK. *Veterinary helminthology.* Durban, IL: Butterworths Publishers, 1983.
13. Hoberg EP, Lloyd S, Omar H. *Libyostrongylus dentatus* n. sp. (Nematoda:trichostongyloidae) from ostriches in North America with comments on the genera *Libyostrongylus* and *Paralibyostronglus*. *J Parasitol* 1993;81:85–93.
14. Malan FS, Gruss B, et al. Resistance of *Libyostrongylus douglassi* in ostriches to levamisole. *J So African Vet Assoc* 1988;59:202–203.
15. Fockema A, Malan FS, Cooper GG, et al. Anthelmintic efficacy of fenbendazole against *Libostrongylus douglassi* and *Houttuynia struthionis* in ostriches. *J So African Vet Assoc* 1985;56:47–48.
16. Soulsby EJL. *Helminths, arthropods and protozoa of domesticated animals.* 7th ed. Philadelphia: Lea & Febiger, 1982.
17. Teixeria de Freitas JF, Lent H. "Spiruoridea" parasitos de "Rheiformes" (Nematoda). *Memorias do Instituto Oswaldo Cruz* 1947;45:743–760.
18. Davis JW, Anderson RC, Korstadh, Traever DO. *Infections and parasitic* diseases of wildbirds. Iowa State University Press, 1971.
19. Shaw JL, Moss R. Factors affecting the establishment of the caecal threadworm *Trichostrongylus tenuis* in red grouse (Lagopus lagopus scoticus). *Parasit* 1989;99:259–264.
20. Hudson PJ, Dobson AP. Population biology of *Trichostrongylus tenuis,* a parasite of economic importance for red grouse management. *Parasit Today* 1989;5:283–291.
21. Schmidt GD. *CRC handbook of tapeworm identification.* Boca Raton: CRC Press Inc, 1986.
22. Gruss B, Malan FS, Roper NA, et al. The anthelmintic efficacy of resorantel against *Houttuynia struthionis* in ostriches. *J So African Vet Assoc* 1988;59:207–208.
23. Klos H, Lang EM. *Handbook of zoo medicine diseases and treatment of wild animals in zoos, game parks, circuses and private collections.* New York: Van Nostrand Reinhold Co, 1982.
24. Bartlett CM, Anderson RC. *Paronchocerca struthionus* n.sp. (Nematoda:Filarioidea) from ostriches (*Struthio camelus*), with a redescription of *Paronchocerca ciconiarum* Peters, 1936 and a review of the genus. *Can J Zool* 1986;64:2480–2491.
25. Noda R, Nagata S. *Struthiofilaria megalocephala* gen. et sp. n. (Namatoda:Filarioidea) from the body vacity of an ostrich. *Bull. Univ. Psaka Pref. Ser. B.* 1976;28.
26. Kazacos KR, Winterfield RW, Thacker HL. Etiology and epidemiology of verminous encephalitis in an emu. *Avian Dis* 1982;26:389–391.
27. Kazacos KR, Fitzgerald SD, Reed WM. *Baylisascaris procyonis* as a cause of cerebrospinal nematodiasis in ratites. *J Zoo Wildlife Med* 1991;22:460–465.
28. Law JM, Tully TN, Stewart TB. Verminous encephalitis apparently caused by the filarioid nematode *Chandlerella quiscali* in emus (*Dromaius novaehollandiae*). *Avian Dis* 1993;37:597–601.
29. Granath WO Jr. Fate of the wild filarial nematode *Chandlerella quiscali* (Onchocercidae: Filarioidea) in the domestic chicken. *Poult Sci* 1979;59:996–1000.
30. Robinson EJ Jr. *Culicoides crepuscularis* (Malloch) (Diptera:Ceratopogonidae) as a host for *Chandlerella quiscali* (VonLinstow, 1904) comb. n. (Filarioidea: Onchocercidae). *J Parasitol* 1971;57:772–776.
31. Dunn AM. *Veterinary helminthology* 2nd ed. London: William Heinemann Medical Books Ltd, 1978.
32. Watters CE, Joyce KL, Heath SE, Kazacos KR. *Cyathostoma* infection as the cause of respiratory distress in emus (*Dromaius novaehollandiae*), in *Proceedings.* Assoc Avian Vet 1994;151–155.
33. Greve JH, Harrison GJ. Conjunctivitis caused by eye flukes in captive-reared ostriches. *J Am Vet Med Assoc* 1980;177:909–910.
34. West AF. Studies on the biology of *Philophthalmus gralli* Mathis and Leger, 1910 (Trematoda:Digenea). *The American Midland Naturalist* 1961;66:363–383.
35. Nolen PM, Murry HD. *Philophthalmus gralli:* Identification, growth characteristics, and treatment of an Oriental eyefluke of birds introduced into the continental United States. *J Parasit* 1978;64:178–180.
36. Bennett GF, Huhzermeyer FW, Burger WP, Earle RA. The leycocytozoidae of Southern Africa birds, redescription of *Leucocytozoon struthionis.* Walker 1912. *The Ostrich, South Africa Ornithological Society* 1992;63:83–85.
37. Fantham HB, Porter A. *Plasmodium struthionis* sp. n., from Sudanese ostriches and *Sarcocystis salvelini* sp. n., from Canadian speckled trout (*Salvelinus frontinelis*) together with a record of a *Sarcocystis* in the eel pout (*Zoarces angularis*), in *Proceedings.* Zool Soc London 1943;113:25–30.
38. Gajadhar AA. *Cryptospoidum* species in imported ostriches and consideration of possible implications for birds in Canada. *Can Vet J* 1993;34:115–116.
39. Gajadhar AA. Host specificity studies and oocyst description of a *Cryptosporidium* sp. isolated from ostriches. *Parasitol Res* 1994;80:316–319.
40. Penrith M-L, Burger WP. A *Cryptosporidium* in an ostrich. *J So African Vet Assoc* 1993;64:60–61.
41. Penrith M-L, Bezuidenhout AJ, Burger WP, Putterill JF. Evidence for cryptosporidial infection as a cause of prolapse of the phallus and cloaca in ostrich chicks (*Struthio camelus*). *Onderstepoort J Vet Res* 1994;61:283–289.
42. Hanley RS, Woods LW, Stillian DJ, Dumonceaux GA. *Serpulina*-like spirochetes and flagellated protozoa associated with necrotizing typhlitis in the rhea (*Rhea americana*), in *Proceedings.* Assoc Avian Vet 1994;157–162.
43. Dhillon AS. Histomoniasis in a captive great rhea (*Rhea americana*). *J Wildl Dis* 1983;19:274.

44. Borst GHA, Lambers GM. Typhlohelpatites b., struisvogles (*Struthio camelus*) veroorzarakt door een Histomonas-infectie. *Tijdschr Diergeneeskd* 1985;110:536.
45. McMillan EG, Zellen G. Histomoniasis in a rhea. *Can Vet J* 1991;32:224.
46. Hegner R. Specificity in the genus *Balantidium* based on size, shape of body, and macronucleus with descriptions of six new species. *Am J Hygiene* 1934;19:38–67.
47. Yamada M, Yoshikawa H, Tegoshi T, et al. Light microscopical study of *Blastocystis* spp. in monkeys and fowls. *Parasitol Res* 1987;73:527–531.
48. Stenzel DJ, Cassidy MF, Boreham PFL. Morphology of *Blastocystis* sp. from domestic birds. *Parasitol Res* 1994;80:131–137.
49. Orosz SE, Mullins JD, Patton S. Evidence of toxoplasmosis in two ratites. *J Assoc Avian Vet* 1992;6:219–222.

Chapter 13

Infectious Diseases

Simon M. Shane and Thomas N. Tully, Jr.

INTRODUCTION

Infectious diseases are responsible for losses in the three species of commercial ratites at all ages. Due to the relatively small numbers of birds and their extensive distribution in small flocks throughout North America, it has not been possible to study and document the etiology, epidemiology, and pathogenesis of infectious diseases of ratites. Most of the investigations conducted by diagnostic laboratories have been based on small numbers or individual moribund or dead birds. The relatively high value of ratites during the past 5 years has impeded studies on the susceptibility of the three species to various pathogens and their response to therapy. Lack of experimental subjects, nonavailability of suitable embryo, tissue culture, and other *in vitro* systems to propagate pathogens of ratites have restrained progress and delayed an understanding of disease processes. Much of the research work reported from the Republic of South Africa, Europe, and Israel, although informative, is not directly applicable to the United States due to differences in climate, management, systems, prevailing diseases, and their reservoirs and vectors.

Most of the knowledge concerning the diseases of ratites has been adapted from commercial and exotic birds. It is evident that ratites are susceptible to many diseases of poultry and free-living avian species. The emergence of diseases in ratite populations will be a function of contact with commercial chickens and turkeys, free-living avian reservoirs, vectors, and vehicles of indirect transmission.

Paralleling the development of the poultry industry, the spectrum of diseases affecting ratites will change in response to increased intensification from small groups of multiplier birds to large commercial flocks raised for slaughter. Economic and logistic factors will promote cluster populations of emus and ostriches in integrations located in relatively limited geographic areas. This will create an opportunity for adaptation of existing avian pathogens to ratites.

Trading and speculation leading to extensive and frequent movement of emus, rheas, and ostriches characterized the early growth of the industry. This created the opportunity for dissemination of chronic diseases such as tuberculosis and salmonellosis. Proximity of emus and ostriches to commercial poultry in the future will result in adaptation and transmission of these pathogens to ratites. Velogenic Newcastle disease has been documented in ostriches in Israel, and infectious bursal disease has been diagnosed in ostriches in California and Florida.

Growth of the industry and the development of large flocks will favor emergence of conditions or syndromes caused by more than one pathogen or interaction of pathogens and the environment. Immunosuppressive viruses such as infectious bursal disease and adenoviruses may exacerbate bacterial infections, including colibacillosis and salmonellosis. Environmental stress and confinement in feed lots will in all probability give rise to pasteurellosis and erysipelas analogous to the turkey industry. Clostridial enteritis will occur with greater frequency as a result in changes in diet and interactions among immunosuppressive viruses, systemic and gastrointestinal infections, and parasites. New disease syndromes will be an inevitable result of increased flock and population density.

Recommendations in the chapter on biosecurity relating to single-age placement and hygiene will reduce the probability of infection. The introduction of vaccination programs against specific diseases has ameliorated losses due to equine encephalitis and clostridial myositis. In the future, infectious bursal disease, pox, and Newcastle disease may have to be prevented by immunization. Following the emergence of new diseases such as avian influenza and pasteurellosis, it will be necessary to prepare inactivated autogenous vaccines to control specific infections. Fortunately, the resources of the veterinary profession and the biologics industry will be available to producers. Extensive experience gained in poultry and exotics will be applicable to ratites.

It is only during the past 5 years that pathologists and epidemiologists have documented infectious diseases in ratites. Readers are referred to standard texts on avian diseases [1] and microbiology [2] for general information on specific infectious agents, their classification, and methods of isolation and identification. The following outline highlights infectious agents responsible for specific conditions described in the three commercial ratite species.

BACTERIAL INFECTIONS

Salmonellosis

Etiology and Occurrence.

A variety of *Salmonella* species have been isolated from emus, ostriches, and rheas. Paratyphoid *Salmonella* species, including *S. typhimurium*, reflect contamination of the environment and feed, contact with rodent and wild bird reservoirs, and defects in biosecurity. Emus can be infected with *Salmonella pullorum* and develop antibodies against this pathogen [3], although there have been no documented field cases resulting in mortality. The pathogen has been eradicated from the commercial poultry industry but *S. pullorum* persists in backyard flocks and gamefowl. Isolation of *Salmonella arizonae* from the ovary of a mature ostrich with reproductive dysfunction suggests that this organism may be present in U.S. flocks.

Transmission.

Salmonella can be transmitted directly by contact with clinically normal or asymptomatic infected carriers and their droppings or indirectly through contaminated equipment, trailers, or the environment. Vertical transmission by fecally contaminated eggs has been described. The possibility of transovarial or transoviductal infection to embryos is possible and this mechanism may become apparent with intensification of the industry.

Epidemiology.

Because *Salmonella* sp. are ubiquitous and persist for long periods in soil and feces, infection on a specific farm may result in losses during incubation, brooding, and rearing. Birds can remain as clinically normal carriers infecting penmates and the environment.

Rodents, free-living birds, and mammals serve as reservoirs of salmonellosis, and appropriate pest control and implementation of strict biosecurity will limit the introduction of infection. Deficiencies relating to contamination of eggs and hatchery hygiene will promote transmission of salmonellosis.

Clinical Signs.

Salmonellosis should be considered in cases characterized by elevated embryonic death or posthatch mortality. Affected juveniles and adults may show depression, diarrhea, or die without prodromal signs. It is emphasized that salmonellosis has no specific clinical presentation.

Postmortem Examination.

Dead chicks may show omphalitis, enlargement of the spleen and liver, and generalized vascular congestion. In more advanced cases, nonspecific enteritis may be evident. There are no specific or pathognomonic lesions of salmonellosis.

Diagnosis.

Routine microbiological screening of yolk sac swabs, intestine, liver, and spleen will reveal *Salmonella* sp., providing appropriate enrichment culture and isolation procedures are followed. Cloacal and oviduct swabs from mature breeders and culture of the yolk sac of dead embryos may reveal the presence of *Salmonella* in reproductive tract infections.

Treatment.

Young chicks with intestinal or systemic salmonellosis should be isolated and provided with supportive therapy. Administration of quinolone antibiotics will suppress clinical signs and reduce mortality in flocks but will result in chronic carriers.

Prevention.

Appropriate biosecurity procedures are required to prevent introduction of infection. A whole blood *Salmonella pullorum* plate agglutination test should be performed as a component of the pre- or postpurchase examination or before interstate transport. Routine screening of newly acquired birds by cloacal swab during quarantine is strongly recommended. Microbiological monitoring of incubators, feed, and the soil from pens and routine culture of tissues from all postmortem submissions will indicate the presence of infection and expedite quarantine and decontamination.

The emergence of *Salmonella* infection in the ostrich or emu industries could have profound implications with regard to food-borne infection. Any documented, proven outbreak of salmonellosis in consumers of emu or ostrich meat would be deleterious to the industry. It is incumbent on producers to identify and eliminate salmonellosis from commercial flocks at an early stage in the development of the industry to maintain the image of high quality, wholesome meat.

Erysipelas

Etiology and Occurrence.

Erysipelothrix rhusiopathiae erysipelas has been diagnosed in a large flock of emus showing acute mortality [4]. The causal organism is widely distributed in domestic food producing animals and is a major problem in commercial turkeys. The pathogen is recovered on occasions from backyard chickens, waterfowl, quail, and exotic birds. The organism causes erysipeloid in humans characterized by cellulitis at the site of infection or, in more severe cases, generalized septicemia.

Transmission.

Infection occurs following ingestion of the organism in contaminated soil. Susceptible birds may be infected through small skin lacerations caused by fences or trailers or as a result of injury caused by penmates. Insects have been implicated in transmission of erysipelas in turkeys and waterfowl.

Epidemiology.

It is presumed that free-living birds or vermin may introduce the infection which contaminates the soil of pens and feeders. The pathogen can remain viable in biological material for many months, and sporadic losses will occur among susceptible birds on an affected farm.

Clinical Signs.

Death usually occurs acutely but may be preceded by a short period of depression.

Postmortem Changes.

Generalized septicemia characterized by enlargement of the spleen and liver and congestion of the major vessels. Occasionally, petechiae on the serosal surfaces of viscera may be observed on postmortem examination.

Diagnosis.

Isolation and identification of *Erysipelas* sp. from a heart-blood swab or from spleen or liver tissue.

Treatment.

Penicillin or quinolone antibiotics by the intramuscular route.

Prevention.

Flocks on farms with a history of erysipelas should be immunized with a commercial, formalin inactivated aluminum hydroxide adjuvant bacterin licensed for turkeys. It is noted that administration of this vaccine represents extra label use and a written clearance based on informed consent should be obtained from the owner and insurance company before initiating a vaccination program. No adverse reactions have been noted following administration of the inactivated turkey vaccine in a single flock of emus. It is recommended that emus and ostriches on affected farms should receive the first dose by the subcutaneous route in the dorsum of the neck at approximately 6 weeks of age followed by subsequent doses at 20 and 40 weeks of age. An annual booster prior to onset of the breeding season is recommended.

Colibacillosis

Etiology and Occurrence.

Diverse *Escherichia coli* serotypes are widely distributed in the environment and in domestic animals and birds. Generally, *E. coli* is an opportunist organism, and only a few serotypes are primary pathogens in humans, poultry, and domestic species. *E. coli* is a normal inhabitant of the avian digestive tract.

Transmission.

E. coli contamination of drinking water frequently occurs when a flock is supplied from a shallow or uncased well. Water supplied to ratites should be chlorinated at a level of 1–2 ppm. Water should be monitored at frequent intervals to confirm freedom from mineral impurities and microbial contaminants.

Transmission.

Mechanical transmission of *E. coli* following fecal contamination of the shell or deficiencies in egg and hatchery hygiene can result in congenital omphalitis (Figure 13.1).

E. coli in contaminated water can be responsible for generalized septicemia following exposure to an immunosuppressive infection, a mycotoxin, or under conditions of environmental stress.

Clinical Signs.

Affected chicks are depressed, huddle, and are disinclined to eat and drink.

Lesions.

Affected chicks show an unhealed navel or the presence of an enlarged yolk sac which contains viscous and frequently malodorous exudate.

E. coli septicemia in juvenile or adult birds is characterized by enlargement of the liver and spleen and generalized venous congestion [5]. Necrotizing enteritis has been recorded in immature emus [6].

Peritonitis associated with *E. coli* infection may occur following foreign body penetration of the proventriculus in ostriches, or the ventriculus or intestine in emus. Peritonitis is characterized by fibrinous exudate and, in chronic cases, adhesions of the peritoneum to the serosal surface of viscera. Airsacculitis may be present by extension from the primary lesion.

E. coli may also be isolated from the oviduct of mature breeders, but caseous salpingitis as observed in mature chickens has not been diagnosed in ratites.

Omphalitis

In addition to *E. coli*, a number of other pathogenic bacteria be isolated from the yolk sac in cases of omphalitis. *Pseudomonas* sp., *Klebsiella* sp., *Proteus* sp., *Salmonella sp.*, and *Citrobacter* sp. are associated with embryonic mortality, reduced chick viability, and neonatal death.

Diagnosis.

Isolation and identification of the organism from yolk sac and heart blood swabs is generally diagnostic. Isolation of *E. coli* from cloacal swabs is nonspecific.

Treatment.

An appropriate antibiotic should be included in supportive therapy of birds infected with *E. coli*. Treatment of omphalitis is generally unproductive, since survivors are invariably stunted.

Prevention.

Water should be chlorinated at a level of 2 ppm and should be free of *E. coli* and other bacterial pathogens.

Egg handling and hatchery hygiene and decontamination of the incubators and brooding areas should conform to acceptable industry practice.

Pasteurellosis

Etiology and Occurrence.

Pasteurella multocida, *Pasteurella hemolytica*, and possibly other *Pasteurella* spp. Although there is only limited documentation of pasteurellosis in ratites [7], the organism is widely distributed in avian and mammalian species and will in all probability occur as an opportunist pathogen following intensification of the industry.

Transmission.

Direct contact with infected or recovered carriers. Exposure to contaminated equipment, feed bags, and housing. Vermin and free-living birds serve as reservoirs of infection.

Epidemiology.

By analogy with pasteurellosis in turkey and broiler breeder flocks, pathogenic *Pasteurella* spp. are probably introduced into the environment of flocks by rodents and infected avian species. The organism may then be disseminated during handling and routine management. Environmental stress, immunosuppressive agents, and movement of birds may precipitate clinical pasteurellosis.

Clinical Signs.

In commercial poultry, pasteurellosis may range from an inapparent infection with negligible flock morbidity to episodes of peracute mortality. Severity of pasteurellosis is influenced by climatic and nutritional factors. Clinical signs include depression and death in acute outbreaks. In chronic cases, affected birds may show arthritis, sinusitis, or cellulitis. Otitis interna may result in torticollis.

Postmortem Lesions.

Poultry and exotics which die of acute septicemia show enlargement of the spleen and liver and generalized vascular congestion. In consistency with clinical signs, chronic infection may be characterized by purulent arthritis. Affected turkey and broiler breeders show focal hepatic necrosis and hemorrhagic or caseous oophoritis. Pneumonia is a frequent observation in growing turkeys infected with *Pasteurella* sp. and this change may occur in young ratites.

Diagnosis.

Isolation and identification of the causal organism from grossly affected tissues or heart blood.

Treatment.

Intramuscular administration of tetracyclines in acute cases followed by drinking water medication. There is no legal provision for medicating ratite diets, but turkey feed containing tetracycline at 400 g/ton can be fed for 2 weeks to suppress mortality.

Prevention.

High standards of biosecurity are required to prevent introduction of infection into flocks or farms.

In the event that pasteurellosis becomes a problem in a specific component of the industry or in a localized area, it will be necessary to administer autogenous inactivated bacterins. Live attenuated *Pasteurella* vaccines, as used in the poultry industry, should not be administered to ratites without extensive preliminary evaluation under strictly monitored laboratory and field trials.

Campylobacteriosis

Etiology and Occurrence.

Campylobacter jejuni is ubiquitous and widely distributed in commercial and free-living avian species [8]. *Campylobacter* spp. occur in the mucous layer surrounding enterocytes. Generally, *Campylobacter jejuni* is not pathogenic in poultry, but specific toxigenic strains may cause enteritis in turkey poults, probably in combination with enteric viruses and cryptosporidia. *Campylobacter* spp. are extremely susceptible to desiccation and cannot survive for longer than a few hours outside the host or unless protected by moist biological material.

Campylobacter spp. are pathogenic to humans, and poultry meat is a recognized source of infection. This organism may emerge as a significant contaminant of ratite meat, as the industry concentrates on end-products.

Transmission.

Campylobacter spp. can be introduced into a flock by infected carriers, vermin, free-living birds, and fecally contaminated water and moist soil.

Clinical Signs.

No specific disease entity of ratites has been attributed to *Campylobacter jejuni*. A related organism, *Desulfovibrio*, has been implicated in a single case of cloactitis and intestinal prolapse in young emus on a specific farm. *Campylobacter jejuni* was also isolated from affected 1- to 2-month-old birds in the absence of any other bacterial, viral or protozoal agent.

Clinical Signs and Postmortem Lesions.

No specific documentation is available. Generally, *Campylobacter* sp. infection in commercial and free-living avian species is not associated with obivous lesions.

Diagnosis.

Isolation of *Campylobacter jejuni* from the intestine of clinically normal birds is not considered to be significant. The organism may be regarded as a potential pathogen if isolated from birds showing enteritis in the absence of any detectable bacterial, protozoal, or viral infection. A section of ligated intestine on ice should be submitted to a laboratory for *Campylobacter jejuni* isolation. This requires special selective media and incubation at 40–42°C in a microaerophilic environment. Submission of isolates to a specialized laboratory for serotyping and evaluation of toxin production is recommended if *Campylobacter* is suspected as a primary pathogen.

Treatment.

Campylobacter is sensitive to erythromycin.

Prevention.

Chlorination of drinking water, decontamination of housing and equipment, and maintaining all-in-all-out placement programs may reduce the prevalence of infection.

Mycoplasmosis

Mycoplasmosis is a significant disease in commercial poultry, causing severe depression in growth rate and feed conversion efficiency in both broilers and turkeys. *Mycoplasma gallisepticum* and *M. meleagridis* interact synergistically with *E. coli* and respiratory viruses to cause airsacculitis, resulting in high levels of condemnation at processing. *M. synoviae* in turkeys and broilers results in arthritis. *Mycoplasma cloacale* has been isolated from tracheal swabs of ostriches and emus. The three pathogenic *Mycoplasma* of commercial poultry have not been identified in ratites [9].

It is possible that with intensification of the ratite industry and concentration of birds in feed lots, mycoplasmosis may emerge as a disease of economic significance.

Transmission.

Mycoplasma spp. are transmitted vertically by the transovarian route and by direct contact with clinically affected and previously infected birds. In addition, the organism, although sensitive to environmental exposure, can be readily transmitted on contaminated clothing, feed bags, equipment, and in trailers.

Epidemiology.

Multiple-age placement, deficiencies in biosecurity, maintaining different species in proximity, and trading in live birds contribute to dissemination of infection.

Epidemiology.

Mycoplasmas are associated with erosive disease, exacerbating intercurrent *E. coli* and other bacterial infections. Introduction of pathogenic *Mycoplasma* spp. into commercial ratite flocks will reduce growth rate and reproductive efficiency. Some *Mycoplasma* spp. and the closely related genera *Ureaplasma* and *Acholeplasma* are associated with reproductive infection in turkeys and these organisms have the potential to depress egg production, fertility, and hatchability in ratites. It is emphasized that these pathogens have not been documented in ratites, although there are no appropriate surveillance programs for screening cases of reproductive dysfunction.

Clinical Signs.

No specific syndrome has been associated with mycoplasmosis in ratites. One clinical report has been published on a single emu showing arthritis attributed to *Mycoplasma* infection diagnosed by fluorescent antibody examination of joint exudate.

Postmortem Changes.

No documentation exists relating to mycoplasmosis in ratites. Respiratory infection would be manifested by airsacculitis. Seropurulent arthritis would be observed in the joints of birds infected with pathogenic *Mycoplasma* spp.

Diagnosis.

Isolation and identification of the pathogen from synovial fluid or tracheal swabs.

Despite extensive serologic screening using the serum plate agglutination test, there is no evidence of antibodies to any of the three *Mycoplasma* spp. of chickens or turkeys.

Suitably equipped diagnostic laboratories can perform a highly sensitive polymerase chain reaction-gene probe assay for *Mycoplasma gallisepticum* and *M. synoviae*.

Treatment.

Should treatment of mycoplasmosis be necessary, tylosin or a quinolone antibiotic by the parenteral and oral routes would be effective.

Tuberculosis

Etiology and Occurrence.

Mycobacterium avium has been diagnosed in ostriches and emus in the United States and in ostriches in Canada [10]. The condition occurs in zoo ratites and in commercial operations. Avian tuberculosis is now generally confined to backyard flocks and collections of exotic and ornamental fowl, having been eliminated from commercial poultry. *Mycobacterium avium* can infect pigs and will sensitize cattle to the intradermal tuberculin test, producing a false positive reaction. Immunosuppressed humans are extremely susceptible to *M. avium*.

Transmission.

Infected ratites serve as carriers of infection, disseminating the organism in feces. Free-ranging ornamental birds and backyard poultry are reservoirs of infection. Fecally contaminated clothing, equipment, and trailers can transmit infection, and the organism can remain viable in the soil of pens for up to 12 months.

Epidemiology.

Extensive trading in emu, ostrich, and rhea, maintaining multiage flocks with frequent introduction of different species of ratites and exotic birds, uncontrolled movement of personnel, and deficiencies in biosecurity will promote dissemination of infection. Concentration of large numbers of birds on potentially contaminated soil of auction lots represents an efficient method of transmitting avian tuberculosis.

Clinical Signs.

Tuberculosis remains occult in ostriches and emus. Unlike domestic poultry in which emaciation is noted by atrophy of the pectoral musculature, weight loss is more difficult to detect in ratites. A number of cases presented for diagnosis have shown prolapse of the terminal intestinal tract.

Postmortem Changes.

Characteristic granulomas are present on the pleura, mesentery, intestinal serosa, and peritoneum (Figure 13.2). Enlargement of the spleen and involvement of the liver is evident. Granulomas may be present in the marrow of the tibiotarsus.

Diagnosis.

Demonstration of acid fast organisms in tissues obtained at biopsy or necropsy. The intradermal sensitivity test using *M. avium* tuberculin is considered diagnostic in live birds. Attempts have been made to screen feces for the presence of *M. avium*, but the technique has not been validated under commercial conditions [11].

Treatment.

None recommended.

Prevention.

Maintaining closed flocks or screening newly acquired breeding stock using the intradermal sensitivity test during the quarantine period and maintaining strict biosecurity will prevent introduction of infection.

Tuberculosis has the potential to impede development of the ratite industry. Tuberculosis is a health-related exclusion on most insurance policies. The disease will reduce the reproductive period and life span of ostrich and emu breeders. Currently, there is no program for screening ratites, but the intradermal sensitivity test (Figure 13.3) may have to be introduced to certify flocks free from infection, paralleling the successful eradication of bovine tuberculosis in the dairy industry.

Chlamydiosis

Etiology and Occurrence.

Chlamydia psittaci, a widespread obligatory intracellular pathogen occurs in a wide range of domestic and free-living avian species and domestic animals. Chlamydiosis is responsible for mortality in commercially reared rheas [12], but anecdotal reports suggest that neonatal ostriches are susceptible. In addition to losses in rheas, *Chlamydia psittaci* can infect human contacts, especially individuals with reduced pulmonary capacity or with immune deficiency.

Transmission.

The disease is transmitted by both acutely infected birds and latent carriers. Chlamydiosis is a systemic disease in avian species, and the infective elementary body is excreted in feces. Infection occurs following ingestion of fecal material or consumption of feed or water contaminated with respiratory or ocular exudate of affected birds. Humans and possibly birds can be infected by inhalation of elementary bodies.

Epidemiology.

Serotyping of isolates of *Chlamydia psittaci* obtained from infected rheas in Texas implicated pigeons as the possible source of infection. Free-living birds, ratites, and exotics can serve as latently infected reservoirs which excrete elementary bodies intermittently without showing any clinical abnormality.

Clinical Signs.

Affected rheas show depression and ocular discharge of short duration preceding death. Most cases present as acute mortality without prodromal signs.

Postmortem Changes.

Marked splenomegaly and moderate hepatomegaly. In a number of cases, characteristic lesions of avian chlamydiosis, including peritonitis, pericarditis, and subepicardial hemorrhage, are observed. Individual birds show fibrinous airsacculitis and pulmonary congestion.

Diagnosis.

Gimenez or Machiavelli-stained impressions or sections of spleen and liver reveal characteristic elementary bodies. The organism can be cultured in specific pathogen-free embryonated chicken eggs inoculated by the yolk sac route. *Chlamydia psittaci* can be identified in exudate using fluorescent antibody stain or enzyme immunosorbent assay antigen detection. Antibodies to *C. psittaci* can be assayed using latex agglutination, elementary body agglutination or direct complement fixation. Antibody titer can be used for epidemiologic investigation and to monitor the progress of infection in flocks [13].

Treatment.

Individual birds can be treated with doxycycline by the intramuscular or oral routes. Flocks can receive tetracycline in feed at the level of 400 g per ton. It is noted that current FDA regulations prohibit addition of tetracycline to ratite diets. Medicated turkey diets can be supplied for periods not exceeding 2 weeks.

Prevention.

Maintaining closed flocks and operating with a high level of biosecurity will limit the possibility of introducing chlamydiosis. Elimination of free-ranging birds is recommended, especially pigeons and doves which may co-mingle with ratites at feeders. Newly acquired birds should be quarantined for 60 days. During this period two successive *C. psittaci* antibody titer assays at 30-day intervals will confirm freedom from exposure to the pathogen.

Clostridial Enteritis

Etiology and Occurrence.

A number of *Clostridium* spp., including *C. perfringens* and *C. colinum*, have been isolated from intestines of dead ostriches and occasionally emus. The condition is analogous to clostridial enterotoxemia of commercial poultry [14]. A single case of *Clostridium chauvoei* enterotoxemia has been documented in two zoo ostriches [15]. *Clostridium difficile* was isolated from the liver of one of a batch of 20-day-old ostrich chicks, which died with a history of anorexia and weight loss [16].

Transmission.

Clostridium spp. is a gram positive, spore-forming organism which can remain viable in soil for long periods. There is no information on the frequency of isolation of *Clostridia* from the intestinal tract of normal ratites.

Epidemiology.

Clostridial enterotoxemia occurs following proliferation of a toxin-forming strain in the intestinal tract. Some managemental and nutritional factor predisposing to onset of disease in commercial chickens include abrupt changes in diet, starvation, stress following moving, weighing, or vaccination of flocks. In chickens, *Clostridial enteritis* is often preceded by an intestinal infection, such as coccidiosis which causes gastrointestinal hypomotility. Ingestion of large quantities of soil, substrate or grass clippings may precipitate *Clostridial enteritis* in young ostriches.

Clinical Signs.

Generally, acute mortality is observed, but anorexia and depression of short duration may precede death.

Postmortem Changes.

Intestinal lesions can range from mild hyperemia to extensive pseudomembranous enteritis. In advanced cases, the intestine is distended and the lumen is filled with desquamated epithelium and exudate combined with ingesta. Focal hepatic necrosis may be present.

Diagnosis.

Isolation and identification of *Clostridium* spp. using anaerobic culture.

Treatment.

Administration of zinc bacitracin solution in drinking water to penmates of birds which die of *Clostridial enteritis* should restore normal feeding and activity.

Prevention.

Eliminate managemental deficiencies or nutritional factors which are identified and may be responsible for onset of the condition. Under commercial conditions, frequent outbreaks of necrotic enteritis may be suppressed by inclusion of zinc bacitracin in feed at the level of 30 g/ton. This will require FDA approval based on trials to demonstrate effectiveness and safety appropriate to a minor species. It is anticipated that *Clostridial enteritis* may become a significant problem in ostriches reared under intensive feed lot management.

Infectious Coryza

Etiology and Occurrence.

Haemophilus spp. has been isolated from outbreaks of rhinitis and sinusitis in ostriches in Israel [17]. The condition is analogous to infectious coryza in chickens. Although *H. paragallinarum* is present in the United States poultry population, no cases have been reported in ratites.

Transmission.

Haemophilus sp. is transmitted from clinically infected birds or recovered carriers. Indirect infection through contaminated clothing, equipment or feed bags occurs commonly in commercial poultry. Intraflock transmission in expedited by feeders and waterers.

Epidemiology.

Multiage farms, trading units, or farms where comingling with poultry species occurs are at risk of infection. The organism is extremely susceptible to environmental exposure and does not persist for longer than 24 hours in the absence of a host.

Clinical Signs.

Affected juvenile ostriches show lacrimation and seropurulent exudate from the nares. In advanced cases, sinusitis is evident.

Postmortem Examination.

Seropurulent sinusitis is present.

Diagnosis.

Culture and identification of *Haemophilus* spp. using appropriate microbiological techniques.

Treatment.

Administration of trimethoprim-sulfa or penicillin-dihydrostreptomycin will suppress clinical signs. Recovered birds will in all probability remain carriers of infection.

Prevention.

Appropriate biosecurity, including quarantine and observation, of acquired stock.

If coryza becomes a problem in commercial ratite units, autogenous inactivated bacterins may be required to reduce the severity of infection.

Anthrax

Etiology and Occurrence.

Bacillus anthracis has been diagnosed in ostriches in South Africa, resulting in acute mortality [18].

Transmission.

Ingestion of organisms which can remain viable in the spore form in soil for extended periods.

Epidemiology.

Ostriches and possibly emus are at risk if housed in pens or locations where cattle and horses have previously died of anthrax.

Clinical Signs.

The disease is characterized by peracute mortality without prodromal signs.

Postmortem Changes.

It is necessary to examine a blood smear from a peripheral vessel to exclude anthrax in ostriches submitted for postmortem examination with a history of peracute mortality. Absence of any obvious cause of death, including trauma, and submission from an area or location where anthrax has been previously diagnosed should raise concern.

Postmortem changes will include a grossly enlarged spleen and liver, and generalized venous congestion.

Diagnosis.

Demonstration of characteristic 3–6 micron nonsporulated square-ended bacilli in a blood smear. Isolation and identification of the organism from liver, spleen, and blood.

Treatment.

None.

Prevention.

Ostriches should not be housed on farms with a history of anthrax. Carcasses of birds which die of anthrax should be disposed of by deep burial or incineration. Healthy contact birds should be moved from the affected farm and treated with penicillin-dihydrostreptomycin as a prophylactic measure. If anthrax becomes a problem under commercial conditions, it will be necessary to administer inactivated vaccines.

Necrotizing Typhlitis of Rheas

Etiology and Occurrence.

This condition is attributed to mixed infection with spirochetes and trichomonad-like protozoa [19]. The relative contribution of the two agents is unknown but their effect is apparently synergistic.

Transmission.

It is presumed that immature birds are exposed to the pathogens by ingestion of contaminated soil or feed. Based on experience with *Trichomonas* sp. in *Columbiformes* and *Galliformes*, a chronic carrier state probably exists.

Epidemiology.

Multiage farms, high stocking density, and defects in hygiene predispose to the condition.

Clinical Signs.

Juvenile rheas aged one month and older are susceptible to infection. Birds may show depression and anorexia for a short period before death. Flock mortality may exceed 50%, especially following introduction of older birds or mixing of flocks.

Postmortem Changes.

Distension of the ceca and colon with obvious hyperemia. Fibronecrotic and pseudomembranous changes are present in the mucosa of the colon and the cecum. In advanced cases, ceca cores are present at the apex of the cecum.

Diagnosis.

Histological examination confirming erosion and ulceration of the mucosa with the presence of filamentous bacteria. Elliptical protozoa can be visualized using periodic acid-Schiff stain or by scanning electron microscopy.

Treatment.

Oral metronidazole in combination with parenteral lincomycin reduces mortality in affected flocks.

Prevention.

No specific measures can be suggested other than separation of different age groups, maintaining high standards of hygiene, and complete disinfection of brooder pens before restocking with a group of younger birds.

FUNGAL DISEASES

Mycosis results from an active infection with a fungal pathogen [20]. This condition should be differentiated from mycotoxicosis, which follows ingestion of preformed phytotoxins of fungal origin.

Respiratory mycosis occurs most frequently in neonates and juveniles and is a consequence of shell surface contamination after oviposition, handling, or during storage or incubation. Latent mycoses of the gastrointestinal and respiratory tracts can be activated by environmental or climatic stress, malnutrition, immunosuppression, or concurrent disease.

Aspergillosis

Etiology and Occurrence.

Aspergillus fumigatus and other *Aspergillus* spp.

Transmission.

Exposure of the developing embryo to *Aspergillus* spores, especially during the period from internal pipping through 48 hours after hatch [21].

Epidemiology.

Contaminated nest substrate or mishandling of eggs can result in penetration of the eggshell by *Aspergillus* spores. Cooling of the egg contents after oviposition creates a slight negative pressure which favors penetration through the pores and the outer shell membrane. Generally, spores remain dormant in the air cell and infect chicks at the time of internal pipping. When the shell is broken during hatching, spores are released into the incubator and hatchery environment, resulting in infection of chicks hatching from noninfected eggs. Deficiencies in air flow and hygiene and the practice of retaining chicks in hatchers for periods exceeding 24 hours will promote lateral infection.

Clinical Signs.

Affected chicks are generally anorexic, fail to grow, and demonstrate dyspnea characterized by gasping within two days of hatch. Death usually occurs within the first 2 weeks. In severe cases usually caused by profound contamination of incubators, virtually all chicks hatched in a season may be infected or die of aspergillosis.

Postmortem Lesions.

Aspergillosis is characterized by the presence of 1–2 mm yellow to green colored granulomata of the air sacs and lungs (Figure 13.4).

Treatment.

Therapy is not recommended based on high cost and poor response.

Prevention.

Hygiene in the chain of egg collection and incubation should be evaluated and appropriate improvements effected. Application of disinfectants with fungicidal action in the hatchery and incubators and thorough cleaning of brooder pens will suppress infection. The hatchery environment including setters and hatchers can be monitored using petri dishes containing Sabouraud dextrose medium exposed for 15 minutes.

Sporadic aspergillosis involving the air sacs and lungs may occur in mature ostriches maintained as breeders or in ratites in zoo enclosures.

Clinicians in Australia have described pulmonary aspergillosis in immature ostriches from 6 months of age onwards. Cases reveal a history of overcrowding, dusty feed, or exposure to high ambient temperature and humidity. It is possible that affected birds acquired infection during the late incubation or early brooding period with activation of latent infection by stress. Affected birds show dyspnea, cyanosis, and decreased physical endurance. The condition can be diagnosed by demonstrating *Aspergillus* antibodies in serum using an *Aspergillus* extract as an antigen in an agar gel diffusion test. Treatment of severely affected cases with ketoconazole was attempted with variable results [22].

Aspergillus mycosis can occur as focal granulomata in mature ostriches [23] and emus [24].

Zygomycosis

Ventriculitis and proventriculitis and systemic mycosis can occur following infection with the zygomycetes, including the genera *Basidia*, *Mucor*, and *Rhizopus*. Cases have been described from Israel and the United States.

Transmission.

Ingestion of the infective agent in feed or substrate.

Epidemiology.

Zygomycosis is regarded as an opportunistic infection which occurs with intercurrent, protozoal or bacterial infection or in immunocompromised subjects.

Clinical Signs.

Affected birds show depression, anorexia, weight loss, and decreased fecal output.

Postmortem Changes.

Multifocal ulceration of the mucosa of the distal esophagus, proventriculus, and ventriculus. Occasionally, focal stomatitis is observed [25].

Diagnosis.

Histological examination confirms the presence of hyphae in the mucosa, submucosa, or muscularis of the proventriculus and ventriculus. The fungus can be cultured on Sabouraud's agar. Comprehensive biological examination of tissues generally reveals intercurrent infection with *E. coli*, *Staphylococcus aureus*, *Pseudomonas aeruginosa*, *Klebsiella pneumoniae*, or *Campylobacter jejuni*.

Treatment.

Supportive therapy, including parenteral fluids, broad-spectrum antibiotic, and metronidazole, to suppress concurrent bacterial and protozoal pathogens, respectively.

Prevention.

Identify and prevent risk factors, including exposure to immunosuppressive viruses and eliminate fungal contamination of feed and substrate.

Candidiasis

Etiology and Occurrence.

Candida sp., including *C. albicans* and *C. mucor*.

Epidemiology.

Candidiasis, a fungal infection of the upper digestive tract, usually reflects immunosuppression or malnutrition or may result from prolonged antibiotic therapy.

Clinical Signs.

Anorexia, depression, and loss in weight. On examination, stomatitis characterized by white raised lesions is observed beneath the tongue or on the oropharyngeal mucosa.

Postmortem Examination.

Mild cases reveal focal stomatitis and involvement of the mucosa of the esophagus or proventriculus. Severe cases present with pseudomembranous stomatitis and pharyngitis.

Diagnosis.

Isolation and identification of the causal organism. Microscopic examination of direct scrapings obtained from lesions will show characteristic budding yeast forms. The organism can be isolated and identified using appropriate media. Histological confirmation can be obtained from tissues obtained at postmortem examination.

Treatment.

Ketoconozole or nystatin, together with supportive therapy, is indicated. Concurrent antibiotic therapy, especially long-term tetracycline in drinking water, should be discontinued.

Prevention.

Investigate and resolve possible malnutrition or immunosuppressive infections. Upgrade hygiene and ensure that water is chlorinated to a level of 2 ppm. Isolate affected birds from the remainder of flock. Addition of copper sulphate to drinking water at a level of 1:2,000 may be attempted in large-scale feedlot operations.

VIRAL DISEASES

Eastern Equine Encephalitis

Etiology and Occurrence.

A specific arbovirus which results in a viscerotropic form of the disease in emus [26].

Transmission.

The virus is transmitted from passeriform reservoirs to susceptible birds by mosquito vectors, including *Culiseta melanura*, *Aedes sollicitans* and possibly by *Culicoides* spp. (midges).

Epidemiology.

Emus are a recent introduction to the new world and are highly susceptible to togaviridae, including eastern equine encephalitis virus. Concentration of birds in pens adjacent to swampy areas or exposure in seasons with excessive rainfall can result in a high population of vectors which will transmit the virus to susceptible flocks.

Clinical Signs.

Eastern equine encephalitis is characterized by acute mortality. Birds show depression of 10–14 hours duration followed by recumbency. Terminally, hemorrhagic diarrhea or emesis of blood stained ingesta occurs.

Postmortem Examination.

The lumen of the intestine may contain up to 500 ml of unclotted blood. Extensive petechiation and ecchymoses are observed in the serosa of viscera and pericardium (Figure 13.5).

Diagnosis.

The causal virus can be isolated from spleen, liver, blood, and intestine. Specialized laboratories are equipped to receive and process specimens to diagnose eastern equine encephalitis. A state or federal veterinarian should be consulted for information on submission protocol, packing, and shipping. Because humans are susceptible to eastern equine encephalitis, appropriate precautions should be taken during postmortem examination and subsequent handling of specimens, since birds are viremic at the time of death.

Treatment.

Treatment is not practical in advanced cases. A few individual birds have responded to supportive therapy, including fluid replacement and administration of antibiotics to prevent secondary bacterial infection.

Prevention.

Emus should not be housed in swampy areas or in regions of high rainfall or where local factors lead to large populations of vectors. Emus can be protected against eastern equine encephalitis with an inactivated commercial equine vaccine. It is suggested that a bivalent vaccine containing antigens against both eastern and western equine encephalitis be administered according to the following schedule:

- 6–8 weeks
- 10–12 weeks
- 16–18 weeks
- at 6-months and thereafter in April and September

Administration of the equine product represents extra label use and practitioners should obtain a clearance from owners before vaccinating flocks. A complete equine dose should be injected intramuscularly using appropriate restraint and aseptic technique.

Studies conducted in Louisiana and Florida have shown that seroconversion occurs in 50% of recipients after two consecutive vaccinations. Three vaccinations will result in protective titers in 80% of recipients, although individual birds do not seroconvert. The prevalence of clinical eastern equine encephalitis in an area should be decreased by general application of vaccine. No adverse effects have been recorded on either reproduction or livability following administration of the inactivated equine product.

Western Equine Encephalitis

Etiology and Occurrence.

Western equine encephalitis virus [27].

Transmission.

Mosquito vectors transmit virus from passeriform reservoirs.

Epidemiology.

As for eastern equine encephalitis.

Clinical Signs.

Affected birds show depression, progressing to paresis and in advanced cases, recumbency and paralysis. Flock morbidity ranges from 10–50%, and mortality is usually below 20% in affected birds. Recumbent birds die within 48 hours, but mildly affected emus recover within 2 weeks with supportive therapy.

Postmortem Changes.

In chronic cases, cachexia and hydropericardium are evident. Affected birds may show evidence of gastrointestinal stasis, and secondary bacterial infection resulting in pneumonia.

Diagnosis.

Selection and identification of the specific virus from homogenates of heart, brain, lung, and spleen tissue. Recovered birds demonstrate high antibody titers to western equine encephalitis virus. Acute and recovery phase serum can therefore be used for retrospective diagnosis.

Treatment.

Supportive therapy, including intravenous and subcutaneous 5% dextrose lactate, B-complex vitamins, steroids, and feeding using an esophageal tube.

Prevention.

Administration of bivalent eastern and western equine encephalitis vaccine, according to the schedule for eastern equine encephalitis.

St. Louis Encephalitis

Etiology and Occurrence.

St. Louis encephalitis virus.

Transmission.

Mosquito vectors, including *Culex pipiens* and *Culex tarsalis*.

Epidemiology.

The disease has been reported in the vicinity of Tampa, Florida, where periodic epidemics occur in the human population following heavy rainfall and warm weather which promote large numbers of vectors.

Clinical Signs and Postmortem Examination.

As for Western equine encephalitis.

Treatment.

Supportive therapy.

Prevention.

No vaccine is available. Susceptible birds should be moved from areas where St. Louis encephalitis is endemic.

Avian Influenza

Etiology and Occurrence.

Type A orthomyxoviruses which are classified according to hemagglutinating (HA) and neuraminidase (NA) activity of glycoproteins in the viral envelope. Avian influenza viruses are widespread in free-living and migratory birds which serve as reservoirs for infection. The disease represents a severe threat to commercial poultry. Mild strains suppress growth rate, depress reproductive efficiency, and sensitize chicken and turkey flocks to secondary bacterial infection of the respiratory tract. Highly pathogenic strains may kill up to 90% of susceptible commercial flocks.

H7N1 strain influenza A caused up to 80% mortality in young ostrich chicks in South Africa and Israel [28]. An H5N2 virus in rheas produced a moderate to severe upper respiratory infection. Serologic evidence confirms that emus and rheas are susceptible to infection with a wide range of type A influenza viruses [29].

Transmission.

Influenza virus may be introduced from wild birds which co-mingle with ratite flocks. Direct infection can occur by introducing infected carriers which may or may not be clinically affected at the time of transfer. Indirect infection follows movement of contaminated equipment, feed bags, or transport of birds in improperly cleaned trailers which previously carried infected stock.

Prevention of influenza in ratite flocks requires strict isolation from backyard and exotic poultry. Because of the risk of infection, managers and owners of ratites should not hunt ducks or come into contact with zoo or exotic species. For mutual protection, there should be complete separation of equipment and personnel involved in commercial poultry farming and production of ratites.

Evidence from South Africa suggests that periodic outbreaks of influenza occur in the Klein Karoo Valley in South Africa, the location of the world's most intensive ostrich-rearing industry. It is anticipated that with growth of the industry in the United States, influenza may become a serious restraint to production efficiency, especially with concentration of birds of different ages on the same or adjacent farms.

All avian influenza viruses obtained from ratites have demonstrated relatively low pathogenicity for commercial poultry, but recent studies suggest that the 1994 isolates are more pathogenic than AI virus from wild ducks. Unfortunately, small changes in the amino acid sequence of the carboxy terminus of the HA1 protein can enhance virulence. Isolation of any H5 or H7 serotype is regarded with concern by U.S. regulatory authorities. Quarantine and restriction of interstate transport will invariably follow isolation of potentially pathogenic strains of avian influenza.

Serologic surveys in the United States have shown exposure of ratites to numerous HA and NA serotypes, confirming widespread exposure to influenza virus. Epidemiologic investigations of outbreaks of influenza in ratites during 1993 implicated two auction locations in Texas in dissemination of virus. One auction had 86 in-state buyers and 33 out-of-state buyers, including one from Canada. Follow-up studies revealed antibodies against avian influenza, *Chlamydia*, and *Salmonella* in birds passing through the auction barn.

Clinical Signs.

Ostriches aged 5 days to 12 months and infected with H7N1 strain showed acute mortality associated with respiratory distress and copious green discolored urine. Juveniles in feed lots showed up to 100 morbidity and 80% mortality. Older birds were generally less susceptible to infection, possibly because of previous exposure to H7 virus. Mortality up to 20% was recorded in some feed lot flocks aged 8–14 months old.

Affected emus and rheas in the United States showed depression, ocular and nasal discharge and diarrhea. These signs are not specific for influenza.

Postmortem Changes.

Young ostriches in South Africa and Israel infected with the highly pathogenic H7N1 strain showed multifocal hepatic necrosis, enlargement of the spleen and, in advanced cases, nephrosis and fibrinous airsacculitis.

Diagnosis.

The virus can be isolated from tracheal and cloacal swabs using specific pathogen-free chicken eggs inoculated on the tenth day of incubation via the chorioallantoic route. Viral isolates causing embryo death and shown to hemagglutinate avian erythrocytes are subtyped using a panel of antisera. Isolation and identification of influenza virus is conducted by designated state laboratories and the National Veterinary Services Laboratory in Ames, Iowa.

Serum of live birds can be screened for influenza antibodies using the agar gel diffusion procedure.

Treatment.

Mild influenza can be treated by supportive therapy, including fluids and antibiotics to prevent secondary bacterial infection.

Highly pathogenic influenza does not respond to treatment. In the event of a declared disease emergency, compulsory depopulation and disposal of carcasses may be mandated by U.S. federal authorities.

Prevention.

The risk and consequences of infection require operation of units as closed flocks with strict control over entry of personnel and equipment. Newly purchased birds should be quarantined and screened for antibodies to avian influenza. Contact with free-living birds should be prevented.

Farmers and managers should restrict visits to other ratite units or auction barns to a minimum. Appropriate protective clothing and decontamination procedures should be used to prevent introduction of infection.

In some areas, inactivated autogenous vaccines are used to protect commercial poultry flocks from clinical influenza. Immunization does not prevent infection or shedding of virus and inactivated vaccines may promote selection of more pathogenic variants. A recent change in USDA regulations allows the manufacture of H5 and H7 strain influenza vaccines in the United States. Acceptable control of endemic highly pathogenic H5N2 influenza is being achieved in Mexico in commercial chickens using an inactivated vaccine.

Infectious Diseases

Newcastle Disease

Antigenically related paramyxovirus type 1 strains are responsible for lentogenic (mild), mesogenic (moderate), and velogenic viscerotropic (severe) forms of Newcastle disease in domestic poultry. Velogenic viscerotropic Newcastle disease is exotic to the United States but occurs in Mexico and other countries in Central and South America and in Africa and Asia. Outbreaks of velogenic Newcastle disease with nervous manifestations have been reported in commercial ostrich units in Israel and in rheas in a zoo in Brazil. The literature documents infection of 236 avian species, confirming the widespread susceptibility of birds to Newcastle disease virus.

Transmission.

Newcastle disease virus can be spread by the aerogenous route for up to 2 miles. Ratite flocks are susceptible to indirect infection by transfer of contaminated equipment or personnel. Backyard poultry, free-living, and migratory birds serve as reservoirs of Newcastle disease virus.

Epidemiology.

Ostriches and possibly other ratites may become infected with Newcastle disease due to proximity to large concentrations of diseased poultry. This situation led to velogenic Newcastle disease in ostriches in Israel where the disease is endemic [30]. In the context of the United States, it is more probable that widespread lentogenic Newcastle disease may affect ratites in areas where large numbers of poultry and exotic species are held in proximity. Breeding units and feed lots in the arid parts of the United States with sparse poultry populations will probably remain free of Newcastle disease in the foreseeable future.

Clinical Signs.

Velogenic Newcastle disease in immature ostriches results in high morbidity and up to 50% mortality. Affected birds show neurologic signs characterized by torticollis, incoordination, and recumbency. Up to 80% mortality occurs following experimental infection of 3-month-old ostrich chicks. Although there are no records of infection in adult birds, it may be presumed that cessation of mating and egg production occurs following exposure to velogenic Newcastle disease virus.

Postmortem Changes.

No specific gross lesions are present in acutely affected cases. Histopathology includes edema, neuronal degeneration, and perivascular lymphocytic cuffing in the brain.

Diagnosis.

Isolation and identification of paramyxovirus in homogenates of brain tissue of affected birds using 10-day-old embryonated specific pathogen-free eggs inoculated via the yolk sac route.

Surviving birds develop antibodies within 10 days, which can be assayed using the hemagglutination inhibition procedure.

Treatment.

Supportive therapy, including fluids and antibiotics to suppress secondary bacterial infection.

Prevention.

In areas where velogenic Newcastle disease is endemic, Hitchner B1 strain, live attenuated Newcastle vaccine can be administered to ostriches by the intraocular route. Birds should be vaccinated at approximately 14 and 30 days of age followed by an inactivated oil emulsion product administered by the subcutaneous route at 45 days. It is emphasized that ostriches should not be vaccinated against either Newcastle disease or influenza in the United States, as there is currently no risk of infection with the highly pathogenic forms of these diseases. The presence of antibodies stimulated by vaccination will complicate epidemiologic studies which may have to be conducted to determine the distribution of infection in the event of an outbreak of either velogenic Newcastle disease or highly pathogenic influenza. Ratites with antibodies to these diseases as a result of vaccination may be quarantined or killed by federal authorities in the event of an exotic vvND eradication campaign.

Adenovirus

Adenoviruses are responsible for a number of severe and economically significant diseases in commercial poultry, including hemorrhagic enteritis in turkeys, quail bronchitis, egg drop syndrome, and inclusion body hepatitis-hydropericardium syndrome in chickens. Adenoviruses have been isolated from young ostriches showing high mortality in Oklahoma and adjoining states [31].

Transmission.

Adenoviruses can be transmitted vertically by the transovarian route. In addition, adenoviruses can be transmitted laterally from infected chicks and possibly asymptomatic carriers. In chickens, adenovirus infection can remain latent until onset of reproduction resulting in both vertical and horizontal transmission.

Maintaining open flocks with extensive trading, purchase of adult breeders without specific knowledge of reproductive history and livability of progeny, and defects in biosecurity contribute to dissemination of virus.

As with adenoviral infections of poultry, coinfection with other agents may influence the severity of infection. Immunosuppressive agents, including infectious bursal disease virus and mycotoxins, and as yet undiagnosed pathogens, may exacerbate adenoviral infection and contribute to morbidity and mortality.

Clinical Signs.

Acute mortality occurs in ostrich chicks commencing at approximately 2 months of age. Birds show marked depression and anorexia with chalky-gray malodorous diarrhea. Up to 90% mortality has been recorded in affected rearing units.

Postmortem Changes.

Multifocal enteritis, fibrinous airsacculitis and pulmonary congestion are observed. Histopathology reveals necrotizing enteritis and hepatitis and lymphoid depletion of the spleen. One case involving many hundred ostrich chicks revealed hypoplasia of bile duct epithelium, suggesting previous exposure to aflatoxin as a risk factor or progenitor of adenovirus infection.

Diagnosis.

An adenovirus was isolated from the liver of affected chicks in a specific U.S. case in addition to a wide range of bacterial pathogens including various enterobacteriaciae from the intestine, liver and respiratory tract.

Treatment.

No specific treatment has been devised, although supportive therapy is recommended. Surviving chicks will probably remain carriers of the virus.

Prevention.

Epidemiologic studies show that chicks from specific hens can introduce infection into growout units. Subsequent horizontal infection may then result in high mortality in unrelated chicks. It is therefore necessary to operate flocks on a closed basis and to implement appropriate biosecurity precautions.

Adenovirus infection represents a potential threat to the development of the ostrich industry. It will be necessary to develop procedures to identify infected parents and to rear their chicks in isolation. Development of an inactivated vaccine may be necessary to suppress clinical disease in ratites raised under intensive conditions. Inactivated adenoviral vaccines are used commercially in chickens to reduce the economic impact of egg drop syndrome and inclusion body hepatitis. Considerable work will be necessary to define the etiology, pathogenesis and control of adenoviral infection in ostriches and possibly other ratites.

Infectious Bursal Disease

Etiology and Occurrence.

The avibirnavirus responsible for infectious bursal disease in chickens has been isolated from cases of "ostrich chick fading syndrome" in California [32] and Florida. Infectious bursal disease virus is immunosuppressive and is responsible for extensive losses in the commercial egg and broiler industries worldwide.

Transmission.

Direct contact between infected poultry and ratites may have promoted adaptation of the virus to young ostriches. Infectious bursal disease virus is resistant to environmental exposure and can persist for 90 days in the earth floors of incompletely decontaminated barns and pens.

Clinical Signs.

Birds show depression, anorexia and diarrhea over a 3- to 4-day period. Terminally, birds assume sternal recumbency and may show coarse muscular tremor and abnormal movement of the head.

Postmortem Changes.

These are nonspecific and include pulmonary congestion, enteritis and, on occasions, airsacculitis reflecting secondary bacterial infection. Histological examination of the bursa of Fabricius shows severe atrophy consistent with the pathognomonic lesion in chickens.

Diagnosis.

Isolation and identification of infectious bursal disease virus. Serologic surveys have not been conducted to determine the extent of infection in U.S. flocks.

Treatment.

Only supportive therapy is recommended.

Prevention.

Biosecurity may prevent introduction of the virus onto a farm and is currently considered the most important preventive measure. It is emphasized that producers should not administer commercial chicken IBD vaccines to ratites. The mode of action of a live attenuated vaccine in ostriches is unknown and there is a distinct possibility that more virulent viruses can evolve by passage of the live attenuated chicken vaccine strain through an unnatural host. The control of infectious bursal disease in commercial poultry is extremely complicated and involves the use of vaccines with different levels of antigenicity and pathogenicity in coordinated programs monitored by comprehensive serologic flock profiling [33].

Ostrich Pox

Etiology and Occurrence.

Avipox viruses occur extensively and have been recorded in a wide range of domestic and exotic avian species. The disease has been diagnosed in ostriches in the Republic of South Africa [34], Israel [35] and the United States.

Transmission.

Avipox viruses are transmitted by mosquitoes. The occurrence of the disease is a function of climatic conditions, including temperature and rainfall and the presence of reservoir hosts and the mosquito vector. There is evidence that horizontal transmission can occur by physical contact between infected and susceptible birds.

Epidemiology.

Contact between high concentrations of poultry and ratites and the presence of vectors contribute to outbreaks of pox.

Clinical Signs.

Chicks can be affected from 2 weeks of age onward. The cutaneous form is characterized by the appearance of 5 mm to 2 cm proliferative lesions on the eyelids and adjacent to the nares (Figure 13.6). The diphtheritic form which produces tracheitis and stomatitis in poultry results in dyspnea characterized by gasping and open-beak breathing. Affected chicks become anorexic and dehydrated. Flock losses of up to 50% in a season have been recorded in areas where the disease is endemic.

Postmortem Lesions.

Characteristic diphtheritic lesions may be observed in the trachea and oropharynx. Proliferative cutaneous lesions up to 2 cm in diameter may be present on the margin of the beak and eyelids (Figure 13.7). Histological examination shows characteristic intracytoplasmic inclusion bodies.

Diagnosis.

Histopathology is diagnostic. If required, the virus can be isolated and identified by inoculating the chorioallantoic membrane of 10-day-old embryonated eggs derived from a specific pathogen-free flock.

Treatment.

Supportive therapy, including application of antibacterial compounds to skin lesions.

Prevention.

In areas where the disease occurs ostrich chicks should be vaccinated using a commercial fowlpox vaccine by the intradermal route in accordance with the manufacturer's recommendations relating to reconstitution and administration of vaccine. Control of mosquitoes may be attempted in brooding units.

Viral Enteritis

Etiology and Occurrence.

A number of viral agents, including adenoviruses and coronaviruses [36], have been isolated from the intestinal tract of young ostriches showing high mortality within 1–3 weeks of hatch. By analogy with turkeys, it is apparent that a number of viral agents may act synergistically to produce a clinical syndrome characterized by anorexia, diarrhea, dehydration, and elevated mortality.

Transmission.

Viral agents are probably transmitted from clinically normal reservoirs in the flock or alternatively environmental infection may occur. Some viruses, including adenoviruses and reoviruses, may be transmitted by the vertical route. Intraflock transmission is usually rapid and may be accentuated by high stocking density or defects in biosecurity and hygiene.

Viral pathogens may interact with cryptosporidia or enteric bacteria, including *Aeromonas* spp., *E. coli*, and *Clostridium* spp., which serve as opportunistic pathogens. Lack of documentation and structured investigation of field outbreaks of posthatch mortality precludes a more detailed understanding of the epidemiology of enteric infections.

Clinical Signs.

There is a history of high mortality in successive breeding seasons in stock aged 1–4 weeks. All three ratite species show depression, anorexia, diarrhea, and dehydration. Terminally, birds may show incoordination or coarse muscle tremor.

Postmortem Examination.

Dehydration is evident on examination. Mild enteritis characterized by hyperemia of the jejunum is usually present, but there are no specific or pathognomonic lesions. In advanced cases, clostridial infection may present as ulcerative enteritis.

Diagnosis.

Routine aerobic or special anaerobic bacteriology may identify secondary bacterial agents. Electron microscopy or attempts at viral isolation may provide an indication of the range of viral pathogens present. Histological examination may reveal the presence of inclusion bodies, suggestive of specific viral agents.

Treatment.

Supportive therapy, including fluids, tube feeding and antibiotics, may suppress mortality.

Prevention.

In the absence of a specific etiology, it is not possible to provide preventive recommendations. Correlating high mortality with individual matings may identify vertically transmitted agents. Eliminating obvious deficiencies in management and hygiene may improve livability.

Borna Disease

Etiology and Occurrence.

Borna disease virus is responsible for high mortality in 2- to 8-week-old ostrich chicks reared under intensive conditions in Israel [37].

Transmission.

It is presumed that insect vectors are responsible for transmission of the causal virus.

Epidemiology.

The disease appeared in brooding units operated by collective farms after initiation of intensive rearing operations in the late 1980s. Mortality exceeded 50% of the annual crop on one farm in 1990, but a decline to 20% followed introduction of preventive immunization in 1992. It is emphasized that Borna disease has not been diagnosed in the United States, but neurotropic viruses, including the alphaviruses, Newcastle disease and avian encephalomyelitis may be introduced into ratite populations in the future.

Clinical Signs.

Affected chicks show paresis characterized by a disinclination to move. This progresses to paralysis within 4–8 days and affected birds die of dehydration.

Postmortem Examination.

No specific diagnostic gross lesions are present.

Diagnosis.

Microscopic examination of the brain shows neuronal degeneration and lymphocytic perivascular cuffing. A specific enzyme linked immunosorbent assay has been used in Israel to demonstrate the presence of virus in brain tissue. The disease can be reproduced in young susceptible ostriches by intramuscular injection of brain tissue from affected birds.

Treatment.

Early cases can be provided with supportive therapy. Hyperimmune serum from recovered birds containing specific Borna virus antibody is therapeutic in the early stages of the disease.

Prevention.

Should this disease emerge in the United States, or in countries other than Israel, inactivated viral vaccines will have to be developed to immunize parent stock to ensure solid maternal immunity in progeny. Chicks may also have to be immunized at approximately 3 to 4 weeks of age following waning of acquired antibody. This approach will require extensive epidemiologic investigation and the development of a suitable antibody assay.

MISCELLANEOUS VIRAL INFECTIONS

Investigations of cases of fading chick syndrome have revealed various viral agents consistent with studies on commercial broilers and turkeys with stunting syndrome. The significance of these viral isolates is unknown, and it is doubtful whether any of the agents as a single entity is capable of reproducing a specific disease. It is possible that potential pathogens, including rotaviruses, reoviruses, parvoviruses, coronaviruses, and picornaviruses, serve as opportunistic pathogens. Their presence in the intestinal and respiratory tracts may result from immunosuppression or intercurrent infection with bacterial pathogens. The isolation of a wide range of viruses is a function of diagnostic resources, including electron microscopy, tissue culture, and isolation using SPF chick embryos. It is possible that many viral infections of emus and ostriches which currently occur and will emerge as clinical entities as the United States industry matures will be diagnosed using conventional egg and tissue culture systems used for poultry pathogens. The viruses responsible for disease in ostriches, emus, and rheas will in all probability originate in commercial and backyard chickens and turkeys.

In the rapidly developing field of ratite diseases, it must be noted that isolation of a virus from one or more birds or cases of a specific disease or syndrome does not necessarily confirm the etiologic relationship with the condition. The high value of emus and ostriches up to the present time, has precluded structured trials on sufficient numbers of susceptible birds to establish Koch's postulate regarding causation. Experiments carried out in Israel and South Africa, have demonstrated the infectivity of avian influenza and Bornavirus. With the emergence of an end-products market and the expansion of the ostrich and emu industries, research workers will require access to experimental subjects to develop pathogen-free flocks for disease studies. It is important that the ratite industry fund research activities directed to identify significant pathogens affecting the three commercial species.

References

1. Calnek BW, Barnes HJ, Beard CW, Reid WM, Yoder HW. *Diseases of poultry*, 9th ed. Ames, IA: Iowa State University Press, 1991.
2. Purchase HG, Arp LH, Domermuth CH, Pearson JE, eds. *Laboratory manual for the isolation and identification of avian pathogens*, 3rd ed. Kennett Square, PA: American Association of Avian Pathologists, 1989.
3. Tully TN, Shane SM. Salmonella pullorum seroconversion in emus (*Dromius novaehollandiae*), in *Proceedings*. Assoc Avian Vet 1993;315–317.
4. Griffiths G, Buller N. *Erysipelothrix rhusiopathiae* infection in semi-intensively farmed emu. *Austr Vet J* 1991;68:121–122.
5. Rao MS, Ramachandra RN, Ziauddin S, Raghavan R. Colibacillosis in an ostrich (*Struthio camelus*). *End J Comp Micro Imm & Inf Dis* 1981;2:40–41.
6. Baldwin CA, Hines ME, Styer EL, Cole JR. A mixed infection in an emu with necrohemorrhagic enteritis (Abst), in *Proceedings*. 37th Ann Meeting, American Association of Veterinary Laboratory Diagnosticians, Grand Rapids, MI, 1994;70.
7. Akoha A. An outbreak of pasteurellosis in a Kano zoo. *J Wildl Dis* 1980;16:3–5.
8. Shane SM. The significance of *Campylobacter jejuni* infection in poultry: A review. *Avian Path* 1992;21:189–213.
9. Mohan R. Mycoplasma in ratites, in *Proceedings*. Assoc Avian Vet 1993;294–296.
10. Shane SM, Camus A, Strain MG, Thoen CO, Tully TN. Tuberculosis in commercial emus (*Dromaius novaehollandiae*). *Avian Dis* 1993;37:1172–1176.
11. Clark SL, Colllins MT. Detection of *Mycobacterium avium* in fecal samples of birds using a modified Bactec system, in *Proceedings*. Annual Meeting, American Association of Veterinary Laboratory Diagnosticians, Grand Rapids, MI, 1994;60.
12. Camus AC, Cho DY, Poston RP, Paulsen DP, Oliver JL, Law JM, Tully, TN. Chlamydiosis in commercial rheas (*Rhea americana*). *Avian Dis* 1994;38:666–671.
13. Grimes JE, Arizmendi F. Case reports of ratite chlamydiosis: An update on the chlamydias, in *Proceedings*. Assoc Avian Vet 1994;133–140.
14. Shane SM, Gyimah JE, Harrington KS, Snider III TG. Etiology and pathogenesis of necrotic enteritis. *Vet Res Comm* 1985;9:269–287.
15. Lublin A, Mechani S, Horowitz HI, Weisman Y. A paralytic-like disease of the ostrich (*Struthio camelus masaicus*) associated with *Clostridium chauvoei* infection. *Vet Rec* 1993;132:273–275.
16. Shivaprasad HL. Necrotizing hepatitis associated with *Clostridium difficile* in an ostrich chick, in *Proceedings*. 37th Annual Meeting, American Association of Veterinary Laboratory Diagnosticians, Grand Rapids, MI 1994;67.
17. Perelman B. Upper respiratory disease, in *Proceedings*. 3rd Annual Ostrich Conference, College of Veterinary Medicine, Texas A&M University 1991.
18. Snoeyenbos GH. Anthrax. In: Biester HE, Schwarte LH, eds. *Diseases of poultry*, 5th ed. Ames, IA: Iowa State University Press, 1965;432–435.
19. Hanley RS, Woods LW, Stillian DJ, Dumonceaux GA. Serpulina-like spirochetes and flagellated protozoa associated with a necrotizing typhlitis in the rhea (*Rhea americana*), in *Proceedings*. Assoc Avian Vet 1994;157–162.
20. Perelman B, Kuttin ES. Fungal infections in ostriches, in *Proceedings*. Western Poultry Disease Conference, Davis, California 1994;21.
21. Marks SL, Stauber EH, Ernstrom SB. *Aspergillus* in an ostrich. *J Am Vet Med Assoc* 1994;204:784–785.
22. Black D. Aspergillosis in ostriches in Australia, in

Proceedings. Annual Symposium for Veterinarians, Texas A&M University 1993.
23. Fitzgerald SD, Moisan PG. Mycotic rhinitis in an ostrich, in *Proceedings*. N Central Avian Dis Conf, Lansing, MI, 1993;89.
24. Chakravarty IB. A case history of mycotic infection (aspergillosis) in an emu (*Dromiceius novaehollandies*) in Delhi zoo. *Indian Vet J* 1976;53:881–882.
25. Jeffrey JS, Chin RP, Shivaprasad HL, Meteyer CU, Droual R. Proventriculitis and ventriculitis associated with zygomycosis in ostrich chicks. *Avian Dis* 1993;38:630–634.
26. Tully TN, Shane SM, Poston RP, England JJ, Vice CA, Cho DY, Panigrahy B. Vicerotropic eastern encephalitis in a flock of emus (*Dromaius novaehollandiae*). *Avian Dis* 1992;36:808–819.
27. Ayers JR, Lester TL, AB Angulo. An epizootic attributable to Western equine encephalitis virus infection in emus in Texas. *J Amer Vet Med Assoc* 1994;205:600–601.
28. Allwright DM, Burger WP, Gayer A, Terblanche AW. Isolation of an influenza A virus from ostriches (*Struthio camelus*). *Avian Path* 1993;22:59–65.
29. Slemons RD, Fischbach WL, Goclan SA. Pathogenesis of ratite-origin influenza virus infection in chickens (ABST), in *Proceedings*. Amer Vet Med Assoc 1995;153.
30. Samberg Y, Hadash D, Perelman B, Meroz M. Newcastle disease in ostrich (*Struthio camelus*): Field case and experimental infection. *Avian Path* 1989;18:221–226.
31. Raines AM. Adenovirus infection in the ostrich (*Struthio camelus*), in *Proceedings*. Assoc Avian Vet 1993;304–312.
32. Chin RP, Woolcock PR. Identification of birnavirus-like particles from the intestines of 8-week-old ostriches (Abst), in *Proceedings*. Western Poultry Disease Conference, Sacramento, CA 1994;110.
33. Lasher HN, Shane SM. Infectious bursal disease. *World's Poul Sc J* 1994;50:133–166.
34. Allwright DM, Burger WP, Gayer A, Wessels J. Avian pox in ostriches. *J S Afr Vet Ass* 1994;65:23–25.
35. Perelman B, Gur-Lavie A, Samberg Y. Pox in ostriches. *Avian Path* 1988;17:735–739.
36. Frank RK, Carpenter JW. Coronaviral enteritis in an ostrich (*Struthio camelus*) chick. *J Zoo Wildl Med* 1992;23:103-107.
37. Weisman Y, Malkinson M, Perl S, Ashash E, Meir R, Nir A, Ludwig H. Borna disease in ostriches (Abst), in *Proceedings*. Western Poultry Disease Conference, Sacramento, CA 1994;22.

Chapter 14

Developmental Problems in Young Ratites

Brian L. Speer

INTRODUCTION

Developmental problems in growing ratites are commonly encountered. Unfortunately, the etiology and epidemiology of these conditions are inadequately understood from a veterinary perspective, restraining effective therapy and preventive management. Nutritional factors (see Chapter 2), genetic disease, and environmental factors often combine to produce a specific developmental problem. These conditions tend to be inconsistently linked with common etiologies in the United States and elsewhere. As an example, a nutritional specification or management practice associated with excellent performance in one location may produce suboptimal results under different circumstances. This chapter considers common categories of noninfectious developmental problems ranging from abnormal embryonic positions through abnormalities encountered during the juvenile period.

MALPOSITIONS

There are several types of embryonic malpositions encountered in ratites. In malposition I, the head is located between the thighs, and the chick fails to lift and rotate the head to the right during the last days of incubation. Excessively high incubation temperature has been suggested as a possible explanation for the high prevalence of type I malposition. Although this abnormality is regarded as lethal in some avian species [1], it is not invariably fatal in ratites. Early recognition of this malposition allowing prompt insertion of an air hole in the shell to allow the chick to breathe may permit normal hatch. Labored or muffled breathing by malpositioned embryos may be auscultated by an experienced incubationist. Malpositioned chicks often show weaker than normal movement and hatching efforts. The sounds and movement of the chick before anticipated hatch may identify malposition type I prior to death of the embryo.

In malposition II, the head is located at the small end of the egg, and the chick is in the "breech" position. Hatchability of domestic chickens is reduced by approximately 50% with this abnormality. The occurrence of this problem is believed to be related to position of the egg and low incubator temperature [1]. Embryonic mortality in ratites may be lower than in poultry with this malposition. Recognizing the problem in a timely manner and providing an air hole which allows the chick to breathe will be helpful in reducing embryonic mortality. Embryos with type II malposition frequently demonstrate an abnormal silhouette, fail to pip internally, and show reduced movement and vocalization. Early recognition of malpositions in ratite eggs is a function of experience guided by regular candling and examination of eggs during incubation and necropsy of unhatched eggs.

RENAL MINERALIZATION

Renal mineralization is a frequently observed as an abnormality at necropsy in embryos and unhatched eggs submitted for examination. Mineral or crystalline deposits comprise either calcium or uric acid. The commonly known causes of renal mineralization in chicks include hypervitaminosis D_3 and in embryos incubation-related dehydration with consequential urate accumulation in the tubules. It is likely that other factors contribute to this problem, but no other specific etiologies have been identified in ratites. Hypovitaminosis A can contribute to the development of renal disease and hyperuricemia in immature chickens, and presumably ratites, but this has not been correlated with renal calcification [2].

Hypervitaminosis D_3 can induce soft tissue mineralization in chicks if diets fed to hens or chicks are oversupplemented. Vitamin D_3 toxicosis has been reported in handfed chicks [3]. It is likely that hypervitaminosis D_3 occurs in ratites, although this has not been documented. Clinical impressions suggest that ratite species are significantly less susceptible to vitamin D_3 toxicosis than other immature avians such as macaws (*Ara* spp). Dystrophic calcification of renal and other tissues associated with hypervitaminosis D_3 may be seen at necropsy concurrently with visceral gout.

Uric acid is actively secreted into the tubules of the kidneys, independent of the state of hydration of the bird or embryo. As a result, dehydration without an adequate tubular flow rate results in accumulation of urates in the renal tubules of affected chicks. This accumulation of urate crystals can lead to obstruction and, eventually, mechanically induced tubular necrosis. Mineral deposits in renal and other soft tissues is recognized at necropsy as visceral gout. Dystrophic calcium deposition is not associated with renal malfunction caused by dehydration. A careful review of hatchery records relating to weight loss of eggs may denote whether dehydration is a potential cause of kidney or soft tissue mineralization with uric acid in embryos or weak chicks which die during the early posthatch period.

YOLK SAC INFECTION AND RETENTION

Omphalitis and yolk sac retention is common in ratites (Figure 14.1). Most yolk sac material should be absorbed by 14 days of age in the ostrich and emu and presumably in other ratites. The causes of yolk sac retention in chicks include high humidity or high temperature during incubation. Simple yolk sac retention is a noninfectious condition. Omphalitis, or inflammation of the yolk sac and umbilicus, may be caused by a variety of bacterial pathogens. Diagnosis is based on the number of affected chicks and the frequency of isolating specific pathogens. Obviously, single or occasional cases of retained yolk sac would not warrant changes in flock management or incubation. Regular review of records during and at the end of the breeding season will denote the prevalence of yolk sac retention and omphalitis. Consultation between the flock manager and attending veterinarian will reveal problems and expedite preventive action.

Chicks that are weak at hatch will be vulnerable to omphalitis.

Contamination of eggs during storage will promote bacterial penetration from the surface of the egg.

Reference is made to the relevant chapters on egg handling and incubation for appropriate recommendations on hygiene.

Incubation at high humidity will result in "wet chicks" characterized by suboptimal loss in egg weight during incubation. Affected chicks invariably demonstrate an increased prevalence of retained yolk sacs which may contain either infected or sterile contents.

Chicks which receive inadequate oxygen during the last week of incubation may also have a high prevalence of unabsorbed yolk sacs and unhealed navels. Both of these abnormalities will result in omphalitis.

High hatcher temperature, excessive hatcher humidity and inadequate hatcher sanitation can all be correlated with an increase in omphalitis over a specific period, corresponding to the duration of malfunction or incorrect operation of incubators. Noninfected, uncomplicated yolk sac retention can be attributed to mismanagement, although a high proportion of chicks with problems will progress to infection of the yolk sac. Cases attributed to a single hatcher should be expected to consistently demonstrate a common or narrow range of bacterial pathogens.

Excessive or inadequate brooder temperature, poor sanitation, and high environmental humidity will exacerbate omphalitis. Since the primary cause of infection is associated with the brooding environment, usually one or two pathogens will predominate in isolates.

Transovarian or transoviductal infection from the hen is rare, based on published reports. Isolation of a common organism from unhatched eggs or from weak or dead chicks and from the oviduct of the hen suggests vertical transmission. *Salmonella* may be transmitted through the ovum in ratites. In commercial poultry, *Mycoplasma* spp. are significant pathogens transmitted vertically. Retroviruses, picornaviruses, adenoviruses, and reoviruses can be transmitted to progeny by viremic hens.

Oviductal infections, especially *Salmonella* spp., penetrate the egg during formation, resulting in embryonic infection. Breeding and incubation records generally reveal persistent yolk sac infection from progeny of a specific hen and invariably involves a single organism.

Infection of the surface of the egg as it passes through the cloaca during oviposition represents an important mechanism of nonoviductal, vertical transmission. Heavy colonization of the terminal intestinal tract with *Salmonella*, *E. coli* or *Pseudomonas* spp. may result in a high prevalence of omphalitis in the progeny of a specific hen or mating. In one documented case, *Salmonella* was isolated from cloacal swabs from a pair

Developmental Problems in Young Ratites

of high-producing emus and from their unhatched embryos. Other emu pairs on the same farm were unaffected and their eggs showed normal hatchability and chick viability.

Once yolk sac disease has been established in a mating pair or trio, chicks should be isolated and provided with antibiotic therapy and, where indicated, surgical removal of the yolk sac. Regardless of the cause of yolk sac retention, prompt diagnosis and treatment will significantly improve livability. In addition to diagnosis and treatment, the flock manager should be encouraged to review potential causes of yolk sac disease relevant to the facility. Possible routes of infection or mismanagement should be investigated to prevent future losses. The transition from diagnosis and therapy of individual chicks to a flock approach is necessary, as ratite units increase in size.

Most frequently, signs of yolk sac disease are noted during the first few days after hatch extending to three weeks of age. Occasionally, older chicks may be presented with infected or retained yolk sacs. Birds with enlarged or infected yolk sacs or which show decreased absorption of yolk frequently are depressed and anorexic. Abnormal or altered behavior, including decreased activity and abnormal neck posture, is commonly noted.

Numerous diagnostic procedures can be applied to diagnose yolk sac problems. Clinical examination, including a detailed history and records, is helpful in assessing the extent of the problem. Complete blood counts (CBC) may have some value in evaluating yolk sac disease in individual cases. Birds with infected yolk sacs may demonstrate toxic changes in heterophil morphology. Radiology has been suggested as a diagnostic procedure for yolk sac retention, because an enlarged and distended sac can be easily visualized. Arguments against radiology include cost effectiveness, impracticality for use at a rearing facility, and inability to document whether the sac requires surgical removal. The size and consistency of the yolk sac can be determined by abdominal ultrasound examination which can be performed at the farm and may be useful for rapid screening of a batch of young birds. The specific value of ultrasound examinations for retained yolk sacs has yet to be documented but the technique shows promise in the hands of experienced practitioners.

Surgical removal of an infected or retained yolk sac is recommended [28] (see chapter 10). The chick is anesthetized and maintained with isoflurane. After routine surgical preparation of the abdomen, a single incision is made in the skin along the midline of the abdomen, circling the navel. The abdominal wall is entered with extreme care not to rupture the yolk sac, which should lie directly below the abdominal wall. The sac is gently elevated and removed from the abdominal cavity (Figure 14.2). Attachment to the intestine at Meckel's diverticulum is ligated and the sac is removed *in situ*. Routine closure of the abdomen and skin concludes the procedure. When performed by an experienced surgeon, yolk sac removal should require only a few minutes and should be minimally stressful to the chick. Yolk should be submitted to a laboratory for both aerobic and anaerobic culture. Patients may be treated with antibiotics, especially if omphalitis is suspected based on the appearance of the yolk sac or in response to a previous history of infection.

MUSCULOSKELETAL SYSTEM

Neck Deformities at Hatch

Torticollis, or "wry neck" is seen sporadically in newly hatched ratite chicks. In general, the emu is most commonly affected, although the ostrich, rhea, and cassowary (in descending order of frequency) can demonstrate this abnormality. Torticollis has multiple causes which are related to muscle and tendon abnormalities rather than primary central nervous system dysfunction. Vitamin E and selenium deficiency, the presence of teratogens in the egg, mineral and micronutrient imbalances or deficiencies in the hen's diet, excessive handling or turning of eggs in the incubator, and developmental anomalies of the skeleton are all considered to be potential causes of torticollis. Although affected chicks sometimes have severe deformities, many cases respond favorably to conservative splinting and multivitamin and mineral supplementation. Careful review of records and diligent necropsy of hatched and nonhatched chicks can often indicate a cause, facilitating management changes which reduce or eliminate neck deformities and other developmental problems at hatching.

Tumbling Chick Syndrome

A problem in young emu chicks commonly referred to as "tumbling chick syndrome" has been described by growers. This clinical presentation is different from torticollis in that these birds consistently flex the head (Figure 14.3) and often roll over onto their back. When excited or stimulated, this behavior is more pronounced than when resting. A specific etiology for this problem has not been defined, although prenatal nutritional deficiencies, including vitamins B, E, and copper and selenium, have been implicated without scientific justification. Incubator or hatcher-induced hypoxia at the time of hatching, septicemia and bacterial omphalitis have all been suggested as potential causes. Some veterinarians believe that the clinical signs observed are related to an abnormality of the central nervous system rather than the musculoskeletal system. Most chicks, if left to eat on their own, become normal within a few days. As a result, comparatively few birds are available for necropsy and histological examination. Severely affected chicks may require parenteral supplementation with multivitamins or antibiotic therapy [4]. Tumbling chicks, when observed in any significant frequency, should be viewed as an indication of a management-related problem on a nutritional deficiency or imbalance. Further investigation should proceed through integrated veterinary flock management to eliminate or reduce the occurrence of the problem.

Spraddle Leg

Spraddle leg in ratite chicks is a deformity of the coxofemoral joints preventing normal adduction of the limbs [5]. This condition is usually bilateral and symmetrical. In contrast, angular limb deformities are usually unilateral, but if bilateral, the changes are asymmetrical. Edematous or "wet" chicks at hatch may be more predisposed to spraddle than normal hatchlings. A chick with a large yolk sac may be unable to adduct its legs until the yolk sac is reduced in size by resorption. Emu growers have recognized a predisposition to "spraddle" when chicks are deprived of adequate exercise during the first week of life. This occurence has little correlation with loss in egg weight during incubation. Some edematous and spraddle-legged ostrich chicks can be held in a narrow box that prevents abduction of the legs. Hobbling chicks with a self-adhesive or elastic bandage between the tarso-metatarsi maintains adduction [6,34]. Emu chicks should be allowed adequate exercise soon after hatching. An early tendency to spraddle should be recognized by chick managers and corrected by hobbling.

Angular Leg Deformities

Limb deformities of multiple and specific types are commonly encountered in young or growing ratite chicks (Figure 14.4). "Leg abnormalities," "splay leg," or "leg problems" are popular all-inclusive terms used to describe conditions such as twisted or bowed leg bones and swollen or deformed hock joints with or without cartilage abnormalities or displacement of the gastrocnemius tendon. Leg deformities usually precipitate secondary joint and bone abnormalities over time. Deformities of the legs usually result in death and represent a significant economic loss to ratite breeders.

Angular limb deformity can result from abnormal development of any bone (osteodysplasia) including the femur, tibiotarsus, metatarsus, or the phalanges. These bones may be individually deformed, or the condition may involve 2 or more bones. The coxofemoral, femoral-tibiotarsal, tibiotarsal-metatarsal, metatarsal-phalangeal, and interphalangeal joints may be affected. In future studies of leg problems, it will be particularly important to characterize specific pathology. Classification should include an accurate description of the deformities noted, the environment of the flock (temperature, light, substrate, and air movement), genetics of the animals, health status of the breeder birds and chicks, and their nutrient status.

The most commonly encountered leg deformities are characterized by progressive lateral rotation of the tibiotarsus and the tarsometatarsus (Figure 14.5), although other deformities may be noted. The specific etiology of tibiotarsal or tarsometatarsal rotation is not known, although genetic, management, and nutritional factors are regarded as causal factors [6,7,8,9]. This deformity may be progressive or develop slowly. In general, the provision of adequate exercise reduces the prevalence of most deformities. This lends support to the concept that deformities have a primary genetic causation exacerbated by lack of exercise. It is generally accepted that tibiotarsal rotations and other leg deformities should be prevented rather than treated. Maintaining an appropriate growth rate, providing acceptable substrate, and adequate levels of exercise are recommended. Derotational osteotomy may correct the deformity in selected cases, especially with tibiotarsal rotation, but the prognosis is generally regarded as poor [4,5]. There are comparatively few documented cases of normal adult birds that were successfully derotated as chicks compared to the total number of attempts at surgical correction. Intermedullary pinning using hemicerclage wires and plates has been reported, principally to correct lateral rotation of the tibiotarsus [5]. Periosteal stripping, or hemicircumferential periosteal transection, has been described to correct tibiotarsal rotation and angular limb deformities in ratites [4,6]. Wedge osteotomy has been discussed as a potential correction for severe bowing and resultant valgus deformity of the tibiotarsus but success rates have not been quantified [9]. Conservative orthotic management of angular limb deformities has been reported in a limited number of birds [13].

Other developmental abnormalities in commercial broilers and turkeys include tibial dyschondroplasia, chondrodystrophy, and osteochondrosis dissicans [7]. The etiology of these abnormalities has defined in poultry and these abnormalities involve genetic factors interacting with nutrition and management. It is likely that parallels exist with ratite species in which leg abnormalities are multifactorial in etiology [7,8].

Developmental Problems in Young Ratites

Rickets

Rickets is a pediatric condition precipitated by a dietary deficiency or imbalance in calcium, phosphorous or vitamin D_3 or as a result of malabsorption of these nutrients. The absolute quantity of these nutrients as well as their relative inclusion in the diet are important as predisposing factors. Rickets resulting from calcium, phosphorous, or vitamin D_3 deficiency or imbalance is characterized by widening of the epiphyseal cartilage and decreased apposition of bone. The bones become soft and flexible, resulting in a disinclination to walk. Marked enlargement of the joints occurs and the long bones of the legs (femur, tibiotarsus, and tarsometatarsus) are bowed. Imbalances in the calcium to phosphorous ratio occur when chicks are erroneously fed breeder diets or are given free access to oyster shell. Alfalfa hay with no supplemental phosphorous may precipitate rickets. Ideally, chick diets contain a calcium to available phosphorous ratio no greater than 2:1. A more detailed understanding of vitamin D_3, calcium, and phosphorous metabolism of ratite species will be developed with greater intensification of the industry and transition to end-products.

Osteochondrosis Dissicans

Osteochondrosis has been documented as occurring in all leg joints of the ostrich and emu, including the two interphalangeal joints [5] (Figure 14.6). Successful repair of a subchondral bone cyst in a 4-year-old male ostrich has been reported [10]. A primary underlying etiology and the relative incidence of this problem have yet to be established.

"ROLLED TOES"

Rolled or curled toes are commonly seen (Figure 14.7) in ostriches but less frequently in emus and rheas [4,5,6]. The deformity results from rotational instability of the interphalangeal joints, usually resulting in medial rotation of the tip of the digit. In more severely affected chicks, subluxation of the interphalangeal or the metatarsophalangeal joints may occur. Suggested causes include nutrition, defects in incubation, and genetics [5]. Improper substrate or lack of exercise may also be related to the development of this abnormality. Curled toe paralysis due to riboflavin deficiency in commercial poultry is not related to rolled toes in ratites [5]. Correction of the problem requires corrective splinting (Figure 14.8). In more severe cases, surgical correction by cross pinning may be required [4] (see Chapter 10). Subluxation of the metatarsophalangeal joint tends to carry a much lower

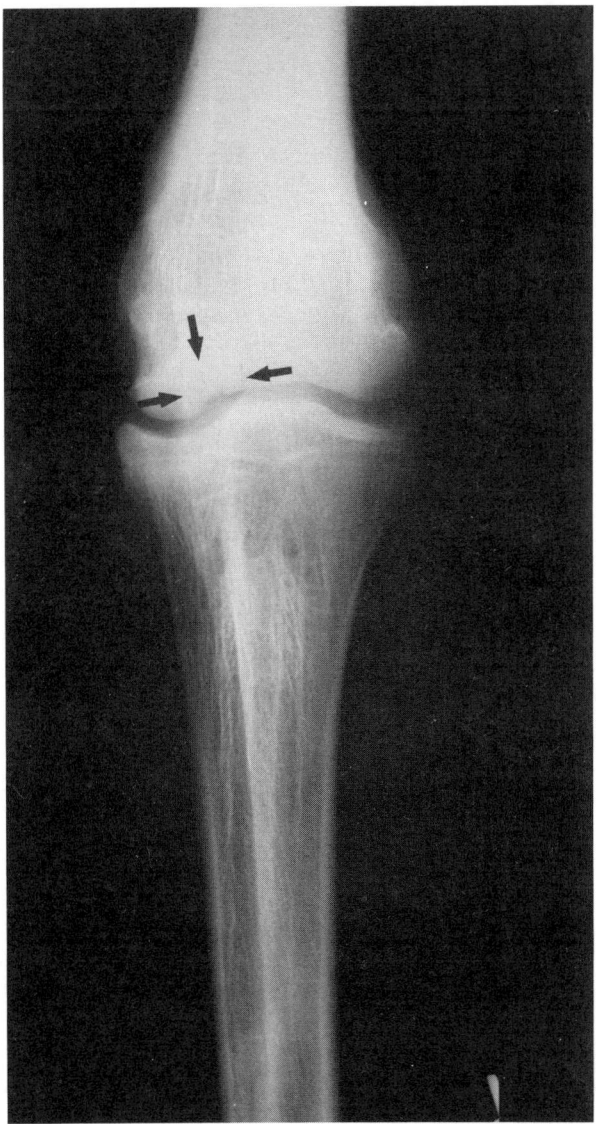

Figure 14.6 Radiograph showing osteochondritis dissicans.

prognosis for resolution compared to simple involvement of the interphalangeal joints.

Slipped Tendons

Slipped tendons occur in all ratite species, but the ostrich and emu are most frequently affected. "Slipped tendon" is a lay term used to describe the displacement of the gastrocnemius tendon laterally from the normal anatomical location between the condyles of the distal tibiotarsal and the promixal tarsometatarsal bones. The retinacular sheath that maintains the position of the tendons generally tears on the medial aspect of the hock, allowing lateral displacement of the tendon [4]. Birds with slipped tendons are unable to stand, and the foot is deviated laterally from the midline of the body. This problem is most frequently encountered in chicks, although unilateral displacement of the gastrocnemius tendon can occur in juveniles and subadults with varus and valgus abnormalities. Causes include genetic predisposition, malnutrition and improper substrate [6]. Trauma can result in a slipped tendon in subadult or obese ostriches. Manganese, copper, biotin, and choline deficiency cause chondrodystrophy of the distal tibiotarsus and proximal tarsometatarsus in chickens and turkeys, resulting in displacement of the tendon due to deformation of the hock joint.

Conservative treatment which comprises corrective bandaging to maintain the tendon in the anatomical position but still allow normal movement of the bird is generally unrewarding [13]. Surgical repair of the medial aspect of the retinacular sheath has been described [4], but involvement of the joint capsule or decubitus represents a guarded prognosis.

Myopathies

Degenerative myopathies are commonly observed in ostriches, emus, and rheas submitted for necropsy [8]. Clinical signs in ratites include depression and reluctance to rise or move with a rapid progression to death. Postmortem findings are generally unremarkable but histopathologic diagnosis is confirmatory.

Angular limb deformities, particularly in the emu, are believed to result from myopathies with the subsequent secondary development of osteopathy. The majority of affected birds are less than 6 months of age. Several causes have been suggested in ratites, including capture myopathy, vitamin E, or selenium deficiency, or furazolidone toxicity [8,11,12,14]. Cardiomyopathy has been reported in three emu chicks with Vitamin E deficiency as a possible etiology [15]. Two rheas with limb abnormalities were tentatively diagnosed with hypovitaminosis E. This was based on mean plasma Vitamin E levels of 1.34 mcg/dl compared to 11.5 mcg/dl in unaffected rheas [16]. Normal organ and tissue levels of Vitamin E or selenium have not yet been documented for ratites.

Acute selenium toxicity has been diagnosed in four cases in ratites. Chicks receiving 3 mg selenium by the parenteral route to prevent myodystrophy died acutely with pulmonary edema and congestion [8]. Selenium levels in the liver of birds in one case ranged from 6.97 ppm to 9.8 ppm. This report suggests that empirical administration of parenteral Vitamin E and selenium is contraindicated unless a specific diagnosis of selenium deficiency is based on analysis of tissues.

GASTROINTESTINAL SYSTEM

Impaction

Impaction of the proventriculus and ventriculus is commonly encountered in the ostrich [17,18,19]. In decreasing order of occurrence, impactions may occur in the rhea, emu, and cassowary. The difference in relative prevalence and predisposition of impaction among these species is possibly related to the anatomy of the upper digestive tract including the proventriculus and ventriculus [20]. Impactions should be viewed as a pediatric problem, as adults are less frequently involved [19]. Impaction is frequently diagnosed incorrectly in ostrich practice based on history and clinical examination. Frequently, a depressed ratite will be presented by owners, believing "impaction" to be the primary complaint, based on the assumption that all ratite species are equally predisposed to the problem. The attending veterinarian should confirm the diagnosis of impaction by radiography, ultrasound, or laparotomy and initiate appropriate investigations to determine the underlying causes of the disorder if successive cases are presented from a flock.

Ratites are geophagic and will pick at and consume any small foreign object within reach. Foreign body ingestion due to mismanagement of substrate is often incriminated as a primary cause. Hypocontractility of the smooth muscle of the gastrointestinal tract may be a possible contributory factor related to primary nutritional and managemental causes. Unthrifty and malnourished birds will demonstrate a higher prevalence of impactions than flocks in ideal conditions [21].

Impactions may be diagnosed by clinical signs apparent on physical examination and can occasionally be confirmed by radiology or ultrasonography. Clinical signs in the ostrich may include abdominal distention and loss of muscle mass. Affected birds show partial or complete anorexia and small hard fecal pellets. Palpation of the abdomen in young birds may reveal a proventriculus distended with rocks, hay, or other foreign material. Grass or hay impaction is more difficult to diagnose (Figure 14.9), but affected chicks can be diagnosed by palpation after an overnight fast. Radiography may be used to document the presence of foreign bodies and enlargement of the proventriculus. Ultrasonography may identify hay, rocks, metallic objects or confirm hypocontractility of the proventricular wall. This technique is becoming an important on-site diagnostic aid in ostrich medicine.

Treatment for impaction involves either conservative medical or surgical approaches or a combination of treatments. Mineral oil and psyllium additives may aid the passage of sand and fine particular material [4,17,19]. More extensive or complete impactions require surgical intervention. The surgical techniques of proventriculotomy and ventriculotomy in the ostrich are described in the relevant chapter of this text and have been documented by several authors. Midventral and left lateral approaches [22,23,24,25] have been used. A ventriculostomy was reported for the removal of foreign bodies in the kiwi [26].

Intusseception

Intusseception is frequently seen in ostrich chicks secondary to impaction but is not confined to chicks with this condition [4,27]. Chronic diarrhea or other enteritides may contribute to the occurrence of this problem. Prompt surgical intervention is necessary when the condition is diagnosed and resection and anastomosis of affected tissues may be required. Most intussceptions are not diagnosed clinically and are identified only at post mortem examination.

Intestinal Torsion

Torsion of the large bowel occurs infrequently. Abrupt changes in feed may be a contributory factor. Clinical signs may include dehydration, depression and diarrhea. Surgical treatment requires resection and anastomosis [19]. Most cases of intestinal torsion are diagnosed only at necropsy.

Cloacal Prolapse

Cloacal prolapse is common in ostrich chicks but may occur in emus [4] (Figure 14.10). Underlying causes such as diarrhea, impaction, or nutritional imbalances should be considered. Prolapsed tissues should be gently repositioned, and either a purse string or two vertical stay sutures are inserted to prevent tissues from re-prolapsing. Anesthesia may be required in some cases, and anti-inflammatory therapy may be necessary to aid in the reduction of edema and swelling of the affected tissues. Young male ostriches with chronic cloacal prolapse may suffer mechanical trauma to the phallus, resulting in potential impairment in future breeding performance.

Fading Chick Syndrome

During the past several years, veterinarians and ostrich growers have recognized a condition in ostrich chicks that is commonly referred to as "fading chick syndrome" [21]. This condition has been characterized by a significant mortality rate with progressive weight loss despite normal food consumption until the later stages of disease. In general, antibiotics and other medication have not yielded any improvement in affected flocks. Postmortem diagnoses include isolation of *Clostridium* and *Salmonella* spp., impaction, bacterial hepatitis, gram-negative and positive bacterial enteritis and septicemia, fungal proventriculitis, and pneumonia. Frequently, no identifiable cause of death has been determined in cachexic birds presented for postmortem examination. This condition is most likely multifactorial in etiology and may involve immunosuppresive agents including viruses and mycotoxins.

Early clinical signs of a "fading chick" may include diminished volume of paralumbar musculature and thin legs. Appetite is generally normal. Most birds are stunted or demonstrate a delayed or absolute failure to grow. A progressive decline in activity occurs, but this is usually difficult to detect, especially by managers having daily contact with the flock. A decrease in activities such as "waltzing" may occur, and affected birds bear weight abnormally on the metatarsal pads rather than the digital cushion. More advanced cases may demonstrate a pronounced decrease in dorsal musculature, decreased appetite, increased susceptibility to stress, and terminal hypoglycemia and dehydration. Hypoglycemic crises occur in the early morning following a cold night or when the bird is under stress. An absolute increase in incidence of impaction, bacterial septicemia, ascites, and clostridial and bacterial enteritis occurs. Chicks sold or shipped in poor condition have significantly higher mortality rates compared to unaffected birds. Mortality is relatively high in the more advanced stages of disease.

The causes of the fading chick syndrome are rather poorly understood. Vertically transmitted viral agents, such as adenoviruses or arenaviruses, which cause a similar syndrome in turkey poults and chicks may be primary pathogens. Infectious bursal disease virus of chickens has been isolated from cases of fading chick syndrome in California and Florida. This virus is responsible for severe immunosuppression in chickens and may precipitate losses due to secondary bacterial and opportunistic fungal infection.

Treatment is based on identification of pathogens present with supportive therapy for the clinical and metabolic abnormalities displayed by affected chicks. Antibiotic treatment is indicated for secondary bacterial infections.

CARDIOVASCULAR SYSTEM

Aortic rupture was reported in four ostriches and attributed to nutritional copper deficiency [28]. Death in these cases was peracute and profound hemorrhage without premonitory signs occurred. Liver copper levels averaged 3.3 ppm (ranging from 2.0–5.0 ppm). These levels were considered deficient based on liver levels of 8.8 ppm copper in 17 birds necropsied for unrelated causes. Dietary levels of copper were not reported. Young adult ostriches aged 6 to 12 months are believed to be the most commonly affected. Aortic rupture was a common occurrence in commercial turkeys, during the 1960s and 1970s. The condition is associated with inherited hypertension in addition to abnormal metabolism of micronutrients. Postmortem diagnoses of aortic rupture have been recorded in emus [8].

MISCELLANEOUS DISORDERS

Lysosomal storage disease of genetic etiology has been documented in a 6-month-old female emu [29] and in two unrelated pairs [30]. Clinical signs include head tremor and ataxia. Affected birds have defects in the clotting mechanism and may show agonal extravasation from the ears and nares. Gross necropsy shows no significant lesions, although some cases may show subcutaneous or intramuscular petechial hemorrhage. Histopathology of the brain reveals vacuolation and granulation of the cytoplasm of neurons. Similar changes are noted in the spinal cord.

REFERENCES

1. Joyner KL. Theriogenology. In: Ritchie BR, Harrison GJ, Harrison LR, eds. *Avian medicine, principles and application*. Lake Worth, FL: Wingers Publishing, 1994;748–804.
2. Siller WG. Kidney diseases in the fowl. In: Gordon RF, Jordon TTW, eds. *Poultry diseases*, 2nd ed. London: Balliere Tindall, 1982;247–259.
3. Takeshita K., et al. Hypervitaminosis D in baby macaws, in *Proceedings*. Assoc Avian Vet 1986;341–345.
4. Hicks KD. Ostrich pediatrics. In: Fudge AM, Speer BL, eds. *Seminars in avian and exotic pet medicine*. Philadelphia: WB Saunders, 1993;2:136–141.
5. Gilslider E. Ratite orthopedics. In: Redig PT, Fudge AM, eds. *Avian orthopedics, seminars in avian and exotic pet medicine*. Philadelphia: WB Saunders, 1994;3:81–91.
6. Stewart JW. Ratites. In Ritchie BR, Harrison GJ, Harrison LR, eds. *Avian medicine, principles and application*. Lake Worth, FL: Wingers Publishing, 1994; 1284–1326.
7. Angel R. Selected problems in ostriches and how they are affected by nutrition, in *Proceedings*. Ostrich Med Conf, Amer Ostrich Assoc. Las Vegas, 1992.
8. Shivaprasad HL. Neonatal mortality in ostriches: An overview of possible causes, in *Proceedings*. Assoc Avian Vet 1993;282–285.
9. Bezuidenhout AJ, et al. Serum-mineral and bone-mineral status of ostriches with tibiotarsal rotation. *Onderstepoort J Vet Res* 1994;61:203–206.
10. Tully TN, et al. A subchondral bone cyst in the distal tibiotarsal bone of an ostrich (*Struthio camelus*). *J Avian Med & Surgery* 1994;9:41–44.
11. Philbey AW, et al. Anasarca and myopathy in ostrich chicks. *Austr Vet J* 1991;68:237–240.
12. Rae M. Degenerative myopathy in ratites, in *Proceedings*. Assoc Avian Vet 1992; 328–335.
13. Vanhooser SL, et al. Aortic rupture in ostrich associated with copper deficiency. *Vet Human Toxicol* 1994; 36:226–227.
14. Vorster BJ, et al. Nutritional muscular dystrophy in a clutch of ostrich chickens. *J So African Vet Assoc* 1984; 55:39–40.
15. Van Der Heyden N. Cardiomyopathy in three emu chicks, in *Proceedings*. Assoc Avian Vet 1994;127–129.
16. Dierenfeld ES. Vitamin E deficiency in zoo reptiles, birds and ungulates. *J Zoo Wildl Med* 1989;10:3–11.
17. Blue-McLendon A. Pediatric disorders of ostriches, in *Proceedings*. Assoc Avian Vet 1993;269–271.
18. Deeming DC, et al. Observations of the commercial production of ostrich (*Struthio camelus*) in the United Kingdom: Rearing of chicks. *Vet Rec* 1993;132:627–631.
19. Wade J. Ratite pediatric medicine and surgery, in *Proceedings*. Assoc Avian Vets 1992;340–356.
20. Fowler ME. Comparative anatomy of ratites. *J Zoo Wildl Med* 1991;22:204–227.
21. Speer BL. The "fading chick syndrome," in *Proceedings*. Western Region Amer Ostrich Assoc 1993;29–34.
22. Gamble K, Honnas C. Surgical correction of impaction in ostriches. *Compendium* 1993;15:235–245.
23. Honnas CM, et al. Proventriculotomy to remove foreign body impaction in ostriches. *JAVMA* 1991;199:461–465.
24. Jacobsen E, et al. Ventriculotomy for removal of multiple foreign bodies in an ostrich. *JAVMA* 1986; 189:1117–1119.
25. Stewart JS. A simple proventriculotomy technique for the ostrich. *J Assoc Avian Vets* 1991;5:139–141.
26. Gasthuys F, DeMeurichy E. Successful ventriculostomy for removal of foreign bodies in a kiwi (*Apteryx australis mantelli*). *J Zoo Animal Med* 1987;18:166–167.
27. Keffen RH. Intussception in an ostrich chick. *J So African Vet Assoc* 1984;55:77.
28. Kenny D, Cambre R. Indications for the surgical removal of the avian yolk sac. *J Zoo Wildl Med* 1992;23:55–61.
29. Swerida DF. Lysosomal storage disease in an emu (*Dromaius novaehollandiae*), in *Proceedings*. Assoc Avian Vet 1994;447.
30. Bermudez AJ, Johnson GC, Suzuki K. Epidemiology of gangliosidosis in emus (Abst), in *Proceedings*. 132nd Am Vet Med Assoc Mtg 1995;141.
31. Van Heerden J, et al. Suspected vitamin E-selenium deficiency in two ostriches. *J So African Vet Assoc* 1983;54:53–54.

Chapter 15

Therapeutics

Thomas N. Tully, Jr.

A knowledge of therapeutics in ratites is limited by documented field observation and a lack of subjects for experimental and clinical research. Studies have been confined to a narrow range of drugs using small groups [1,2,3,4,5]. Therapeutic dosages for ratites have been derived from other avian species by extrapolation and practical clinical experience. Specific information has not been obtained from pharmacodynamic studies. Due to differences among avian and exotic species, allometric scaling derived from basal metabolic rate has been used to develop drug dosages which are dependent on cardiac output for absorption, distribution and metabolism. The computer program Zoodose 2.0 [6] has been developed for ratites. Allometric scaling, although an approximate method to derive drug dosages, is a practical technique in current use for avian species including ratites. Inappropriate dosing may result in either inadequate treatment or toxic levels of administration, creating potentially fatal complications.

Because many ratites producers have limited experience in medicating flocks, veterinarians have the obligation to enlighten and educate owners on the need for proper diagnosis and treatment of disease. A common misconception exists among owners that broad spectrum antibiotic therapy will cure all problems. A specific diagnosis and appropriate treatment is required for a successful clinical outcome.

Ratites present special therapeutic problems to the veterinarian. Although ostriches and emus are relatively large, there is no pectoral muscle to serve as an injection site. Continual administration of intramuscular agents into the gluteal muscles by the owner may lead to injury. The large muscle masses of the legs represent the only practical and safe site for intramuscular administration despite the active renal portal system in ratites. Injection sites should be alternated between the tibiotarsal and thigh muscles of both legs. The lumbar muscles may be used in ostriches if the patient is well fleshed.

The success of oral medication depends on age and species. Ostriches and emus will generally swallow capsules or tablets directly from the hand, but sometimes medication must be hidden in a grape or food ball. Liquid medication is difficult to give to ratites *per os*, and as with all exotic and avian patients, improper administration may lead to aspiration. If medication is added to feed or water, the owner must know it the patient is receiving the desired dose or the consequences of inadequate treatment become apparent. Powder or liquid medication added to drinking water may result in an abnormal color or flavor. Adding a sugar substitute or flavored electrolyte powder may enhance intake of medicated water. Careful observation of consumption by the owner is critical to successful oral therapy. If the bird does not drink voluntarily, an alternative method of administration such as esophageal intubation must be selected.

Label and mixing instructions must be strictly followed to avoid incorrect dose rates. Many water-soluble antibiotics degrade rapidly in solution or when exposed to sunlight. This dictates daily preparation of water-soluble medication.

Subcutaneous administration of therapeutic agents in ratites is not recommended except when prolonged absorption of oil emulsion products is desired. Subcutaneous tissue is poorly vascularized, impeding absorption of drugs.

Intravenous and intraosseous administration contributes to rapid uptake of drugs in all ratites. Catheterization is necessary for prolonged intravenous or intraosseous administration and is only practical in a hospital setting. A venous catheter may be placed in the right jugular vein (ostriches) or medial metatarsal vein. An intraosseous catheter may be used in neonates and requires insertion into the proximal aspect of the tibiotarsus or ulna of the ostrich.

Nebulization is limited to neonates and young juveniles, which are especially susceptible to respiratory infection. Equipment required includes an air compressor, a source of oxygen, an enclosed chamber and a suitable nebulizer. An inexpensive, commercial unit which should satisfy most applications [7] can deliver antimicrobial drugs to the lungs and proximal air sacs providing the particle size of the agent is less than 0.5 microns in diameter.

Topical medication should be used sparingly to avoid matting of feathers. Aerosol antibiotic sprays are effective for minor skin lacerations and abrasions and may be applied quickly from a short distance, reducing stress in handling.

Ointments are preferred over liquids for ocular treatment. The number of applications per day can be reduced due to the slower release of antibiotic. Misting the eye with a water soluble topical spray may be necessary to treat a fractious bird [8].

Nasal flushing by irrigation of the infraorbital sinus or intratracheal administration of antibiotics may be attempted in immature ratites which can be firmly restrained.

Gram-positive rods [9] predominate in the normal flora of the ostrich gastrointestinal tract of ratites. In young birds or older patients subjected to stress or prolonged antibiotic treatment, inoculation of the gastrointestinal tract with a ratite-specific *Lactobacillus acidophilus* culture may promote intestinal function and inhibit mycosis.

FLUID THERAPY

Fluid replacement for ratites can be calculated, applying the Zoodose 2.0 program. If this program is not available, the following formula can be used [10]:

a. fluid maintenance requirement of 50 ml/kg/day
b. deficit = estimated weight × % dehydration

Fluids should be replaced using intravenous or intraosseous routes. The estimated deficit should be replaced over a 48-hour period.

FEEDING THE DEBILITATED PATIENT

Anorexic adult ratites should be force fed using a tube placed directly into the esophagus with care to bypass the glottis. The tube must be positioned distal to the thoracic inlet. Alternatively, a tube may be inserted by esophagostomy using a percutaneous approach. An incision should be made approximately midway between the head and thoracic inlet on the left side [1]. After the end of the tube is positioned in the proventriculus, it is fixed by insertion of a skin suture. Ratites should be fed at 6-hour intervals [1].

Many different therapeutic supplements have been administered to ratite patients. Hills Canine Maintenance Diet, or a slurry of ratite diet, can be offered to adults at a level of 100 g/5 kg body weight, three times daily. For young birds, neonates, and juveniles, Emeraid I and II™ and Nutrastart™ [11] have been successfully used. The Zoodose 2.0 program can be used to calculate daily requirements. Total parenteral nutrition may be attempted, although cost will restrict this approach to immature birds for short periods.

VITAMIN AND MINERAL SUPPLEMENTATION

Vitamin and mineral supplements should be administered to ratites with extreme care and only following diagnosis of a specific deficiency. Indiscriminate supplemental selenium and copper will result in toxicity in the absence of precise diagnosis of a deficiency, resulting in gastrointestinal, musculoskeletal, or reproductive dysfunction. Most commercial diets are adequately supplemented with macro- and micronutrients including selenium, and there should be no need for routine or even therapeutic supplementation. Vitamin and electrolyte solutions may be administered to birds which have been subjected to transport stress, trauma, or infection.

VACCINATION

At this time, only two vaccination protocols are recommended for emus. This species should be immunized against both eastern or western equine encephalomyelitis virus. Despite the relatively confined area where these infections have been recorded, actual experience with insect-borne infections is limited. The fact that emus are often transported over considerable distances and infection is fatal, justifies vaccination.

Equine Encephalitis

Vaccine — Tissue culture propagated (nonchick embryo) inactivated bivalent eastern and western encephalomyelitis vaccine prepared and licensed for equine use. The product may contain equine tetanus antitoxin. Under no circumstances should an equine vaccine incorporating influenza antigen be administered.

Protocol — Initial vaccine: Full equine dose by the intramuscular route at 6 weeks of age. Boosters: 10 weeks of age and at 5 and

6 month-intervals thereafter, before and after the breeding season (April and September). A booster should be administered in the event of an outbreak within 10 miles of the farm or before transfer to an endemic area.

The recommended vaccine protocol for blackleg (*Clostridium chauvoei*) is as follows:

Vaccine "Blackleg" ruminant vaccine with or without tetanus antitoxin.
Protocol Initial vaccine: at 2 months. Boosters: at 3 months and then annually after the end of the breeding season (April).

Avian Pox

Live attenuated fowl pox vaccine may be administered to immature ostriches at 2 and 3 months of age in areas where the disease is endemic. It is noted that the occurrence of pox is dependent on climatic factors which promote high populations of mosquito vectors together with proximity to commercial or backyard poultry.

Special Vaccines

Inactivated autogenous bacterins have been recommended by ratite veterinarians to prevent specific enterobacter infections, usually involving the gastrointestinal tract and as an adjunct to antibiotic therapy. These vaccines should be made from specific isolates derived from affected birds on the farm. Vaccines should be prepared in a state or federally licensed plant using approved manufacturing practices. Autogenous vaccines are expensive because minimal quantities of 1,000 doses are required and their effectiveness has not been proven.

ANTHELMINTIC THERAPY

As with all livestock species, antiparasitic medication should be used appropriately in ratites. Parasites should be identified, and the proper treatment administered in accordance with the prescribed time intervals, as indicated in Chapter 12.

The only current routine antiparasitic program for emus involves monthly treatment of birds aged 1 to 12 months with ivermection (200 mcg/kg) [12] to prevent *Chandlerella quisquali*.

ANTIBIOTIC THERAPY

Antibiotics are used extensively to treat infections of companion and food animal species. Appropriate treatment facilitates recovery and has undoubtedly benefitted producers. Appropriate antibiotics should be administered following identification of a pathogen and determination of the spectrum of antibiotic sensitivity. Misuse of antibiotics is obviously a problem in the emerging ratite industry as indicated by the frequency of resistance to synthetic penicillins and quinolone compounds. The U.S. Food and Drug Administration recognizes the dangers of indiscriminate use of antibiotics with the emergence of drug resistance and occurrence of residues. Many drugs with the potential for misuse are restricted to prescription by a veterinary practitioner. Extensive use of over-the-counter antibiotics has contributed to the problem of emerging drug resistance.

Veterinarians must be aware of the pharmacological action of antibiotics, including spectrum of activity, potential toxicity and possibility of residues (Appendix 15.1).

Gentamicin should be used only with strict veterinary supervision due to the nephrotoxicity of the compound.

Reports of adverse drug or reaction should be forwarded to the Food and Drug Administration and the manufacturer of the drug. Pharmacists and clinical faculty at colleges of veterinary medicine can advise on the availability and source of antibiotics and other therapeutics.

Appendix 15.1

Ratite Formulary

ANTIMICROBIALS

Drug	Route	Dosage	Comment
1. Amikacin Amiglyde(Aveco) Amikin(Bristol Labs) injectable solutions- 50mg/ml 250 mg/ml	I.M.⇒ Air Cell Egg⇒ Egg Dip⇒	7.6-11 mg/kg BID 10-25 mg/egg 2,000mg/gallon distilled water pH6	
2. Amoxicillin Amoxi-Drops, Amoxi- Inject (Smith Kline) suspension(drops)-50mg/ml inject-250mg/ml	P.O.⇒ Drinking water⇒	15-22 mg/kg BID 250 mg/gallon drinking water for 3-5 days	
3. Ampicillin Polyflex(Fort Dodge) 100 mg/ml	P.O.⇒ I.V.,I.M.,S.Q.⇒	11-15 mg/kg TID 4-7 mg/kg TID	
4. Carbenicillin GeoPen(Roerig) injectable solution-200 mg/ml	I.V.	11-15 mg/kg TID	
5. Cefataxime Claforan(Hoechst-Roussel)	I.M.	25 mg/kg TID	young birds
6. Cephalexin Keflex Pedicatric Suspension(Dista) oral suspension-25-100 mg/ml	P.O.	15-22 mg/kg TID	
7. Cephalothin Sodium Keflin(Lilly) injectable solution-100 mg/ml	I.M.,I.V.	30-40 mg/kg QID	
8. Chloramphenicol (Parke Davis);(Fort Dodge) injectable solution-100 mg/ml oral suspension- 30mg/ml	P.O.,I.M.,I.V.,S.Q.	35-50 mg/kg TID for three days	not for use in food animals
9. Chlortetracycline Aureomycin(Cyanamid) tablet-25mg	P.O.	15-20 mg/kg TID	
10. Ciprofloxicin Cipro(Miles) tablets-250,500,750mg	P.O.	3-6 mg/kg BID	
11. Clavamox (SmithKline Beechmam) tablets-62.5,125,250)	P.O.	10-15 mg/kg BID	
12. Doxycycline (Pfizer);(Henry Schein);(Roerig) suspension-5mg/ml syrup-10mg/ml capsules-100mg	P.O.	2.0-3.5 mg/kg BID	
13. Enrofloxacin Baytril(Haver/Diamond) tablet-5.7,22.7,68mg	P.O.,S.Q.⇒ I.M.⇒ Drinking Water⇒	1.5-2.5 mg/kg BID 5 mg/kg BID q 2 days 10% solution,	I.M. injections cause se- vere muscle necrosis

Therapeutics

	injectable solution-22.7 mg/ml oral solution-100mg/ml		10 mg/kg q 3 days
14.	Erythromycin (Sanofi);(Lextron) tablets-250,500mg	P.O.	5-10 mg/kg TID
15.	Flucytocine Ancobon(Roche) capsule-250,500mg	P.O.	80-100 mg/kg BID
16.	Gentamicin (Butler);(Schering) injectable solution-50 mg/ml	I.M.,S.W.	1-2 mg/kg TID	use only as a last resort and only under strict veterinary supervision, causes visceral gout
17.	Griseofulvin Fulvicin U/F(Schering) powder-2.5 grams/packet	P.O.	35-50 mg/kgSID
18.	Itraconizole Sporanox(Janssen) capsules-100mg	P.O.	6-10 mg/kg SID	if neurologic signs develope reduce or discontinue
19.	Ketoconizole Nizoral(Janssen) tablet-200 mg	P.O.	5-10 mg/kg SID
20.	Bactitracin Methylene Disaliylate Solu-tracin-200(A.L.Labs) soluble powder-51.2g/4.1 oz pack	drinking water	4.2oz/gallon drinking water=200mg/gallon	for *Clostridium perfringens*. mix fresh daily and use as sole source of drinking water.
21.	Oxytetracycline LA 200, Liquamycin(Pfizer) Injectable solution-200 mg/ml	I.M.	2.5ml/50kg body weight q 3 days	long acting causes severe muscle necrosis, use as last resort
22.	Ceftiofur Sodium Naxcel(Upjohn) injectable soution-1,4 gram vials	I.M.⇒ air cell(egg)⇒	10-20 mg/kg BID 0.25/egg
23.	Norfloxacin Noroxin(Merck & Co.) tablets-400mg	P.O.	3-5 mg/kg BID
24.	Nystatin Mycostatin(Apothecon) Myco 20(Squibb) suspension-100,000 units/ml	P.O.	250,000-500,000 IU/kg BID
25.	Pipercillin Pipracil(Lederle) injectable solution-200 mg/ml	I.M.	25 mg/kg	for young birds less than six months of age
26.	Polymyxin B Sulfate Aerosporin(Burroughs Wellcome) rubber stopped glass vial with flip off cap—500,000 units (1mg=10,000 units)	I.M.	1.5-2 mg/kg BID
27.	Trimthoprim-sulfa Bactrim(Roche) Tribrissen(Coopers) suspension-8 mg trimethoprim/40 mg sulfamethoxazole per ml	P.O., I.M., S.Q.	10-15 mg/kg
28.	Tetracycline soluble powder capsules-250mg suspension or solution-100mg/ml	drinking water	1–2 teaspoons/liter drinking water	make fresh twice a day
29.	Tylosin (Butler);(Elanco) injectable solution-50 mg/ml 200 mg/ml	P.O.⇒ I.M.,I.V.⇒	5-10 mg/kg TID 3-5 mg/kg BID soluble powder

ANTIPARASITIC AGENTS

Drug	Route	Dosage	Comments
1. Albendazole Valbazen(SmithKline Beecham) oral suspension-113.6 mg/ml	P.O.	1ml/22kg body weight BID for three days repeat in two weeks	For flagellate parasites and tapeworms
2. Dimetridazole Emtryl(Salsbury Lab) soluble powder-36.4gm/6.42oz	drinking water	Miz 1oz/10 gallon drinking water. Treat for 5 consecutive days	antiprotozoal agent. very effective.withdraw 5days before slaughter. do not use on laying hens. not currently available in U.S.
3. Fenbendazole Panacur(Hoechst-Roussel) suspension-100mg/ml	P.O.	5-15 mg/kg SID q 3 days	
4. Ivermectin Ivomec,Eqvalen(Merck,AgVet) injectable solution-10 mg/ml	P.O.,I.M.,S.Q.	200μg/kg	
5. Levamisole Levasole(Pittman Moore) Tramisole(American Cyanamide) L-spartakon(Janssen) injectable solution-136.5mg/ml tablets-20mg	I.M.,P.O.	30 mg/kg	For *Libiostrongylus douglassi*, give 1 month of age then once a month for 7 treatments, then 4 times a year
6. Mebendazole Telmintic, Telmin(Pittman Moore) soluble powder-40 mg/g suspension-33mg/ml	P.O.	5-7 mg/kg	
7. Metronidazole Flagyl(Searle) tablets-250,500mg injectable solution-5 mg/ml	P.O.	20-25 mg/kg BID	
8. Pyrantel pamoate Strongid T(Pfizer) oral suspension-4.5 mg/ml	P.O.	5-7 mg/kg	

VITAMIN AND MINERAL SUPPLEMENTS

Drug	Route	Dosage	Comments
1. B-complex (Butler);(Lextron);(Vedco)	I.M.	2mg/kg-based on thiamine	curley toes in ostrich tumbler emu chicks
2. Vitamin E/Selenium Seletoc(Schering)	I.M.	0.06mg/kg	
3. Copper Sulfate	I.M.	1mg/10kg body weight for 5 days	
4. Vitamin D_3	I.M.	15,000 IU/bird	
5. Vitamin E	P.O.	200-300 IU/bird	
6. Vitamin A&D_3 Injacom(Hoffman-LaRoche)	I.M.	0.25-2cc/bird	young birds
7. Vitamin K_1 (Butler);(Phoenix);(Vet-A-Mix) injectable soln-10mg/ml	I.M.	5mg/kg	clotting disorders
8. $MgSO_4$	P.O.	¼teaspoon/juvenile 2 tablespoons/adult	obstipation

MISCELLANEOUS THERAPEUTICS

Drug	Route	Dosage	Comments
1. Adequin (Luitpold Pharm) injectable solution: 500 mg polysulfated glycosaminoglycan	I.M.	500 mg q 4 days for 28 days	Only use if there is no evidence of infection
2. Aminophylline (Searle) injectable solution: 25 mg/ml tablets-100, 200 mg	P.O., I.M., I.V.	8–10 mg/kg TID	
3. Atropine (Butler);(Vedco); (Phoenix) injectable solutions: 10.5 mg/kg, 15 mg/ml	I.M., I.V., S.Q.	.03–.05 mg/kg TID	
4. Flunixin meglumine Banamine (Schering) injectable solution: 50 mg/ml	I.M.	0.2 mg/kg	For pain
5. Benebac (Pet-Ag) gel form	P.O.	1 g/chick	
6. Chlorhexidine Nolvasan (Fort Dodge) disinfectant solution: 2%	Egg Spray	30 ml/gal water 104–108°F	
7. Cimetidine Tagamet (Smith Kline Beecham) tablets: 200, 300, 400, 800 mg liquid: 60 mg/ml	P.O., I.V.⇒ I.M.⇒	3–5 mg/kg TID 5–10 mg/kg BID	
8. Demerol Hydrochloride (Winthrop Pharm) supplied in syringe/needle units dose range 25–100 mg/ml	I.M.	1 mg/kg	Anesthesia recovery
9. Depomedrol (Upjohn) sterile aqueous suspension: 20, 40 mg/ml	I.M.	200 mg repeat as needed	
10. Dexamethazone Azium (Schering) injectible solution-2 or 4 mg/ml	I.M.	4 mg/kg BID q 2 days 2 mg/kg BID q 2 days	⇐ shock ⇐ trauma
11. Dipyrone Novin (Haver-Bayvet) injectable solution	I.M.,I.V.,S.Q.	20–25 mg/kg	
12. Doxapram HCL (Fort Dodge) injectable solution-20 mg/ml	I.V.	4–8 mg/kg every hour	respiratory stimulent
13. Boldenone Undecylenate Equipose(Squibb) injectable solution- 25,50 mg/ml	I.M.	0.5 mg/pound weight repeat in three week intervals	
14. Flumethasone Flucort solution(Syntex) Flucort tablets(Syntex) injectable solution-0.5 mg/ml tablets-0.5 mg/tablet	P.O.,I.M.,I.V.,S.Q.	1–1.5 mg/kg	
15. Furosemide Lasix(Hoescht-Roussel) injectable solution-50 mg/ml	I.M.	0.1 ml BID-QID q 2 day	edematous chicks
16. Glycipyrrolate Robinal-V(Robins) injectable solution-0.2 mg/ml	I.V.	0.04 mg/kg	

17. Hydrocortisone Hydrocortisone Sodium Succinate (Elkins Sinn) injectable solution- 100,250,500mg and 1 gm multidose vials Hydrocortisone tab- lets(Danbury) tablets-20mg	I.V.⇒ P.O.⇒	40–50 mg/kg SID 3.0–4.5 mg/kg BID	⇐ shock
18. Lomotil (Searle) tablet-2.5 mg oral liquid-0.5mg/ml	P.O.	2.0–2.5 mg/kg TID	
19. Mannitol (Webster);(Vedco) injectable solutions-20 mg/ml, 180mg/ml	I.V.	1500 mg/kg QID	
20. Hemicellulose Psyllium Metamucil(Searle)	P.O.	1 tablespoon with 60 mls water per chick	up to 4 oz, daily for impaction
21. Metoclopramide Reglan(Robins) tablets-10mg syrup-1mg/ml	I.M.,P.O.	1 liter/80kg body weight	gastrointestinal disorders
22. Mineral Oil	P.O.⇒	15 ml/kg-adult	impaction
23. Oxytocin (Butler);(Lextron) (Vedco) injectable solution-20 units/ml	I.M.⇒	1–1.5 ml twice 24 hours apart	adult, egg binding
24. Phenylbutazone Butazolidin(Coopers) injectable solution-400 mg/ml tablets-100,400mg	P.O.	10–14 mg/kg BID	
25. Prednisolone Sodium Succinate Cort-sol(Butller) Solu-Delt-Cortef(Upjohn) tablet-5mg injectable solution- 10–15mg/ml	P.O.,I.M.⇒ P.O.⇒ I.V.⇒	1.5–2 mg/kg BID 1–1.25 mg/kg q 48 hrs 5–8.5 mg/kg q hour	⇐ immune suppression ⇐ prolonged therapy ⇐ shock
26. Azaperone Stressnil(Pitman-Moore) injectable solution-40 mg/ml	I.M.	1 ml/55kg body weight	sedation
27. Butorphanol Tartrate Torbugesic(Fort Dodge) injectable solution-40 mg/ml	I.V.	.005–0.25 mg/kg	
28. Vegetable Oil	P.O.	15 ml/kg-adult ½tablespoon/bird-chicks	impaction

References

1. Jensen J. Ratite pharmacology and therapeutics, in *Proceedings*. Advances in Ratite Health Seminar.
2. Bush M, et al. Preliminary pharmacokinetic studies of selected antibiotics in birds, in *Proceedings*. Amer Assoc Zoo Vet Conf 1979; 45–57.
3. Bush M, et al. Pharmacokinetics of cephalothin and cephalexin in selected avian species. *Amer J Vet Res* 1981;42:1014–1017.
4. Fockema A, et al. Anthelmintic efficacy of fenbendazole against *Libyostrongylus douglassi* and *Houttuynia struthionis* in ostriches. *J So African Vet Assoc* 1985; 56:47–48.
5. Gruss B, et al. The anthelmintic efficacy of resorantel against *Houttuynia struthionis* in ostriches. *J South African Vet Assoc* 1988;59:207–208.
6. Wildlife and Exotic Animal Teleconsultants, P.O. Box 10541, College Station, Texas 77842.
7. Devilbiss Health Care Inc., Somerset, Pennsylvania.
8. Flammer K. Antimicrobial therapy. In: Ritchie BW, Harrison GJ, Hanison LR, eds. *Avian medicine: Principals and application*. Lake Worth, FL: Wingers Publishing, 1994;434–456.
9. Simpson, RB, Jackson TC, Warren KK, Kolp DL. Bacterial flora of the small and large intestines of normal ostriches, in *Proceedings*. Advances in Ratite Health Seminar. [DATE]
10. Huff DG. Avian fluid therapy and nutritional therapeu-

tics. In: Bauck L, ed. *Seminar in avian and exotic pet medicine*. 1993;13–16. [PUBLISHER]
11. Lafeber Products, Ordel, Illinois.

12. Blue-Mclendon A, Graham DL, Ambrus SI, Craig TM. Cerebrospinal nematodiasis in emus, in *Proceedings*. Assoc Avian Vet 1992;326–327.

Chapter 16

Guide to Examination and Health Certification of Ratites

Amy M. Raines and Simon M. Shane

INTRODUCTION

With the rapid growth in numbers of ratites, food-animal veterinarians are required to provide health certificates for insurance and sale of ostriches, emus, and rheas. In addition, it is necessary to prepare certificates for interstate transport in compliance with regulations imposed by state and federal authorities to limit transmission of diseases, including avian influenza, salmonellosis, and chlamydiosis.

Speculative demand has created high values, exposing practitioners to claims from aggrieved clients denied insurance or who acquire birds with latent defects. It is anticipated that the recent decline in value of ratites may induce some owners to attempt to recoup eroded capital through liability and malpractice claims.

This guide is intended to assist practitioners, especially those not familiar with ratites, to conduct a comprehensive clinical or health certification examination. A model insurance certificate is provided as a guide (Appendix 16.1).

It is extremely important to follow a set protocol when examining ratites. The procedure should include a thorough history a long with a review of the environment and the management of the individual bird or flock. Ratites should be evaluated at a distance, by manual examination, and, if required, by conducting specific diagnostic and laboratory procedures. Comprehensive examination by a veterinarian will contribute to an accurate diagnosis of disease and will facilitate treatment and possibly prevent further losses.

FACILITIES

It is necessary to assess the quality and adequacy of facilities and housing (Figure 16.1). Special attention

Figure 16.1 Efficient emu breeding unit showing adequately fenced pens sheltering against climatic extremes with defined feeding and nesting areas. Pens are well drained and provided with crushed stone substrate in the "running space" adjacent to the perimeter and with mowed grass in the center. Courtesy: American Molded Structures.

should be paid to fencing and security, which should be adequate to confine the stock and prevent predator loss. Noninjurious fencing should be installed.

Pens should be large enough to permit exercise (Figure 16.2). Birds should be accommodated in well drained enclosures, with mowed lawn free of noxious or potentially poisonous plants or debris. Adequate clean, chlorinated water should be available in suitable drinkers in each pen (Figure 16.3).

Acceptable standards of biosecurity should be followed with closed flocks of only one species (emu, rhea, or ostrich) maintained on a single facility. In multispecies operations, owners should establish an

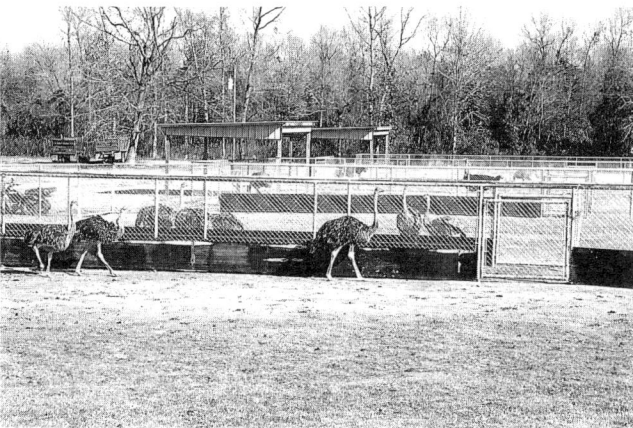

Figure 16.2 Large ostrich pens, well drained and suitably sized to provide exercise.

Figure 16.4 Working pen adjacent to barn and mobile chute for handling, examining, and treating emus.

HISTORY

The source of newly introduced and existing birds should be documented. Introductions should be quarantined for at least 30 (but preferably 60) days, during which health status and function of the digestive tract and locomotory systems can be monitored.

It is necessary to confirm that newly acquired birds were transported in compliance with state and federal requirements.

Previous health certificates and documentation relating to vaccination and treatment should be reviewed. The health status and history of the flock or farm should be determined, and diseases or mortality within the past year should be recorded. Reports concerning the diagnosis and disposition of previous cases, including results of postmortem laboratory procedures, should be noted. Previous history relating to medication, surgery or consultation for congenital or acquired abnormalities in individual birds or the flock should be entered on the veterinary certificate or health records, and details should be provided in an appended explanation.

In the case of mature breeders, a record of egg production and hatchability of females or reproductive performance of a mated pair should be documented.

IDENTIFICATION

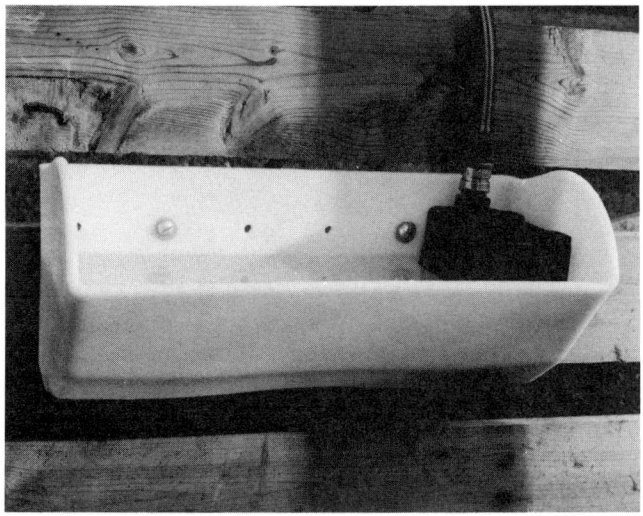

Figure 16.3 Automatic water receptacles which can be regularly cleaned. Courtesy: Hawk Enterprises.

acceptable distance of 1,000 feet between pens and limit contact among ostriches, rheas, and emus. There should be no other commercial or exotic poultry on the facility, as ratites are susceptible to a broad range of pathogens of domestic livestock. Any deviation from acceptable standards of biosecurity or suboptimal conditions should be noted on a separate sheet appended to the insurance or health certificate.

Working pens or chutes should be available on each farm to catch and restrain ratites for inspection, vaccination, and treatment (Figure 16.4).

The training and stockmanship of the owner or farm operator should be evaluated. Owners must be aware of current recommendations concerning preventive immunization of emus against equine encephalitis and routine parasite control in all species.

It is important to record the species, gender, and microchip (Figure 16.5) identification of the subject. Supplementary information, including leg tag number (Figure 16.6), color of plumage (rheas), and height, can be provided. Identification of subjects is critical in the event of subsequent insurance claims or legal action, especially in relation to boarding, syndicate farms, or joint ownership.

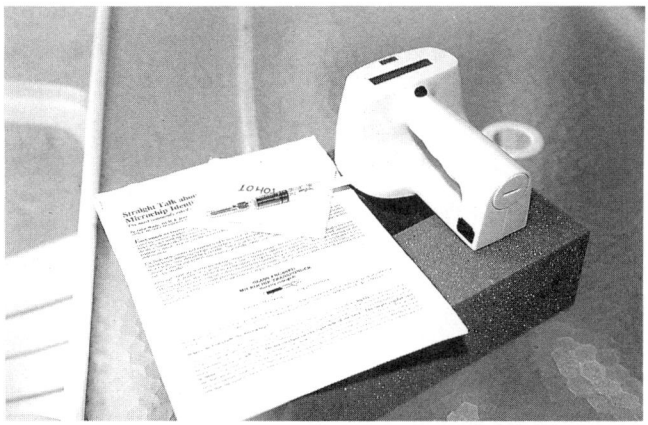

Figure 16.5 Microchip reader is essential to positively identify individual ratites for certification prior to movement or sale. Documentation of observations is necessary to obviate conflicting claims.

Figure 16.7 Examination of emus requires close observation within the group to detect depression or locomotory abnormalities.

Figure 16.6 Leg bands used for farm identification of individual birds.

GENERAL APPEARANCE

Observe the activity of individuals within a group (Figure 16.7). Birds which tend to lag behind penmates may have some locomotory or systemic abnormality.

The size, muscular development, and extent of fat deposit should be consistent with age. Subjects should be alert, active, and respond to stimuli. Aggression, fractiousness, or hypersensitivity, which may indicate a behavioral abnormality predisposing to injury, should be noted on the health certificate or addendum.

LOCOMOTORY SYSTEM

The locomotory system should be evaluated before birds are captured. Observation of gait should confirm that birds do not show incoordination, limping, abnormal position of limbs at rest, or during slow and fast movement. Confirm the normal position of the head and free movement of the neck when standing and during movement. Ensure absence of torticollis, scoliosis, lordosis or kyphosis of the spine, and valgus and varus leg abnormalities. Observe and palpate hock and stifle joints, and examine feet and foot pads to establish freedom from injury, malformations or evidence of surgery. The neck should be examined to exclude possible abessation, foreign body retention, or skeletal malformation.

INTEGUMENTARY SYSTEM

Observe the head with specific reference to the cornea and ocular adnexa and confirm the absence of nasal and ocular discharge. Note symmetry of the head to exclude the possibility of sinusitis or retrobulbar pathology. Examine the oral cavity and in young birds, confirm absence of stomatitis (candidiasis or zygomycosis), geophagia, or malformation of the mandible or maxilla. Examine the plumage and skin for evidence of hyperkeratosis, ectoparasites, abrasions, lacerations, or scars denoting surgery.

CENTRAL NERVOUS SYSTEM

CNS defects will be noted when birds are examined in motion and when standing. Ensure that birds react appropriately to handling and environmental stimuli.

Examine both eyes to confirm pupil reflex and ascertain normal appearance of the eyelids, nictitating membrane, and cornea. Exclude congenital or acquired lenticular opacity (cataracts).

GASTROINTESTINAL SYSTEM

Palpate left ventrolateral quadrant to detect possible impaction of the proventriculus.

Examine the vent area to exclude diuresis or diarrhea denoted respectively by the presence of white crystalline or fecal material on plumage. Confirm that droppings in the pen are of normal consistency and color for the species.

REPRODUCTIVE SYSTEM

If the female is in production, examine the cloacal area for lacerations or discharge. Palpate the ventral abdominal area to determine distention associated with an internal layer. Relate the presence of an egg to the time of previous oviposition.

In male birds, ensure that the intromittent organ (penis) is normal in appearance and sheathed within the cloaca. In breeding birds, the mucosa of the cloaca should be moist in appearance and deep red in color.

ADDITIONAL DIAGNOSTIC PROCEDURES

If required, the following special diagnostic procedures can be carried out:

- Collection of blood for CBC or serum for clinical pathology, serum for avian influenza, or equine encephalitis titer (Figure 16.8).
- Metal detector or radiographic examination (Figure 16.9) for possible ingestion of a foreign body.
- Radiographic or ultrasound examination of breeding hens to assess the status of the female reproductive tract.
- Feces for parasitology.
- Bacterial cultures from the cloaca (Figure 16.10), reproductive tract, and choanae.
- Intradermal tuberculosis sensitivity test (Figure 16.11).

Postmortem Examination

Frequently, dead ratites are presented to practitioners for diagnostic pathology. It is usually more productive to submit carcasses of neonate and young ostriches, rheas, and emus directly to a state diagnostic laboratory rather than perform a postmortem examination under less than ideal conditions. In the absence of

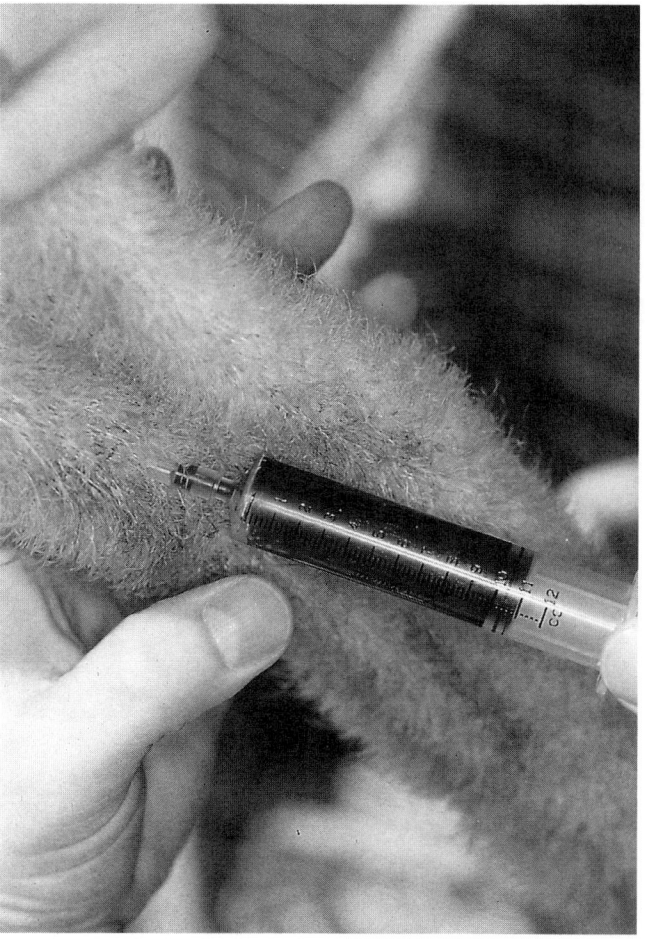

Figure 16.8 Collection of a blood sample from the jugular vein.

experience relating to the normal anatomy and gross appearance of organs and tissues, submission of suitably packed intact cadavers is recommended. This will facilitate diagnosis based on gross examination and subsequent laboratory procedures, including microbiology and histopathology.

Veterinary practitioners can provide owners with referrals to the most convenient avian diagnostic laboratory, or the contact telephone number can be obtained from a state or local emu or ostrich association. Many laboratories will only accept referrals from veterinarians. It is important to adhere to submission requirements, including history, identification of specimens, packaging, method of shipment, and payment. These requirements should be determined in advance.

Carcasses of young ratites and unhatched eggs should be refrigerated, but not frozen, prior to submission. Unhatched eggs or dead chicks should be placed in individual plastic bags, numbered, dated,

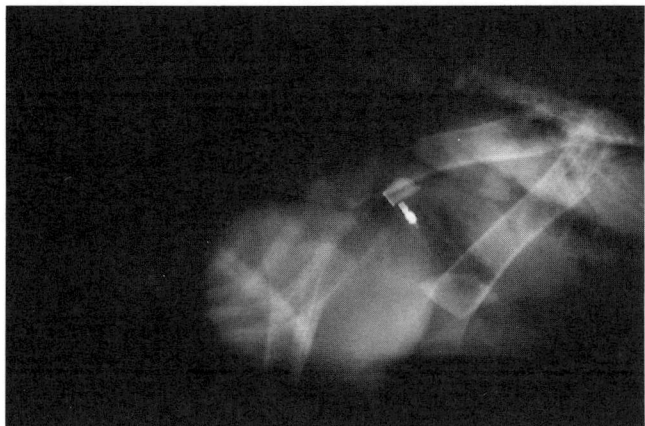

Figure 16.9 Radiograph showing foreign body (cigarette lighter) in ventriculus.

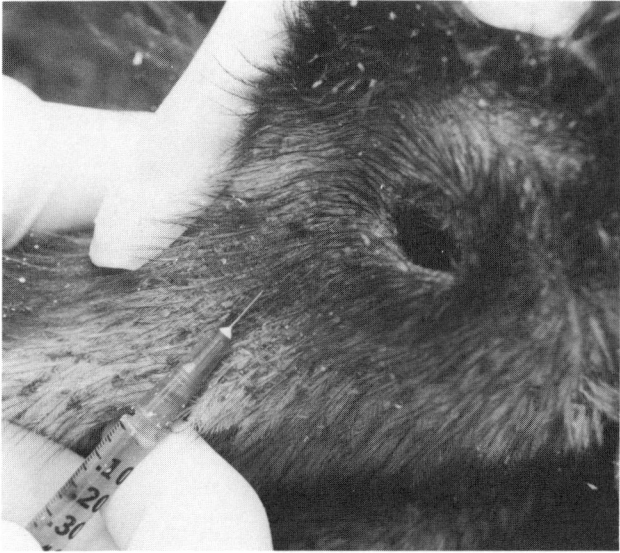

Figure 16.11 Intradermal *Mycobacterium* sensitivity test.

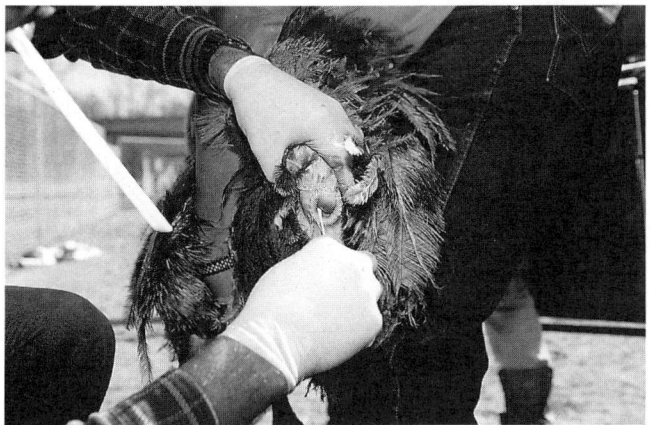

Figure 16.10 Obtaining a cloacal culture for microbiological examination.

and identified according to parents, pen number, or microchip. Individually bagged and refrigerated dead chicks, embryos, or eggs can be placed in a styrofoam chest containing plastic-encased, frozen cool packs. The styrofoam container should be placed in an outer cardboard carton. The submission form detailing the sender's identification and history should be placed in an envelope and sealed to the inside lid of the carton. The outer container should be firmly sealed with adhesive fiber tape.

A brief history describing the background to the problem should be provided. Photocopies of production records, egg weight during incubation, or relevant observations concerning disease conditions, nutrition, or other factors will assist the diagnostician. A full postal address, day and night telephone numbers, and fax number if possible, should be provided to expedite contact. When submitting samples to a laboratory for the first time, a diagnostician or administrator should be consulted concerning submission protocols. Generally, it is inadvisable to dispatch specimens on a Friday or immediately proceeding a holiday weekend, because delays associated with handling may result in deterioration and autolysis, especially during summer.

Where possible, dead ostriches, emus, or rheas should be transported to a diagnostic laboratory for postmortem examination. In the event of an insurance claim or if the possibility of a malpractice claim exists, a comprehensive examination by a suitably equipped and staffed laboratory is essential to obtain an impartial diagnosis and opinion. If it is not practical to submit an entire bird, the veterinary pathologist at the diagnostic facility should be consulted concerning the specimens to be submitted.

The following recommendations are provided to increase the probability of a diagnosis from specimens collected during a field postmortem:

- The examination should be conducted in an enclosed, well-ventilated area with sufficient light to observe and recognize changes in tissues and organs.

- Sterile plastic bags and culturette swabs should be available to collect specimens for microbiological examination. Tissues for histopathology should be placed in 10% buffered saline in adequately labeled, leak-proof containers. A ratio of 1:10 for specimen to fluid should be maintained to ensure adequate fixation of tissue.

- Postmortem examination of adult ostriches and emus requires instruments suitable for equine or bovine

Figure 16.12 Handling biological material with zoonotic potential requires appropriate precautions.

necropsy. Sterile scissors and forceps should be available to collect specimens for microbiological examination.

Prosectors and assistants should be protected from potential zoonotic infection by wearing face shields (Figure 16.12), coveralls, boots, and gloves. It is noted that in eastern equine encephalitis, viremia occurs terminally, and free blood in the body cavity may infect humans. Since ostriches are susceptible to anthrax, blood smears should be examined before commencing postmortem examination of any mature bird which has demonstrated acute mortality, especially from a farm or area where anthrax has been previously diagnosed. Erysipelas has been diagnosed in ostriches and emus, and erysipeloid cellulitis will result from accidental laceration of the skin during necropsy. Rheas frequently die of chlamydiosis and represent a zoonotic danger to human contacts.

- Thorough examination of the exterior of the subject should be carried out, noting abrasions, lacerations, evidence of leg or wing fractures, or cervical dislocation. Cloacal and choanal swabs can be obtained, and evidence of hemorrhage or discharge from the cloaca nares or oral cavity should be noted.

- The oral cavity, trachea, and esophagus should be examined for changes characteristic of specific diseases. It is advisable to collect suitable sections of trachea, esophagus, oral mucosa, conjunctiva, and brain for histologic examination.

- The body cavity can be opened by careful incision of the midline, extending from the cloaca to the xiphoid process. The incision should be in an anterolateral direction extended through the rib cage using shears.

- Sections of liver and spleen (1 cm^3) should be obtained for microbiological examination using aseptic technique. The gastrointestinal tract should be examined with specific reference to possible accumulation of fibrous material, sand, or rocks in the proventriculus and ventriculus, consistent with impaction. The jejunum, large intestine, and ceca should be opened and examined for the presence of endoparasites or evidence of enteritis. The kidneys can be evaluated for the presence of nephrosis or accumulation of urate. The reproductive status of the female tract should be evaluated, and swabs should be obtained from the lumen of the oviduct and from ovules, if present.

- The heart, lungs, and major vessels should be examined, and a blood swab should be obtained from the right atrium for microbiological examination.

- The lungs should be removed from the body cavity and examined for bacterial or mycotic infection.
  Sections of tissue showing any gross abnormality should be submitted for microbiological and histological examination. As a routine, it is advisable to submit a 2 cm × 0.5 cm section of all major organs in addition to skin and skeletal muscle.

- Femorotibial (stifle) and tarsal (hock) joints should be opened and examined for evidence of arthritis. Swabs of exudate should be submitted for microbiological examination.

- Four unstained blood smears obtained from a peripheral vessel should be submitted for hemoparasite examination.

- Ingesta representing the ventriculus, mid-jejunum, cecum, and terminal intestinal tract should be submitted in separate, labeled bags for microbiological or parasite examination. If clostridial enteritis or myositis is suspected, a representative section of ligated intestine or muscle should be submitted with a request for isolation and identification of anaerobic bacteria.

Specimens for microbiological and histological examination must be packed in leak-proof containers conforming to the requirements of the carrier.

Most laboratories will be able to provide a preliminary diagnosis within 24 hours based on gross examination. Bacterial examination requires at least one week for culture and identification. Viral isolation may require as long as 3 weeks if successive passage on eggs or tissue culture is required. Most laboratories will provide a written report within 2 weeks, but special procedures, including serology or comprehensive microbiology, may delay submission of the final report.

Appendix 16.1

Veterinary Certificate for Insurance of Ratites

Species	Emu	Ostrich	Rhea
Sex		Male	Female
Date Hatched		mo / yr	
MicroChip		— —	

OWNER(S) _____
ADDRESS _____
CITY _____ STATE _____ ZIP _____
PHONE ( ) _____ (W) / ( ) _____ (H)
DATE OF EXAM _____ PLACE OF EXAM _____

IDENTIFICATION: Plumage Color: _____ Height of back: ___ ft. ___ in. Weight ___ lbs.

SYSTEM EVALUATION	Normal	Abnormal	Explanation if abnormally detected/observed
Locomotory			
Integumentary			
CNS			
Gastrointestinal			
Reproductive			
If breeding:	No. eggs produced (F)		State time period:
	No. progeny sired (M)		State time period:
VACCINATIONS (past year)	Antigen:		Dates Administered:
*PREVIOUS CLINICAL PROBLEMS OR SURGERY			
*PREVIOUS REPRODUCTIVE PROBLEMS			
*RESULTS OF SPECIAL LABORATORY EXAMS	Fecal	Serology	
	CBC/Clin. Chem.	Radiology & Ultrasound	
*ADDITIONAL COMMENTS RELATING TO SUBJECT, OTHER BIRDS, OR RANCH			
EXAMINING VETERINARIAN (please print)	Name		License
	Address		
	Telephone ( )		

*PLEASE ATTACH SUPPLEMENTARY COMMENTS AND REPORTS

I certify to the best of my knowledge and belief that no clinical or disease-related circumstances exist that would affect the insurability of the identified bird, and that the identified bird was examined in accordance with acceptable standards of diligence and professional judgment.

_____   _____   _____
Signature   Date   Time

THIS CERTIFICATE VALID FOR 15 DAYS ONLY FROM DATE ISSUED

Chapter 17

Jurisprudence

Simon M. Shane

INTRODUCTION

The purpose of this chapter is to acquaint practitioners and owners with the broad principles of law pertaining to ratite medicine. Both veterinarians and owners are strongly advised to obtain professional legal advice on any matter relating to contractual relationships, establishing corporations, acquiring property, or if an apparently unresolvable issue arises. Prompt referral of situations which may result in litigation will facilitate preparation for a civil action or may result in resolution of a dispute through negotiation or mediation.

Ratites are expensive in relation to companion and food animals and there is considerable potential for injury of humans by mature ostriches. Both veterinarians and owners should therefore be aware of their respective obligations under law and conduct their activities in an ethical and responsible manner to minimize the possibility of disputes. Both veterinarians and farmers must comply with federal, state, and local laws which are enacted for the benefit of society in general and ultimately for ratite owners and practitioners.

CONTRACT LAW

A contract is an agreement between parties that creates an obligation to perform. The following elements comprise a valid contract:

- An offer, which may be either a written or oral statement initiated by a buyer or seller, or a provider or consumer of services.
- Acceptance of the offer by the other party. This constitutes a subjective element of a contract referred to as "a meeting of the minds." As far as possible, offers and acceptances should be in writing, as this reduces the possibility of misunderstanding which may later escalate into an unresolvable dispute.
- Consideration, represented by either payment or a promised action or service, is the third objective element of a contract. Generally, courts will require that the consideration be reasonable in relation to the value of the asset or the service provided. Claims based on unreasonable consideration are usually not enforceable.

In the event of a dispute resulting in a legal action, either party has legal recourse to satisfy the provisions of the written or oral contract. If the court establishes that a breach of contract has occurred, it may enforce specific performance by the defendant or award damages to compensate the plaintiff for losses sustained. In addition, the courts may order restitution if the defendant was unjustly enriched as a result of the breach.

There are a number of current ratite production practices which may result in contractual disputes. These include the following:

- preselling of hatchlings and immature stock,
- boarding of mature pairs with compensation derived from a share in the sale of progeny, and
- syndication and other unconventional but expedient arrangements which are initially beneficial to both parties.

Most disputes will occur as a result of actual or perceived negligence, suboptimal performance of stock, or with falling prices due to a decline in demand relative to supply.

PROFESSIONAL LIABILITY

Equine veterinarians have traditionally represented the greatest risk and generated the highest number of claims against professional liability insurance. This situation is due to the value of patients, expectations of clients for recovery and restoration of function and performance, and the probability of injury associated with handling large and sometimes fractious animals. In many respects, ratite practitioners are subject to the same risks, and it is anticipated that claims from this segment of practice will increase as the current young ratite population matures and is served by the profession during the next few years.

NEGLIGENCE

Veterinarians and owners should be aware of the law of negligence, which is a component of civil tort law. Negligence is represented by imprudent actions, commissions, or omissions by an individual which result in harm or injury. The courts will compare the actions of the defendant with those of a rational and careful person under similar circumstances. Injury to a client or bystander by an inadequately restrained ostrich could result in an action against the veterinarian under the law of negligence. An injury to a person which was not due to any imprudent act or omission by the veterinarian would not represent negligence, provided the courts accept that the occurrence could not have been foreseen by a reasonable person.

Few emu and ostrich ranches have adequate facilities to restrain stock for examination and treatment. Currently, most breeders are housed in small units of up to 4 pairs, and neither fixed nor portable chutes are available for restraint. Practitioners will become more aware of the risks of negligence claims associated with injury of owners and their stock. Owners will either have to provide catching and restraining installations and trained manpower, or forego veterinary services. Alternatively, veterinarians may have to provide portable or mobile chutes which will require thorough decontamination after use on each farm. Inevitably, clients will have to compensate veterinarians for additional capital and operating costs in addition to escalation in liability insurance.

MALPRACTICE

Malpractice is an extension of negligence and relates specifically to the provision of professional services. Malpractice is represented by an action by a practitioner at variance with the services performed by a skilled, competent professional exercising due care and diligence.

Disputes between veterinarian and client may arise as a result of diagnosis, treatment, or its outcome. All veterinarians are obliged to carry professional liability insurance to protect against a successful negligence or malpractice claim. Generally, the veterinary profession is less vulnerable than physicians with regard to malpractice, but negligence claims may arise from injury of owners and bystanders during examination or treatment. Although few malpractice claims against veterinarians are successful, the contingency fee system favors defendants initiating claims against veterinarians.

Legal action requires an investigative phase (discovery), detailed oral examinations (depositions), discussions with legal advisors and experts, followed by the actual trial. All these stressful, time-consuming and expensive stages should, as far as possible, be avoided by both plaintiff (client) and defendant (veterinarian) by mediation and compromise.

Claims arise from the following causes:

- Poor communication between veterinarian and owner, especially regarding the outcome of a case.

- Undue expectations on the part of the owner with regard to restoration of function or reproductive efficiency.

- Dispute over fees.

- Improper certification for purchase, insurance, or interstate movement.

- Provision of substandard service by the practitioner.

- Personal disposition of the client or outright avarice.

RESOLUTION OF DISPUTES

Generally, claims arise as a result of a number of factors which have their basis in misunderstanding over diagnosis, treatment outcome, and costs. Rarely does a practitioner fail to apply the skills and diligence expected of a careful professional. Claims and disputes can be resolved by methods other than litigation.

- Negotiation between the practitioner and client, especially in the case of a dispute over a fee, can generally resolve issues if both parties approach the matter in good faith.

- Mediation by an impartial third party can sometimes result in an equitable decision. Mediation is relatively inexpensive and a decision can be obtained fairly quickly. The result usually requires concessions by both parties but is not binding. The mediator's decision, if unacceptable, to either the veterinarian or client can be subsequently referred for legal action.

- Arbitration involves a formal review of the circumstances of a case by an impartial arbitrator or panel. Both parties must agree in advance to be bound by the decision of the arbitrators who function according to an accepted code of practice. Generally, arbitration is cheaper and quicker than a formal court action and can be regarded as midway between mediation and a civil case requiring pleadings, discovery, and a trial.

INFORMED CONSENT

It is generally accepted that informed consent by an owner is necessary before commencing any program of treatment which may range from antibiotic therapy to extensive surgery. For the owner to provide consent for a procedure, it is necessary that the veterinarian fully explain the diagnosis (the cause of the disease or condition), alternative methods of treatment, relative costs, and a prediction of outcome with alternative treatments. If the owner comprehends the information provided by the veterinarian, a rational and informed decision can be made. The legal requirements relating to informed consent vary according to locality within the United States. Some jurisdictions hold that the information normally provided by a reasonable practitioner working in the same community would represent adequate disclosure. Courts and other jurisdictions may decide that the information provided should conform to what any reasonable practitioner would convey to a client anywhere in the United States under similar circumstances.

NEGLIGENCE AND MALPRACTICE CLAIMS

Claims against veterinarians in terms of liability insurance are strenuously defended and complaints without merit do not succeed. The process of initiating or defending an action can be time consuming and expensive to both practitioner and client. In the event of a malicious suit, the practitioner may countersue claiming defamation and reimbursement for expenses involved in case preparation. Although judgments against plaintiffs are rare if malice is proven, or if there was insufficient cause to have initiated the action, the practitioner may succeed in a countersuit and obtain damages.

To successfully pursue a claim for malpractice, it is necessary for the plaintiff (client) to establish the following four elements:

- To demonstrate that the veterinarian had a duty and obligation to the client.
- To prove that the practitioner provided inferior service or committed errors constituting substandard care.
- To show proximate cause, relating the action of the practitioner to the specific outcome of the case.
- To quantify a loss resulting from the veterinarian's action or neglect.

These elements are considered in greater detail below.

Duty and Obligation

Establishing the duty of the veterinarian would be based on medical records confirming agreement by the practitioner to undertake the case. Veterinarians do not have to accept all cases presented. Once having initiated treatment, services should be rendered to the best of the practitioner's ability until either successful resolution of the case or termination of the relationship as requested by the owner. It is expected that practitioners will exercise reasonable care consistent with the standards of the veterinary profession. Emergency treatment outside the conventional practitioner-client relationship does not obligate a veterinarian to continue treatment which may be initiated only to save life or relieve suffering. Generally, veterinarians acting either on their own initiative or at the request of owners, or some representative of the owner, can render emergency treatment and be absolved of liability in terms of samaritan laws which have been enacted in various states. Under specific circumstances, veterinarians who render emergency service but commit acts of gross negligence may be liable for damages directly arising from their actions.

Standard of Care

The courts will generally judge the activities of a veterinarian in terms of what a reasonable and careful practitioner would do under similar circumstances. In some states, a "similar locality" rule exists which would compare the actions of a veterinarian with colleagues in that area. Other jurisdictions would hold to a broader standard reflecting either the veterinary profession in the United States or veterinarians specifically engaged in the practice of ratite medicine. A decision concerning the appropriateness of the actions of the veterinarian and adherence to acceptable standards is based on the submissions of experts, duly accepted by the court, based on their qualifications and experience. Experts review evidence, including clinical records, radiographs, and depositions. They render opinions and submit to cross examination by opposing attorneys and may be required to justify their opinions to the court.

Actions for malpractice may be based on acts of commission. Severing a nerve or a major blood vessel while draining an abscess would not be acceptable practice and would be regarded as an act of commission. Leaving a surgical case before complete recovery from an anesthetic or neglecting to revisit a patient could be regarded as an act of omission, in the event of an unacceptable outcome.

The unprecedented expansion in ratite production has created a demand for the veterinary profession to provide services to owners. A number of veterinarians in private practice and university faculty have acquired experience and training in the broad areas of ratite surgery and medicine or in specific disciplines, based on food animal medicine, anesthesiology, surgery, dermatology, or ophthalmology. Practitioners with advanced training or board certification in a specific field may be regarded as specialists. As such, they command greater remuneration but are held to a higher standard of practice. A veterinarian without formal training or qualifications, who assumes the privileges and status of a specialist and undertakes procedures which would normally be carried out by a suitably qualified veterinarian or team of specialists, is vulnerable to claims for malpractice. The standard of a specialist will be applied when assessing the care and services provided. Practitioners are strongly advised to refer cases to specialists or university clinics and hospitals if special skills or equipment are required.

Proximate Cause

To succeed in a claim for malpractice, the plaintiff must show a direct relationship between the actions of the practitioner and the outcome. The practitioner may successfully claim as a defense that the outcome in question could not have been predicted based on the presentation of the case, including history, clinical examination, and the radiographic and laboratory findings.

Since primary and contributory causes of death of a patient are critical to establishing proximal cause, it is essential that a thorough postmortem examination be conducted in the event of death. Cadavers or tissues from birds subject to euthanasia as a result of severe locomotory or neurologic dysfunction should be submitted for evaluation by a board certified pathologist. Referral of terminal cases to a regional diagnostic laboratory or university department of pathology is strongly recommended to protect both the interest of the client and the practitioner. In addition to a comprehensive gross examination, pathologists can perform histological, toxicological, and microbiological examination of tissues.

Findings are incorporated into a comprehensive case report which addresses the primary and contributory causes of death. It is possible that a bird may die of an intercurrent infection or congenital abnormality not apparent prior to routine surgery. In the event of death, adequacy of surgical technique, including hemostasis and integrity of suture lines, can be confirmed. Often when all the facts are disclosed, the practitioner is vindicated and the aggrieved owner is placated. Thorough postmortem examination resolves conflicts and obviates the need for unnecessary claims based on an incorrect presumption of malpractice.

DAMAGES

The fourth element of a claim for malpractice requires the owner to quantify a loss. The death of a ratite or failure to restore reproductive or locomotory function will be reflected in a specific claim consistent with the prevailing market value of the bird. The courts may award compensatory damages based on either purchase price or fair market value.

Courts may award consequential damages to owners based on a reasonable estimate of losses sustained as a result of mortality or reduced activity. For example, the death of one of a pair of emus bonded and proven as breeders in early September may result in loss of production for the complete season. Even if it is possible to purchase a male or female of equivalent age, incompatibility between the newly mated pair may result in infertility. The producer would forego income from the eggs produced by the hen. The value of the loss could be calculated on the basis of egg production, hatchability, and chick viability of the pair during the previous season and from other pairs housed on the farm during the specific year. Market value of chicks in the given year would be used to calculate the loss.

Under special circumstances, the courts may award punitive damages if the activities of the veterinarian are regarded as willful or wantonly negligent. It is necessary to show malice or fraud in relation to the unprofessional acts on which the successful claim for malpractice is based. Knowingly issuing a false certificate of health or acting for both buyer and seller in providing a prepurchase certificate would be regarded as a willful and fraudulent act. This might render the practitioner liable for punitive damages which may exceed the compensatory and consequential damages awarded.

AVOIDING CLAIMS FOR NEGLIGENCE

Veterinarians should ensure that ratites are suitably restrained.

Only personnel directly involved in examination and treatment should be in the vicinity to avoid possible injury to bystanders.

If owners insist on participating in the handling and treatment of their birds, the veterinarian should obtain a written indemnification in addition to ensuring that procedures are carried out with due care for the health and safety of the owner, farm employees, and the patient(s).

To avoid malpractice claims, practitioners should never guarantee cures or directly or indirectly create a false anticipation of success.

Clients should be made aware of the risks associated with any procedure or treatment.

In the event of restraint, anesthesia, or surgery, the owner's formal, written consent should be obtained.

Euthanasia should not be carried out without the owner's written consent or, in an emergency, from the owner's representative supported by telephoned or faxed permission of the insurer.

Detailed and comprehensive records should be maintained relating to all diagnostic procedures and treatments. All patients should be positively identified, preferably by a previously inserted microchip.

Veterinarians are obligated to refer to specialists, cases which are considered to be beyond their experience, skills or resources available.

Owners should be apprised of risks associated with zoonoses, including but not limited to chlamydiosis in rheas and avian tuberculosis and eastern equine encephalitis in emus.

All prescription medications should be adequately labeled and, where appropriate, dispensed in child-proof containers.

RESTRAINTS ON THE USE OF DRUGS

Pharmaceuticals are controlled by the Food and Drug Administration, Center for Veterinary Medicine. This agency implements the Food, Drug and Cosmetic Act as amended and strictly controls the registration, approval, labeling, and use of all drugs.

The Food and Drug Administration imposes restraints on veterinarians with regard to dispensing and administration of drugs. Since ratites are regarded as minor species, all drugs are used in accordance with "extra-label" provisions. It is possible for a veterinarian to prescribe and administer legally available drugs provided the following conditions are satisfied:

- The drug should be approved for a companion or food animal species.
- The agent should be in common use by practitioners without adverse effect in the species.
- The drug should be administered by a veterinarian in terms of an ongoing professional-client relationship. Therapeutic response and possible side effects should be monitored, and the practitioner should be familiar with the farm and flock.
- The owner should be informed of and acknowledge the extra-label status of the drug and should be apprised of any known or foreseeable side effects or adverse reaction. Some chemotherapeutic agents, such as gentamicin, can severely damage the avian kidney but are generally innocuous in mammals.

BIOLOGICAL PRODUCTS

Biologics are controlled by the United States Department of Agriculture, Animal and Plant Health Inspection Service in terms of the Virus-Serum-Toxin Act as amended. This agency enforces the licensing, manufacture, labeling, and distribution of veterinary biologicals to ensure potency and safety.

When eastern equine encephalitis appeared in emu flocks in Louisiana in 1992, it was necessary to implement a program of preventive immunization using an inactivated equine vaccine. In the absence of available flocks for experimental evaluation, owners of birds considered at risk volunteered their flocks to receive vaccine. A value judgment decision was made to vaccinate birds based on the risks involved and the fact that an inactivated vaccine was to be used. Owners were initially required to sign a consent form and release university faculty and referring practitioners from responsibility for adverse reactions. In addition, the pilot program required flocks to be clinically monitored, and the serologic response to vaccination was determined.

PESTICIDES

Veterinarians and owners should use only insecticides approved by the U.S. Environmental Protection Agency, the Food and Drug Administration, and state regulators where applicable. There is a narrow range of products suitable for prevention and treatment of the ectoparasites of ratites, and application should be carried out strictly in accordance with federally approved labeling. This will prevent environmental contamination and reduce the possibility of toxicity in either applicators or treated flocks.

Index

abdominal air sacs, 96, 119
abdominal procedures, 101
abdominocentesis, 55, 56
abnormal embryonic positions, 63, 67A, 72
abnormalities
　acquired, 20
　congenital, 20
　congenital leg, 20, 150
abrasions, 156, 170
acanthocephala, 119
acaricide, 116
acepromazine, 81
acholeplasma, 132
acid-base balance, 20
acid-base imbalance, 20
acid detergent fiber (ADF), 18
acid fast organisms, 133
Acinetobacter spp., 53
acquiring property, 173
act of commission, 174
act of omission, 174
acute phase serum, 139
adenocarcinomas, 56
adenovirus, 141, 143
adipose tissue, 3
administration, 155, 156
Aegyptianella pullorum, 122t, 122
aegyptianellosis, 122
Aeromonas, spp., 144
aerosol, 141
agar gel diffusion test, 137, 140
aggression, 42, 79
air cell, 51, 62, 65
air sacculitis, 40
　fibrinous, 133, 140, 142
airway occlusion, 86
alanine transaminase (ALT), 111
albumin, 51, 110
alfalfa, 22t
alimentary tract disorders, 110
alkaline phosphatase (ALP), 111, 112
all-in-all-out, 37

allometric scaling, 155
ambient air, 63
Amblyomma spp., 115t
amino acids, 14
aminoglycoside toxicosis, 109
aminopeptidase, 11
ammonia, 38
ampulla, 48
amylase, 12, 112
anaerobic culture, 134
analgesic agent, 92
anaphylactic shock, 116
anemia, 109, 122
　blood loss, 109
　deficiency, 109
　hemolytic, 109
　nonregenerative, 109
　regenerative, 109
anesthesia, 79–94
　complications, 90, 91
　monitoring, 87
　recovery, 99
anesthesiology, 79–94
anesthetic protocols, 90–99
anesthetic recovery, 91
aneurysm, 109
angular limb deformities, 102, 150, 152
anisocytosis, 109
anorexia, 112, 143
antagonist to anesthetic agents, 84
anthelmintic therapy, 115–124, 157
anthrax, 135, 170
antibiotic
　quinolone, 128, 129, 132
　therapy, 157
antibodies, 132, 133, 139, 140, 141, 144
antibody titer, 134
anticholinergic agents, 81
antioxidants (feed), biological, 18, 21
aortic rupture, 21, 153
apnea, 90

arbitration, 174
arbovirus, 138, 144
arginine, 14
Argus persicus, 115t
arrhythmias, 90
arterial blood pressure, 88
arthritis, 130
　purulent, 130
　seropurulent, 132
artificial incubation, 59, 65
arytenoid cartilages, 85
ascarid eggs, 118
Ascaridia spp., 117t, 118
aspartate transaminase (AST), 111
aspergillosis, 22, 38, 107, 136
Aspergillus flavus, 22
Aspergillus fumigatus, 137
aspiration, 80
assisted hatch, 65, 72
ataxia, 154
atropine, 81
auctions, 33
auscultated, 88
autogenous inactivated bacterins, 135
avian
　biochemical analysis, 109–110
　hematology, 106–110
　homeostases, 13, 110
　influenza, 139
　pox virus, 143, 157
　tuberculosis, 107, 132
avibirnavirus, 142
avidin, 51
avipox, 143
award damages, 176
axilla, 10
Ayre's T piece, 85
azaperone, 81

Bacillus anthracis, 135
bacterial infection, 53, 112, 128–136
bactericidal, 35

Baermann apparatus, 124
Bain's nonrebreathing system, 85
bands, 166
barbules, 1
barley, 15, 22t
basal metabolic rate, 115
Basidia, 137
basophilia, 108
basophilic stippling, 109
basophils, 108
Baylisascaris columnaris, 117t
Baylisascaris procyonis, 117t
Baylisascaris spp., 113, 117t, 120
B-carotene, 18
beak, 88, 167
behavioral problems, 52
bile, 11
    acids, 111
    pigments, 4
bile duct hypoplasia, 142
bilirubin, 112
biochemical assay, 105, 109
biological antioxidant, 18, 21
biological products, 127
biosecurity, 31–38, 127, 131, 143, 165
biotin, 19
    deficiency of, 152
bird stocks, 45
biting midges, 116
blackflies, 116, 122
Blasocystis hominis, 123
blood
    -brain barrier, 120
    clotting, 19, 20
    collection, 105
    collection technique, 105
    film, 105
    hemolyzed, 110
    loss related anemia, 109
    pH, 89
boarding, 173
body condition, 80
body temperature, 80t
bonded pairs, 41
bone
    intramedullary, 110
    ischial, 3
    pneumatic, 103
    screws, 103
booming, 7
Bordetella spp., 113
Borna disease, 144
Bornavirus, 144
brachial arteries, 88
brachial vein, 7

bradycardia, 81
breach of contract, 173
"breech" position, 147
breeder nutrition, 18, 23, 60
breeder pen, 42
breeders, 39, 42, 51, 57
breeding cycle, 49
    out of season, 4
    status, 99
brooder temperature, 148
brooding unit, 144
budding yeast, 137
bursal depletion, 122
bursa of fabricius, 6
butterfly catheter, 83

cachexia, 112, 139
caeca, 5
caecal threadworm, 118
calcium, 19, 48, 55, 106, 110, 147, 150
calcium binding protein, 110
calmus, 116
campylobacteriosis, 131
Campylobacter jejuni, 131, 137
Campylobacter spp., 131
Candida albicans, 137
Candida mucor, 137
candidiasis, 137
candling, 52, 64, 147
capnography, 89
carbarsone, 123
carbaryl dust, 115, 121
carbohydrates, 15, 17
carbon dioxide, 15, 60
carbonic anhydrase, 57
carcass analysis, 168–171
cardiac arrest, 90
cardiac compression, 90
cardiac LDH, 111
cardiodepressant, 81
cardiomyopathy, 152
cardiorespiratory depressant, 81
cardiovascular system, 87, 90
carriers, 32
cartilage, 21
    arytenoid, 85
casque, 2
cassowary, 1, 2, 4, 5, 8
cataracts, 167
catheter placement, 87
caudal vascular ligaments, 48
cecum, 5, 118
cellulitis, 40, 130
cellulolysis, 15, 18
cellulose, 16, 17, 18

central nervous system (CNS), 150, 167
centrifugal floatation, 124
cereals, 22t
cerebral nematodiasis, 120
cestodes, 119
chain-link fencing, 39
Chandlerella quiscali, 116, 117t, 120, 157
Chapmania tauricollis, 117t, 119
chemical restraint, 81–83
chemical sedation, 44, 81–83
chemistry panel, 112
chenodeoxycholic acid, 111
chewing lice, 116
chick
    edematous, 60, 62, 67A, 70
    hobbling, 150
    malformed, 67A
    mortality, 122
    small, 67A
    sticky, 67A
    viability, 60
Chlamydia psittaci, 133
chlamydiosis, 107, 133, 170
chloride, 19
chlorine compounds, 35
cholesterol, 11
choline (B6), 19
    deficiency, 152
chondrodystrophy, 20, 150, 152
chondrotomies, 102
chorioallantoic
    inoculation, 140
    membrane, 54
chromium, 21
ciliates, 123
circulatory collapse, 41
citrated blue-top tubes, 105
civil action, 173
civil tort law, 174
clavicle, 3
clitoris, 49
cloacal
    prolapse, 122, 153
    swab, 43, 140
    temperature, 89
cloacal-rectal reabsorption, 110
closed flocks, 116, 134, 140
clostridial enteritis, 134
clostridial enterotoxemia, 134
clostridial myositis, 134
Clostridium chauvoei, 134, 157
Clostridium colinum, 134
Clostridium difficile, 134
Clostridium perfringens, 134
Clostridium spp., 18, 134, 144

Index

clutch, 50, 55
cnemial crest, 3
coccidia, 122
  oocysts, 122
coccidiosis, 122–123, 134
Codiostonum struthionis, 117t, 118
colibacillosis, 129
colon, 12
commissions, 174, 175
compacted dirt, 96, 152
compensatory damages, 176
complete blood count, 112
concrete, 37, 38
conjunctivitis, 38, 121
Contortospiculum rhea, 119
contract, written, 173
contract law, 173
convection ventilation, 38
convulsive behavior, 83
copper, 21, 149
copper sulfate, 138
coprodeum, 5
coprourodeal fold, 5
copulation, 50, 52
coracoid, 3
corn, 15, 22t
cornea, 38
corneal reflex, 88, 90
coronavirus, 143, 145
corticosteroids, 90, 107
corticosterone, 110
countersue, 175
coverslip smear, 105
coxofemoral joints, 150
cranial vascular ligaments, 48
creatine kinase (CK), 111
creatinine, 110
cresols, 35
crop, 3
crude fiber, 17
crura, of the phallic bodies, 9
Cryptosporidium spp., 122t, 122, 144
cuff inflation, 86
culicoides, 32, 116, 138
Culicoides crepuscularis, 116, 120
curled toe, 102, 151
curled toe paralysis, 151
cuticle, 61, 70
cyanoacrylate glue, 101
Cyathostoma variegatum, 117t, 121

damages, 176
  cellular, 111
  decontamination, 35
  defendant, 174

deficiency anemias, 109
dehumidification, 70
dehydration, 109, 112, 144, 148
Deletrocehalidae, 117t
depopulation, 37
depression, 34, 95, 152
dermatitis
  bacterial, 116
  necrotizing, 56
dessication, 131
Desulfovibrio, 131
detergents, 35
detomidine, 81
developmental anomalies, 147–154
developmental orthopedic disease, 149–51
developmental problems in young ratites, 147–155
dew point values, 74
dextrin, 17
diabetes mellitus, 110
diaphysis, 102, 103
diarrhea, 122, 123, 138, 142, 144
diazepam, 81
dicalcium phosphate, 20
Dicheilonema rhea, 117t, 119
Dicheilonema spicularia, 117t, 120
dietary energy, 15, 16
differential counts, 107, 108
digestibility, 16
digestibility coefficient, 11
digestible energy (DE), 15
digestible enzyme, 4, 12
digestive system, 3
digital arteries, 88
digital cushion, 2
1,25-dihydroxycholecalciferol, 18
dimetridazole, 123
diptheric pox form, 143
direct complement fixation, 133
direct life cycle, 116
disease prevention in industry, 33
disinfection, 35
dislocation of cervical vertebra, 37
dissociative combinations, 88
DNA finger printing, 49
DNA sexing, 49
dorsal groove, 8
dosages, of anesthetic agent, 82t
double fencing, 39, 42
doxapram, 90
doxycycline, 55, 133
drainage, 38, 101
drug excretion, 81
drug metabolism, 81, 155
drug resistance, 33, 70, 157

drumming, 50
dry bulb isotherms, 74
dry bulb temperature, 74
ductus deferens, 48
duodenum, 4, 12
duty and obligation, 175
dyspnea, 143
dystrophic calcification, 148
dystrophic calcium deposition, 148

early embryonic mortality, 52, 67A
early hatch, 62, 67A
eastern equine encephalomyelitis, 113, 138, 156, 157, 170
E. coli (see *Escherichia coli*)
ectoparasites, 115
edema, 141, 186
  ischemic, 10, 103
EDTA, 105, 108, 110
egg, 54, 124
  bound, 55
  collection, 60
  disinfection, 35, 36, 61
  drop syndrome, 141
  formation, 51, 57
  peritonitis, 51, 55, 100
  production, 48, 53
  quality, 57, 59
  retention, 98
  sanitization, 61, 70
  size, 57, 60
  storage, 61
  turning, 64, 65, 70
  uniformity, 59
eggshell, 32, 51, 57, 136
egg yolk lipids, 17
ejaculation, 50
ejaculatory ducts, 48
electrocardiogram display, 88
electroejaculation, 50
electron microscopy, 136
elementary bodies, 133
elementary body agglutination, 133
emaciation, 132
embryo, 63
  egg position, 63
  respiration, 51
embryonic malformations, 19, 20, 21, 62
embryonic malpositions, 67A, 72, 147
embryonic malposition II, 63, 147
embryonic mortality, 52, 62, 128
emu, 41–46, 98
  handling, 43
  hatchery management, 69–79

nutrient requirements, 23, 28
encephalomalacia, 19
endoparasites, 116
endoscopy, 95
endotracheal tube, 91
end-tidal carbon dioxide, 89
energy, 15, 16
enteritis, 128, 142
    hemorrhagic, 141
    necrotizing, 130
    pseudomembranous, 134
    ulcerative, 137
Enterobacteriacea, 55
enterotoxemia, 134
enthalpy value, 76
enzyme immunosorbent assay antigen detection, 133, 144
enzymes, 11
eosinophilia, 108
eosinophils, 106, 108
epidemiology, 127
epididymis, 48
epiphyseal endplates, 151
epiphysis, 103
epithelization, 101
equine encephalitis virus (*see* virus)
equine nasogastric tube, 97
erectile tissue, 8
erysipelas, 129, 170
erysipelas bacterin, 129
Erysipelothrix rhusiopathiae, 129
erythrocyte counts, 108
erythrocytic anisocytosis, 109
erythrocytic ballooning, 109
erythrocytic polychromasia, 109
erythromycin, 131
erythropoiesis, 109
Escherichia coli, 32, 53, 129, 131, 132, 137, 144, 148
esophageal lacerations, 102
esophageal lumen, 97
esophageal stethoscope, 88, 89
esophagotomy, 97, 156
esophagotomy tube, 98
esophagus, 3
establishing corporations, 173
estrogen, 53, 110
etiology, 127
etorphine, 81
euthanasia
    written consent for, 172
excitable induction, 83
excitable phase, 81
excitable recovery, 83
excitation, 83, 108
exercise, 38, 42, 102
exploratory celiotomies, 101

extraction ventilation, 38
extraembryonic membranes, 51
extra label provisions, 177
extra label use, 138
extravasation, 154
exudates, seropurulent, 135
eye, 9
eyelids, 143

facilities, 165
fading chick syndrome, 122, 142, 145, 153
farm selection, 33, 34
fasting, 80
fats, 15, 17
fatty acids, 15, 17
feather mite, 116
feathers, 2, 20, 21
fecal flotation, 124
fecal moisture content, 19
fecal route, 116
fecal specimens, 124
feces, 35
feed ingredients, 20
feeding the debilitated patient, 156
feed refusal, 20
feed storage, 19
femur, 3, 150, 151
    fractures of, 103
fenbendazole, 118, 119, 121
fence rubbing, 41
fencing, 38, 41
fermentation, 16, 18
fermentative digestion, 12
fertility, 51, 52
fiber, 17
fibrinous exudate, 130
fibroelastic tissue, 4
fibrolymphatic bodies, 8
Figure-8 hemicerclage wire, 103
fistulas, 98
flagellated protozoan, 122
flagellates, 122
flies, 116
flock, 52
    moribidity, 102
    mortality, 118
flooring, 38
fluid administration, 86, 156
fluid maintenance requirement, 156
fluid replacement, 156
flukes, 121
fluorescent antibody, 132, 133
folacin, 19
Foley catheter, 54
follicle-stimulating hormone (FSH), 49, 50

follices, 50
    F1, 50
    F2, 50
    F3, 50
follicular maturation, 50
folliculocentesis, 56
forages, 22, 118
forced air unit heaters, 38
forced air ventilation, 38
foreign body, 95, 152
formalin, 35, 124
fossa, 56
fracture
    compound, 102
    long bone, 103
    open comminuted, 102
    pelvic limb, 102
frostbite, 56
fructose, 17
fungal infection, 113, 136
fungal spores, 124, 136
fungicides, 136
furazolidone toxicity, 152

gait, 120
galactose, 17
gallbladder, 6, 111
gamma glutamyl transferase (GGT), 112
gamonts, 122
gapeworm, 121
gastrocnemius tendon, 150, 151
gastrointestinal tract, 168
general appearance, 167
genetic disease, 150
genetics, 59
genital duct, 7
genital eminence, 8
genital mound, 8
genital tract, 7
gentamicin, 157
geophagia, 19, 152
germinal disc, 47
Giardia, 122
Gimenez-staining techniques, 133
gizzard, 4, 12, 117
globulins, 110, 111
glomerular filtration, 109
glottis, 85, 156
glucose, 17, 91, 106, 110
glutathione peroxidase, 21
glycine, 14
glycogen, 17
gnats, 116
gonadotropins, 56
grackle, 120
granulation tissue, 101

granulomas, 133, 136
granulosa cell tumors, 56
grasses, 22
gross energy (GE), 15
growth plate, 102
gums, 17

Haemophilus spp., 115t, 134
harvested skin, 101
hatchability rate, 19, 21, 59, 65, 147
hatcher-induced hypoxia, 149
hatchery environment, 32, 50, 65
hatchery hygiene, 66
hatching, 65
hay, 38, 40
health certificates, 34, 176
health records, 176
heart, 88, 122
heart blood swab, 170
heart rate, 88
heated floors, 38
heat increment, 16
heating systems, 38
helminths, 113, 116–121
hemagglutination (HA), 139
hemagglutination inhibition procedure, 139
hematocrit, 108, 109
hematocytometer methodology
   direct manual, 106
   indirect manual, 106
hematology values, 106, 107t
hemicellulose, 17
hemicircumferential periosteal transection, 150
hemoconcentration, 109
hemoglobin, 21, 108
hemolytic anemia, 109
hemoparasites, 116
hemorrhage, 154
   subcutaneous, 154
   subepicardial, 133
hemorrhagic
   diarrhea, 138
   enteritis, 131
   oophoritis, 141
heparinized green-top tubes, 105
hepatic necrosis, 134, 140
hepatocellular damage, 111
hepatomegaly, 128, 129, 131
Heterakis gallinarium, 122
heteropenia, 108
heterophilia, 106, 108
heterophils, 106
   immature, 108
   toxic, 108, 149

Hexamita, 122t
hides, 115
hippoboscid fly, 116
histidine, 14
Histomonas meleagridis, 122, 122t
histomoniasis, 122
histopathology, 169
Hitchner B1 strain, 141
hock joint, 150, 152
hooding, 37, 79
horizontal incubation, 64
horizontally integrated cooperatives, 31
horizontal transmission, 37, 142, 143
hormones, 49, 50
housing, 37, 39, 41, 42
Houttuynia struthionis, 117t, 119
hulls, 22
   oat, 22
   rice, 22
   soybean, 22
humerus, 3, 40
humidity, 62, 70, 100, 148
Hyalomma truncatum, 115t
hydration, 80
hydropericardium, 139
hygiene, 37
hyperemia, 134, 144
hyperglycemia, 110
hyperkeratosis, 20
hyperphosphatemia, 110
hyperthermia, 41, 86, 92
hypervitaminosis A, 18, 167
hypervitaminosis D3, 147, 148
hyphae, 137
hypocalcemia, 110
hypocapnea, 90
hypoglycemia, 81, 91, 110
hypophosphatemia, 110
hypotension, 90
hypotensive crisis, 90
hypothermia, 86
hypovitaminosis A, 147
hypovitaminosis E, 152
hypoxia, 109
hysterotomy, 90

iatrogenic, 109
iatrogenic aminoglycoside toxicosis, 109
identification, 166
ileum, 5
immobilization, 83
immunization, 166
   passive, 61
   preventive, 166

immunoglobulins, 61
immunosuppression, 145
immunosuppressive infection, 145
Immunosuppressive viruses, 142
impaction, 3
   of the proventriculus, 95, 152–153
implant, 103
imprinted, 50
imprudent actions, 174
inactivated autogenous bacterins, 135, 157
inclusion bodies, 144
inclusion body hepatitis, 141
   hydropericardium syndrome, 141
incomplete tracheal rings, 90
incubation
   air circulation, 62, 70
   delayed dehydration, 62, 70
   humidity, 62, 70, 100, 143
   room, 61
   temperature, 61, 62, 70, 100, 147
   trays, 62, 70
incubator induced hypoxia, 70, 149
incubators, disinfection of, 35
indeterminate layers, 50
indirect blood pressure monitor, 88
induction, 83
   inhalation anesthetic, 83, 84
infectious bursal disease, 112, 142
infectious coryza, 134
infectious diseases, 127–146
infertile, 5
infertility, 52, 67A
infestation, 115
informed consent, 175
infraorbital sinus antibiotic, 156
infrared candling, 70
infundibulum, 47, 48, 51, 55
inguinal area, 101
inhalation agents, 54
   suggested concentrations for, 85t
injury, 40
inner shell membrane, 32, 51, 54, 61, 63
insect vectors, 32, 119, 144
insemination, 50
insurance
   claim, 112
   examinations, 112
   health certificates, 35, 112
integumentary system, 167
intermediate host, 119, 121
internal pip, 147
interphalangeal joints, 151
intersexes, 53
interstate transport, 165

intestinal
   anastomosis, 101
   prolapse, 131
   resection, 101
   tortion, 153
intraabdominal ovulation, 99
intracellular bacteria, 133
intracytoplasmic inclusion bodies, 143
intradermal tuberculin test, 132
intramedullary pins, 103
intramuscular injections, 155
intraosseous administration, 155
intraosseous catheter, 87, 155
intratracheal antibiotic administration, 156
intravenous administration, 155
intravenous catheter, 83, 155
intravenous injection, 155
intromission, 50
intromittent organ, 7
intubation, 85
intussception, 153
invasive surgical procedures, 92
investigative phase, 174
iodine, 21, 35
iodophors, 35
ionic homeostasis, 13
iron, 21
isofluorane, 81, 84
isoleucine, 14
Israel, 33
isthmus, 47, 51
ivermectin, 116, 118, 120, 157
ixodid ticks (*see* ticks)

jejunum, 5
jugular vein, 7, 155
jurisprudence, 173–177

kanteling, 50
ketamine, 7, 83
ketoconazole, 137
kibbutzim, 33
kidney, 47
kidney tubules, 47
kiwi, 1t, 7, 8
Klebsiella pneumoniae, 137
Klebsiella sp., 137
Klein Karoo Valley, 23, 28, 33, 140
koilin, 4, 117

lactated Ringer's solution, 86
lactic dehydrogenase, 106, 111
Lactobacillus acidophilus, 156
lactose, 17
laparotomy, 92, 113

large animal circuit, 85
large intestine, 4, 5
laryngotracheitis, 32
larynx, 6
laser flow cytometry, 105, 107
late embryonic mortality, 67A
late hatch, 67A
lateral digital extensor tendon, 102
lateral recumbancy, 86
lavaging, 55
laxatives, 95
lead toxicosis, 109
lead II electrocardiogram, 88
leather, 28
left paramedian approach, 95, 99
legal
   action, 174
   advice, 173
   liability, 173–174
   liability insurance, 173, 174
   services, 173
leg bands, 166, 167
lentogenic, 141 (*also see* Newcastle disease)
Leptospira spp., 32
leucine, 14
leukocytes, 106
leukocytosis, 55, 107, 113
Leukocytozoon spp., 116, 122
Leukocytozoon struthionis, 121, 122
leukopenia, 107
levamisole, 118
Libyostrongylosis, 113, 116, 117
Libyostrongylus dentatus, 117t
Libyostrongylus douglassi, 113, 116
lignin, 18
limestone, 20
lincomycin, 136
linea alba, 2
linoleic acid, 7
lipase, 12, 111
lipemia, 110
lipid micelle complexes, 18
lipomas, 56
litigation, 101
liver, 6, 111
locomotory system, 167
low incubator temperature, 62
lungs, 7, 120
lutenizing hormones (LH), 50
lymphocytic perivascular cuffing, 141, 144
lymphoid
   depletion, 142
   tissue, 6
lysine, 14, 22, 23

lysosomal storage disease, 154
lysozymes, 51

Machiaveli-staining techniques, 133
macrominerals, 19
magnesium, 20
magnum, 47, 48, 51
maintenance
   barbituates, 85
   dissociative combinations, 84
   inhalation, 85
malathion spray, 115
malformations, 60
malice, 175
malicious suit, 175
malnutrition, 13–23
malposition, 147
malposition I, 147
malpractice, 174
   claims, 174
maltose, 17
manganese, 60
   deficiency, 152
manual ventilatory support, 89
mask induction, 83, 84
meal,
   blood, 21t, 22t
   cottonseed, 21t, 22t
   fish, 21t, 22t
   meat, 20, 21t, 22t
   peanut, 22t
   soybean, 21t, 22t
   sunflower, 21t, 22t
mean corpuscular volume, 108
mechanical transmission, 32, 134
Meckel's diverticulum, 100
medial metatarsal arteries, 88
medial metatarsal vein, 7
medicated water, 155
medication, 155
menadione dimethylpyrimidol, 19
menadione sodium bisulfate, 19
mesh graft dermatome, 101
mesogenic, 141 (*also see* Newcastle disease)
mesonephric kidney, 19
metabolic
   basal rate, 155
   water, 13
metabolizable energy (ME), 15, 16, 23
metacercaria, 121
metaphysis, 103
metatarsal, 2, 150
methionine, 14
metomidine, 81
metritis, 53, 57

Index

metronizadole, 123, 136
mice, 32
microaerophilic environment, 131
microbial fermentation, 12
microbial flora, 18
microchip, 45, 166
microchip reader, 167
micronutrient imbalances, 19, 20
midazolam, 81
midembryonic mortality, 67A
milo/sorghum, 22t
mineral
    oil, 152
    supplementation, 156
mineralization, of soft tissue, 148
minimal mean arterial pressure, 89
miracidia, 121
mites, 116
model insurance certificate, 165, 172
molybdenum, 21
monitoring techniques, 87–90
monocytes, 108
monocytosis, 108
mononuclear cell lymphocytes, 108
monosaccharides, 17
mosquitos, 32, 116, 138, 143
mucin, 51
Mucor, 137
mucosa, grandular, 48
mucous membrane color, 81, 88
muscle, 111
    pectoral, 81, 135
    rectus abdominus, 2, 100
    relaxation, 83, 84, 111
    tone, 87, 90
    wasting, 111
muscular development, 149–150
Mycobacterium avium, 132
Mycobacterium spp., 32, 57, 113, 132
Mycoplasma cloacale, 131
Mycoplasma gallisepticum, 131
Mycoplasma meleagridis, 131
Mycoplasma spp., 32, 53, 131, 148
Mycoplasma synoviae, 131
mycoplasmosis, 131
mycosis, 136
mycotoxicosis, 21, 136
mycotoxin, 136, 142
myopathy, 92, 152
    capture, 152
    degenerative, 21, 152
    exertional, 41, 152
myositis, 86

nasal flushing, 156
Natt and Herrick's solution, 106
nebulization, 156
neck, 40, 42, 90
necrotizing typhlitis, of rheas, 135
negligence, 174
    avoiding claims for, 176
negotiation, 174
    claims, 174
nematodes, 116–121
neoplasia, 56, 111
nephrotoxic drugs, 157
nephrotoxicity, 157
nervous system, 10, 167
    central (CNS), 150, 167
net energy (NE), 15, 16
neuraminidase (NA), 139
neuromuscular function, 20
neuronal degeneration, 141, 144
neutral detergent fiber, 11, 12, 16, 17
neutral detergent solubles, 17
Newcastle disease, 33, 141
niacin, 19
nitrates, 14
nitrites, 13
nitrogen-free extract, 16
nits, 116
noncuffed tubes, 85
noninflated cuffs, 85
nonrebreathing system, 85
nonregenerative anemia, 109
normal physiologic values, 80t
nutrient requirements, 23–28
nutrition, 11–30
nutritional
    copper deficiency, 153–154
    deficiencies, 13–23
    disease, 13–23
nystatin, 137

oats, 15
obesity, 52
ocular treatment, 156
oligosaccharides, 17
omphalitis, 128, 129, 130, 148, 149
oocyst, 122
oocyte, 47
oophoritis, caseous, 131
open reduction, 103
opiods, 81, 92
    and neuroleptanalgesic combinations, 81
oral contract, 173
oral examinations, 166, 174
oral medication, 155
organs of special sense, 9
oriental eyefluke, 121
orthomxyoviruses type A, 139

osteoblast, 18
osteochondrosis, 102
osteochondrosis dissecans, 150–151
osteodysplasia, 150
osteomyelitis, 111
osteopathy, 152
osteotomy
    derotational, 102, 150
    wedge, 102, 150
ostrich
    capture, 39
    energy and amino acid requirements, 16, 23, 28t, 29t
    handling, 37–40
    hatchery management, 59–67
    nutrient recommendation for South African, 21, 28–29t
    pox, 143
    restraint, 39
    transport, 40
otitis interna, 130
outer shell membrane, 51, 54, 61
ova, 7, 48
ovarian neoplasms, 55
ovary, 7, 47, 50
    cystic, 55
oviduct, 7, 47, 55
    atony, 99
    spasm, 99
oviposition, 47, 48, 51
ovotestes, 53
ovulation, 50, 53
ovum, 47, 51
oxygen, 63, 148
oxytocin, 48, 50, 55

packed cell volume, 109
padding and positioning, 86
pairs, 39, 52
palpebral reflex, 88, 90
pancytopenic, 108
pantothenic acid, 19, 60
papillomavirus, 55
paracentesis, 53
paralysis, 139, 144
paramyxovirus, 141
paramyxovirus type 1, 141
parasite, 115–126
    control, 115–126
    diagnostic methods, 124
    filariid, 120
    neurotropic, 120
    specimen collection, 124
paresis, 122, 139, 144
Paronchocerca struthionus, 117t, 120
parvoviruses, 145
Pastuerella haemolytica, 130

Pastuerella multocida, 130
pasteurellosis, 130
patella, 3
pectins, 17
pedal reflex, 84, 87, 90
pelvic girdle, 3
penicillin, 135
pens, 38, 41, 42, 116
pen system, 38, 39, 41, 42
pepsin, 4, 12
perianesthetic supportive therapy, 86
pericarditis, 133
pericardium syndrome, 139
periosteal elevator, 102
periosteum, 102
peripheral pulse, 80, 88
peritoneal hernias, 57
peritonitis, 55, 98, 130, 133
peroneal nerve, 2, 10, 86, 103
perosis, 20
pesticides, 177
petechiae, 129, 138
phalangeal pad, 2
phalanges, 150
phallic sulcus, 7
phallus, 7, 49
pharyngitis, 137
pharynx, 3
phenols, 35, 70
phenylalanine, 14
Philophthalmus gralii, 117t, 121
phloxine, 106
phosphorous, 19, 106, 110, 150
  nonphytate, 20
photoperiod, 49
physical examination, 80
phytotoxins, 136
pica, 19
picornaviruses, 145
pip, 63, 65, 70
plaintiff, 174
plasma, 105
Plasmodium spp., 116, 122, 122t
Plasmodium struthionis, 122, 122t
plate agglutination test, 128
plucking boxes, 80
pneumonia, 131
polycythemia, 109
polyionic isotonic fluids, 156
polymerase chain reaction-gene probe assay, 132
polysaccharides, 17
pores, 61
postmortem examination, 168–171, 176
postoperative analgesia, 92
postovulatory follicle, 50

postpurchase examinations, 112
potassium, 19, 105, 110
poultry industry, 33
  commercial industry, 139, 148
practitioner-client relationship, 175
praziquantel, 119
preanesthetic agents, 81
preanesthetic preparation, 79–81
prebreeding physical exam, 112
predators, 42
premature ventricular depolarizations, 90
preoperative examination, 79–83
prepurchase examination, 32, 112
prerenal dehydration, 110
preselling stock, 173
pressure transducer, 88
proctodeum, 6, 49
production data, 69
progesterone, 56
prolapse, 56, 153
prosthetic rings, 102
Prosthorhynchus rhea, 119
protective clothing, 34
protein, 14
  electrophoresis, 11
  levels, 110
  serum, 106, 107, 110
protozoal diseases, 121
protozoan, 121
protozoan cysts, 121
proventriculitis, diptheric, 117
proventriculotomy, 95, 98
proventriculus, 4, 12, 117, 137
proximal physis, 102
proximate cause, 175, 176
pseudoleukopenia, 107
Pseudomonas spp., 32, 53, 137, 148
psitticine circovirus, 108
psychrometric chart, 70, 74–78
psyllium, 98, 152
Pterolichidae, 116
pulmonary congestion, 133, 142, 152
pulmonary edema, 152
pulse
  oximetry, 89
  quality, 80, 88
  rate, 87, 88
  rhythm, 87
punitive damages, 176

quail bronchitis, 141
quarantined, 34, 116, 166
quaternary ammonium compounds, 35, 70
quill mite, 116

radiology, 149
raffinose, 17
ratite
  formulary antimicrobials, 158–159
  formulary antiparasitic agents, 160
  formulary miscellaneous therapeutics, 161–162
  meat, 131
  wrap, 91
rats, 32
rebreathing system, 85
rectal pouch, 5
rectocoprodeal fold, 5
rectum, 5
recumbancy, 103, 139, 141
recurrent laryngeal nerve, 102
red cell distribution with percentage anemia, 109
regenerative anemias, 109
regurgitation, 85, 86, 91
relative humidity, 72, 74
renal calcification, 147
renal disease hyperuricemia, 147
renal disease mineralization, 147
renal portal system, 7, 81
reoviruses, 143, 145
repeatability, 59
reproduction, 47–57
  disease, 53–57
reproductive failure, 52–57
reproductive system, 7, 168
reservoir bag, 85, 89
resolution of disputes, 174
resorantel, 119
respirations, 7, 89
respiratory
  arrest, 89
  collapse, 41
  infection, 156
  mucosa, 38
  rate, 7, 89
restraining stocks, 37, 40, 174
restraint devices, 39, 40, 45
restraints on the use of drugs, 177
rete tubules, 48
reticulocytes, 109
retinoic acid, 18
retinol, 18
retinyl acetate, 18
retinyl palmitate, 18
retrobulbar pathology, 167
reverse peristalisis, 51
rhabdomyolisis, 81, 91, 103
rhea, 2, 4, 8, 12, 18, 118, 119
  nutrient requirement, 23

Index

rhinitis, 134
Rhizopus, 137
riboflavin (B2), 19
    deficiency, 151
ribose, 17
rickets, 151
rickettsia, 122
Robert Jones type bandage, 103
rolled toe, 151
rotational deformities, of the tibiotarsus, 102
rotaviruses, 145
rough induction (*see* induction)
round worms, 118

Sabouraud's agar, 137
salivated, 85
Salmonella arizonae, 128
Salmonella entritidis, 55
Salmonella pullorium, 128
Salmonella spp., 32, 55, 128, 148
Salmonella typhimurium, 128
salmonellosis, 128
salpingitis, 53, 98
    caseous, 130
salt, 19
sand, 61
scapula, 3
schizonts, 122
scolex, 119
scoliosis, 120
seasonal infertility, 49
secretory glands, 51
sedatives, 81–83
selenium, 21
    deficiency, 149, 152
    toxicity, 152
self-trauma, 41, 91
semen, 50
    collection, 50
    extenders, 50
    pH, 50
    storage, 50
    volume, 50
seminiferous tubules, 48, 55
septicemia, 107, 108, 130, 149
serological identification, 32
serologic flock profiling, 34
serosal surface, 138
Serpulina-like organism, 122
serum plate agglutination test, 132
serum potassium (*see* potassium)
sexually mature, 49
shade cloth, 41
shell
    gland, 18, 48
    gut, 54
    membranes, 51, 54, 61
    pores, 51, 61
    quality, 57, 60
    surface, 57, 60
    thickness, 57, 60
shepherd's hook, 39, 80
shipping crate, 39
Similium spp., 32, 116, 122
sinusitis, 134–135
skeletal integrity, 19
slipped tendons, 151
small-animal circuit, 85
small intestine, 4, 12
smudge cells, 105, 107
sodium, 19, 110
sodium chloride, 124
sodium hypochlorite, 2, 35
sodium nitrate, 124
South Africa, 21
specific gravity, 124
spermatozoa, 49
sperm concentration, 50
spermiogenesis, 20, 49
spirochete, 113, 135
spiruids, 118
splayed legs, 102, 150
spleen, 7, 107
splenomegaly, 128, 129
spontaneous ventilation, 89
spores, 124, 134, 135, 136
spraddled legs, 67A, 150
stachyose, 17
St. Louis encephalitis, 120, 139
stanchions, 80
standard of care, 175
Staphylococcus aureus, 137
Staphylococcus spp., 137
starch, 15, 17
starvation, 110, 112
stay sutures, 153
sternal recumbancy, 10, 91, 142
sternum, 3
stifle, 3
stocking density, 38
stomatitis, 137, 143
    focal, 137
    pseudomembranous, 137
straw, 91
stress, 107, 110
stress hemograms, 107
strongylids, 117, 124
Struthiofilaria magalocephala, 120
Struthioliperus nandu, 116
Struthioliperus renschi, 116
Struthioliperus rhea, 116
Struthioliperus struthionis, 116
stunting syndrome, 145
subcutaneous administration, 155
subcutaneous hemorrhages, 154
subepicardial hemorrhage, 133
substrate, 38, 98
suckers, 119
sucrase, 11
sucrose, 17
sulfate, 13
sulfonamides, 57
surgery, orthopedic, 92
surgical conditions, 95–104
sweating, 76
Syngamus spp., 113, 117t, 121
Syngamus trachea, 117t, 121
synsacrum, 3

tapeworms, 119, 124
tarsal bone, 3
tarsometatarsus, 102, 151
    fracture of, 103
tarsus valgus deformity, 102
teratogen, 149
testes, 48
testosterone, 49
tetracycline, 131, 133, 137
therapeutics, 155–163
thermal stress, 40
thermal support, 86
thermoregulation, 86
thiamin (B1), 19
thoracic girdle, 3
threonine, 14
thrombocytes, 108
thrombocytopenia, 108
thyroid gland, 21
thyroxine, 21
tibial dyschondoplasia, 150
tibiotarsus, 37, 88, 150, 151
    fracture of, 103
    rotational deformities, 150
ticks, 115, 122
    argasid (soft), 115, 122
    ixodid (hard), 115, 116
    larval, 115
    nymphs, 115
    paralysis, 115
tissue perfusion, 87
tocopherol, 19
tocopheryl acetate, 19
toenail, 2
toe pinch, 84
togaviridae, 113, 138
tongue, 89
topical medication, 156
torticollis, 120, 130, 141, 149
total dissolved solids (TDS), 13
total erthrocyte count, 108

total leukocyte count, 106
total protein, 110
toxic heterophils, 108, 149
Toxoplasma gondii, 122t, 123
trachea, 85, 102
tracheal cartilaginous rings, 6, 86
tracheal cleft, 6
tracheal ring prosthesis, 102
tracheal swabs, 140
tracheitis, 143
trailer, 34, 42
tranquilizers, 81
transfixation cast, 102, 103
transovarial route, 128, 132, 141
transoviductal transmission, 128
transport, 156
trematodes, 119, 121
trichomonads, 122, 122t, 135
Trichostrongylus tenuis, 117t, 118
tri-iodothyronine, 21
trios, 52
true metabolizable energy (TME), 16
trypsin, 12
tryptophan, 14
tuberculosis, 107, 132
tubular necrosis, 148
tubular secretion, 109
tumbling chick syndrome, 149
tumors, 56
tylosin, 132
type A orthomyxoviruses, 139
typhlitis, necrotizing, 122, 135

ulcerative entritis, 137
ultrasonography, 55
ultrasound, 149
umbilicus, 66, 148, 173
urate gout, 148
urea, 110, 112
Ureaplasma, 132
ureters, 48
urethra, 7
uric acid, 109, 147, 148
urodeum, 5, 48
uroproctodeal fold, 6
uterine segment, 51
uterovaginal sphincter, 48, 51
uterus, 48, 51, 99

vaccination, 127, 156
  protocols, 156
vaccines
  autogenous, 140, 157

bactrin, 131
bivalent, 138, 139
inactivated, 135, 138, 142, 144
live attenuated, 141, 143
vagina, 48
vaginal prolapse, 55
valgus, 20, 102
valine, 14
varus, 20
vasa deferno, 48
vascular congestion, 128, 130
vascular rupture, 41
vas deferens, 48
velogenic Newcastle disease, 141
velogenic viscerotropic, 141
venous congestion, 135
vent, 9, 43, 57
ventilation, 23, 38, 62, 90
ventilation exchange rate, 38
ventilatory support, 89
ventricular arrythmias, 90
ventriculitis, 39, 137
ventriculostomy, 153
ventriculus, 4, 12, 152
vent sexing, 9, 43
vermin, 32
vertically integrated production, 31
vertically transmitted diseases, 32, 142, 148
vertical transmission, 32, 142, 148
veterinary certificates, 34, 176
viral
  diseases, 138–145
  enteritis, 193
  infection, 55, 113
  isolation, 138, 139, 140, 141
virucidal, 35
virus, 138–145
  arena, 153
  circo, 108
  EEE, 113, 138, 156
  WEE, 120, 138, 139, 156
Virus-Serum-Toxin Act, 177
visceral gout, 112, 148
vision, 9
vital signs, 80
vitamin A, 18
vitamin C, 19
vitamin D_3, 18, 150
vitamin D_3 toxicosis, 148
vitamin E, 18, 21
vitamin E deficiency, 149, 152
vitamin mineral supplements, 156, 160

vitamin supplementation, 19, 156
vitelline membrane, 47
voluntary ejaculation, 50
voluntary movement, 90
vrotmaag, 117
VVND (*see* Newcastle disease)

walking halters, 45
warm water heating pad, 86
wasting, 34
water, 13
  intoxication, 13
  metabolic, 13
  vapor, 13, 60
wedge osteotomy, 102, 150
western equine encephalomyelitis, 113, 120, 138, 139, 156, 157
wet bulb isotherms, 74
wet bulb temperature, 74
wheat, 15, 22t
wing, 2, 3, 40, 43
  web, 88
wire fencing, 39, 42
wireworms, 113, 116
working pens, 166
wound debridement, 101
wound management, 101
wry neck, 149

xanthine dehydrogenase, 21
xylan, 17
xylazine, 81
xylazine-ketamine, 81, 83, 84
xylose, 17

yeasts, 137
yolk, 47, 54
yolk sac, 100, 129, 130
  absorption, 148
  deposition, 51
  external, 67A, 148
  infections, 67A, 100, 148, 150
  retained, 149
  retention, 100
yolk sacculectomy, 100

zinc, 20
zinc bacitracin, 134
zinc sulfate, 124
zoonoses, 133, 139
zoonotic infection, 131, 133, 170
zygomycosis, 137